ADVANCES IN MOLECULAR AND CELL BIOLOGY

Volume 6 • 1993

EXTRACELLULAR MATRIX

ADVANCES IN MOLECULAR AND CELL BIOLOGY

EXTRACELLULAR MATRIX

Series Editor: E. EDWARD BITTAR
Department of Physiology
University of Wisconsin
Madison, Wisconsin

Guest Editor: HYNDA K. KLEINMAN
National Institute for Dental Research
National Institutes of Health
Bethesda, Maryland

VOLUME 6 • 1993

JAI PRESS INC.

Greenwich, Connecticut *London, England*

55 Old Post Road, No. 2
Greenwich, Connecticut 06836

JAI PRESS LTD.
The Courtyard
28 High Street
Hampton Hill, Middlesex TW12 1PD
England

ISBN: 1-55938-515-4

Manufactured in the United States of America

CONTENTS

LIST OF CONTRIBUTORS

Monique Aumailley	Institute for Biology and Chemistry of Proteins Université Claude Bernard Lyon, France
E. C. Bergijk	Laboratory of Pathology University of Leiden Leiden, The Netherlands
Thomas C. Blochberger	The Eye & Ear Institute of Pittsburgh Department of Ophthalmology University of Pittsburgh School of Medicine Pittsburgh, Pennsylvania
J. A. Bruijn	Laboratory of Pathology University of Leiden Leiden, The Netherlands
Shukti Chakravarti	The Eye & Ear Institute of Pittsburgh Department of Ophthalmology University of Pittsburgh School of Medicine Pittsburgh, Pennsylvania
David A. Cheresh	Department of Immunology The Scripps Research Institute LaJolla California
J. M. Foidart	Laboratory of Biology University of Liège Liège, Belgium
Rafael Fridman	Department of Pathology Wayne State University Detroit, Michigan

Robert Garrone Institute for Biology and Chemistry of Proteins
Université Claude Bernard
Lyon, France

John R. Hassell The Eye & Ear Institute of Pittsburgh
Department of Ophthalmology
University of Pittsburgh School of Medicine
Pittsburgh, Pennsylvania

Daniel Herbage Institute for Biology and Chemistry of Proteins
Université Claude Bernard
Lyon, France

Philip J. Hogg Department of Hematology
The Prince of Wales Hospital
Randwick, Australia

Katherine A. Lee Dermatology Division
The Jewish Hospital of St. Louis
Washington University Medical Center
St. Louis, Missouri

Robert P. Mecham Departments of Medicine and
Cell Biology and Physiology
The Jewish Hospital of St. Louis
Washington University Medical Center
St. Louis, Missouri

Deane F. Mosher Departments of Medicine and
Biomolecular Chemistry
University of Wisconsin
Madison, Wisconsin

A. Noēl Laboratory of Biology
University of Liège
Liège, Belgium

Douglas Noonan Department of Chemical Carcinogenesis
Istituto Nazaionale per la
Ricerca Sul Cancro (Ist)
Genova, Italy

Vishwas M. Paralkar	Laboratory of Molecular Developmental Biology National Eye Institute National Institutes of Health Bethesda, Maryland
William C. Parks	Dermatology Division The Jewish Hospital of St. Louis Washington University Medical Center St. Louis, Missouri
Richard A. Pierce	Dermatology Division The Jewish Hospital of St. Louis Washington University Medical Center St. Louis, Missouri
Jody A. Rada	The Eye & Ear Institute of Pittsburgh Department of Ophthalmology University of Pittsburgh School of Medicine Pittsburgh, Pennsylvania
A. H. Reddi	Laboratory of Musculoskeletal Cell Biology Johns Hopkins University School of Medicine Baltimore, Maryland
Jane Sottile	Departments of Medicine and Biomolecular Chemistry University of Wisconsin Madison, Wisconsin
Xi Sun	Departments of Medicine and Biomolecular Chemistry University of Wisconsin Madison, Wisconsin
Michel van der Rest	Institute for Biology and Chemistry of Proteins Université Claude Bernard Lyon, France
Slobodan Vukicevic	Bone Research Branch National Institute for Dental Research National Institutes of Health Bethesda, Maryland

PREFACE

Studies on the components of the extracellular matrix have expanded recently and greatly increased our understanding of their biological functions and roles in development and in disease. Because so many new areas have been defined, selection of chapter topics was difficult to limit. Areas with recent developments in both basic knowledge and diseased states have been emphasized. Some of the chapters detail evolving frontiers which are of intense interest, but complete information is not known. For example, several bone morphogenetic factors have been described and others likely exist. The structures of these multifunctional proteins are determined and they appear related. The description of extracellular matrix component receptors is far from complete, but worthy of a progress report since much is already known. Of particular interest is the "promiscuity" of some of these receptors in their interaction with more than one matrix protein and even different ligand specificities on different cell types. Such variability in these interactions challenges conventional ideas on the specificity of ligand-receptor interactions. The collagens were traditionally thought of as structural molecules but now have been found to be biologically active. The number of genetically distinct collagen genes is greater than 27 (still increasing!) and their gene regulation is unique. The demonstration of a bi-directional promoter for the α1 and α2 collagen IV chains has stimulated research on gene regulation and the DNA binding proteins. A number of diseases have also been shown recently to involve collagens. The proteoglycans have been thought of as largely carbohydrate-containing molecules with the research emphasis placed on the sugars. With the recent DNA sequencing of several protein cores, these molecules have been found to be of tremendous interest. The basement membrane heparan sulfate proteoglycan contains domains with homologies to laminin, immunoglobulin, and LDL. Proteoglycans have multiple activities and function as adhesion factors, cellular receptors, growth factor binding sites, etc.

Several chapters are devoted to whole basement membrane and to other matrix components such as fibronectin, thrombospondin, laminin, collagen IV, entactin/nidogen, and elastin which have been shown to have remarkable effects on normal and malignant cell behavior. The structures of these matrix components have been described, and now active sites as well as alternate forms are being defined. The role in tumor growth and acquired and genetic diseases is also reviewed. The potential clinical use of the molecules or active sites is only briefly mentioned but likely to be important in the future.

Hynda K. Kleinman
Guest Editor

COLLAGEN:

A FAMILY OF PROTEINS WITH MANY FACETS

Michel van der Rest, Robert Garrone, and Daniel Herbage

Advances in Molecular and Cell Biology
Volume 6, pages 1–67

ISBN: 1-55938-515-4

I. INTRODUCTION

A. Definition of Collagen

The word "collagen" was coined in the 19th century to designate "that constituent of connective tissue which yields gelatin on boiling" (Oxford Dictionary, 1893). Some aspects of collagen chemistry were obviously known for a much longer time for uses like the production of animal glue from hides and bones or the tanning of leather. The more modern view of collagen as an assembly of protomers forming a supramolecular fibrillar structure was only clearly formulated by Gross in the early 1950s (Gross, 1956). Collagen could thus be viewed as a "substance" composed of individual constituent molecules (termed tropocollagen originally and now referred to as collagen molecules) assembled into cross-linked fibrils that give connective tissues their resistance to tension. These molecules were shown to be made of a particular triple-helical assembly of three polypeptide chains. The most abundant molecule, studied in these early works, was shown to be composed of two identical $\alpha 1$ chains and one distinct $\alpha 2$ chain.

It was not until the late 1960s that the existence of different collagen molecules was recognized (Miller and Matukas, 1969; Miller, 1976). This has given rise to our current nomenclature of collagen types that is based on the assumption that trimeric collagen molecules are assembled from polypeptide chains of the same type number (Bornstein and Sage, 1980). As it will be seen later, recent findings indicate that not all collagen molecules conform to this rule. With one likely exception, all the collagen chains are distinct gene products. At the time of this writing, 14 collagen types have been given type number and these include 27 distinct chains. Several additional candidate collagen genes or gene products have been recently reported or are studied in different laboratories.

Many of the recently identified molecules are very different from the originally described type I collagen molecule (van der Rest and Garrone, 1991). The generally accepted current definition of a collagen is that it is a structural protein of the extracellular matrix which contains one or more domains having the conformation of a triple helix.[1] This definition implies that collagen molecules are multidomain proteins and that some domain(s) are involved in the multimolecular aggregates forming the architecture of the extracellular matrix (van der Rest and Garrone, 1990).

The triple-helical domains of collagens are thus the main characteristic of this family of proteins to which they confer unique properties of aggregation. We will thus, after a brief overview of the collagen family of proteins, discuss that particular protein conformation, how it is formed, and how triple-helical domains can interact to form multimolecular aggregates. The nontriple-helical domains of molecules containing triple helices of the collagen type could be of any nature and they are indeed highly diverse in structure and function. These domains will be discussed separately.

B. The Collagen Family of Proteins

Most of the 27 collagen chains have been characterized at the protein, cDNA, and genomic DNA levels. The analysis of the structure data can be used to establish phylogenic relationships and homologies between these various chains. Table 1 gives a list of the 14 collagen types, grouped by homology, and the main characteristics of the molecules, namely their constituent chains and their tissue distribution.

The subfamily of collagens forming quarter-staggered fibrils (collagen types I, II, III, V and XI) is best known. The nine constituent chains all share the same general features and their genes are all clearly derived from the same ancestral gene (Sandell and Boyd, 1990b). They contain a main triple-helical domain of approximately 1000 amino acids/chain (see Figure 1). Their COOH-terminal propeptides are highly homologous between themselves (Dion and Myers, 1987). Their modes of aggregation also appear to be very similar. The subdivisions into collagen types is based on the assumption that chains of a given type number assemble to form trimeric molecules. Recent data indicate that hybrid molecules of chains of two different types can also be formed. The α3(XI) is almost certainly the same gene product as α1(II) (Morris and Bächinger, 1987). In addition, evidence for the existence of hybrid type V–type XI molecules is building up (Niyibizi and Eyre,

Table 1. The Collagen Types

Structures Formed	*Type Number*	*Constituent Chains*	*Molecular Formulas*	*Described In*
Quarter-staggered fibrils	Type I	α1(I)	[α1(I)]3	Dentin, skin (minor form)
		α2(I)	$[\alpha1(I)]_2\alpha2(I)$	Most connective tissues
	Type II	α1(II)	$[\alpha1(11)]_3$	Hyaline cartilages, vitreous humor
	Type III	α1(III)	$[\alpha1(III)]_3$	Fetal skin, vessel walls, soft connective tissues
	Type V/XI	α1(V)	$[\alpha1(V)]_3$	Chinese hamster lung cells
		α2(V)	$[\alpha1(V)]_2\alpha2(V)$	Most type I collagen containing tissues
		α3(V)	α1(V)α2(V)α3(V)	Placenta
		α1(XI)	α1(XI)α2(XI)α3(XI)	Cartilages
		α2(XI)	$[\alpha1(X1)]_2\alpha2(V)$	Bovine bone, endothelial cells, A204 cells
		α3(XI)		
Basement membranes	Type IV	α1(IV)	$[\alpha(IV)_2\alpha2(IV)$	Basement membranes
		α2(IV)		
		α3(IV)	?	Glomerular basement membrane
		α4(IV)		
		α5(IV)		
Hexagonal lattices	Type VIII	α1(VIII)	?	Descemet's membrane, endothelial cells
		a2(VIII)		
	Type X	α1(X)	$[\alpha1(X)]_3$	Growth plate cartilage
Beaded filaments	Type VI	α1(VI)	α1(VI)α2(VI)α3(VI)	Most connective tissues
		α2(VI)		
		α3(VI)		
Fibril-associated	Type IX	α1(IX)	α1(IX)α2(IX)α3(IX)	Cartilage, vitreous humor
		α2(IX)		
		α3(IX)		
	Type XII	α1(XII)	$[\alpha1(XII)]_3$	Tendon, skin
	Type XIV	α1(XIV)	$[\alpha1(XIV)]_3$	Skin, tendon
Anchoring fibrils	Type VII	α1(VII)	$[\alpha1(VII)]_3$	Mesenchyme-epithelium junctions
?	Type XIII	α1(XIII)	?	Epithelial and endothelial cells

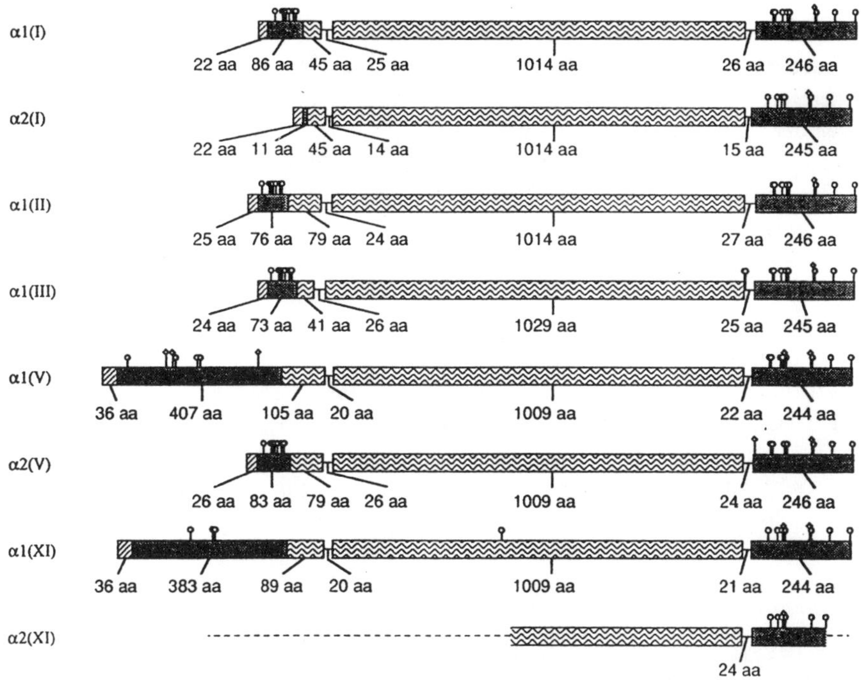

Figure 1. Schematic representation of the structures of the fibrillar collagen chains. The triple-helical domains are represented by (▨), the globular domains by (■), the signal peptide by (▨) and the junction regions containing demonstrated or putative cleavage sites for the N- and C-proteinases by a solid bar. Regions that have not been studied are shown as a dotted line. The positions of cysteine residues (⚲) or putative N-glycosylation sites (⚲) are indicated.

1989; Brown et al., 1991). It might therefore be more appropriate to consider the molecules formed by type V and type XI collagen chains as isoforms.

The other collagen types are distinctly different. They are often called nonfibrillar collagens because of their inabilities to form quarter-staggered fibrils by themselves. Some of them are nevertheless able to form other types of fibrils or to participate in the quarter-staggered aggregates as essential constituents (van der Rest and Garrone, 1991). Several subfamilies can now be distinguished among nonfibrillar collagens, based on comparisons of the protein structures and of the intron-exon gene organizations (Table 1).

Olsen has proposed the abbreviation FACIT (Fibril Associated Collagens with Interrupted Triple-helices) for the subfamily comprising type IX, type XII, and type XIV collagens (Shaw and Olsen, 1991). These collagens are closely associated with the quarter-staggered fibrils and are characterized by the presence of a short triple-helical domain located near the carboxyl end of the molecule. This domain is homologous in this group of collagen types and may play a role in the interaction with the fibrils. Type IX collagen is covalently bound to type II collagen and should therefore be considered as a constituent of the fibrillar assembly (Eyre et al., 1987; van der Rest and Mayne, 1988; Wu and Eyre, 1989). The structural homologies between type XII and type XIV collagens are actually much higher than between these two molecules and type IX collagen (Aubert-Foucher et al., 1992). Type XII and type XIV collagens are characterized by the presence of very large N-terminal nontriple-helical domains with clear similarities to adhesive glycoproteins, a characteristic which is not found in type IX collagen (Dublet et al., 1989; Gordon et al., 1989; Yamagata et al., 1991).

The aggregates formed by the other nonfibrillar collagens are very diverse. Some of them assemble into meshworks that are responsible for the formation of sheets or "protein membranes" that surround tissues and organs. The best known of these sheet-forming collagens is type IV collagen which is now known to include five different chains (Hostikka et al., 1990). This suggests the existence of isoforms in this collagen type. These five type IV collagen chains constitute clearly another distinct subfamily of collagen gene products. Type VIII collagen is responsible for the formation of another sheet structure, the hexagonal lattice of Descemet's membrane (Sawada et al., 1990) and is closely homologous to type X collagen, which is synthesized by hypertrophic chondrocytes in the growth plate (Yamaguchi et al., 1989; Muragaki et al., 1991a). *In vitro*, type X collagen is able to form hexagonal aggregates which are very similar to that observed in Descemet's membrane (Kwan et al., 1991). Type VIII and type X collagen thus constitute yet another subfamily of collagen gene products.

Type VI collagen is characterized by the presence of very large multidomain amino and carboxy terminal nontriple-helical regions (Bonaldo and Colombatti, 1989; Koller et al., 1989; Chu et al., 1989; Chu et al., 1990b; Colombatti et al., 1990) and assembles into beaded filaments that are formed by a complex aggregation process (Engel et al., 1985). While the nontriple-helical regions of type VI collagen contain domains of homology with other proteins (see below), this collagen type with its three constituent chains represents a subfamily of its own. Because the size of its triple-helical domain is similar to that of type VIII and type X collagens, it has been sometimes included in the same subgroup of "short chain" collagens (Ninomiya et al., 1990). We feel that this is not appropriate any more since besides the triple helix no clear homology can be detected between type VI and these other two collagen types.

Type VII collagen also constitutes a subfamily of its own. It is characterized by a very long triple-helical region and by a large N-terminal domain that appears as a three-finger structure by rotary shadowing electron microscopy (Burgeson et al., 1990). This collagen aggregates into structures known as anchoring fibrils which contribute to the cohesiveness of the junction between epithelial basement membranes and the underlying stroma (Keene et al., 1987b).

Type XIII collagen has been described at the cDNA level (Tikka et al., 1988) and shown to be subjected to very complex alternative splicing (Pihlajaniemi and Tamminen, 1990; Tikka et al., 1991). Little is known yet about the protein itself which has been detected only with antibodies prepared against peptides synthesized according to the cDNA sequence. Again, it bears no clear homology with any of the other collagen types.

This quick survey of the collagen family of proteins shows how diverse these molecules are. Their common denominators (i.e., the presence of triple-helical domains in their structures and the fact that they are constituents of the architecture of the extracellular matrix) justify to group them into a unique protein family although it could certainly be argued differently. To better understand this diversity, we will discuss first the properties of the collagen triple helix, the posttranslational modification steps required for its formation, and how this structure has the potential to form the different types of aggregates that are used in a number of different biological situations.

We will then describe the nontriple-helical regions of collagen molecules. They are extremely diverse in size, shape, and properties. Within certain subfamilies of collagen molecules, it can be said that the major differences among the various types reside in their nontriple-helical regions. We will show how the different collagen molecules are assembled into supramolecular aggregates that can, in most cases, be observed by electron microscopy as distinct histological structures. These aggregates are often stabilized by covalent cross-linking between constituent chains and molecules. These cross-links and the biosynthetic steps involved in their formation will then be discussed. Finally, we will show how the collagen diversity is reflected at the gene level.

II. THE COLLAGEN TRIPLE HELIX

A. Conformation

The collagen triple helix has a coiled-coil structure made of three parallel polypeptide chains (Figure 2). Each chain forms a left handed helix with an approximate threefold screw-axis symmetry, closely similar to a polyproline helix (Ramachandran and Ramakrishnan, 1976). The three chains are assembled around a central axis and form a right-handed superhelix; This superhelix is due to the twisting of the chain helices around the central axis by about +30° at every turn.

Every third amino acid is thus in the center of the helix and, for steric reasons, only glycine, with a side chain limited to a single hydrogen atom, can occupy this position without altering the triple-helical conformation. Hence the primary structure of a collagen triple helix can be written as $(\text{Gly-Xaa-Yaa})_n$. The presence of another amino acid in the Gly position or the presence of imperfections in this repetitive structure seriously alters the stability or the conformation of the helix.

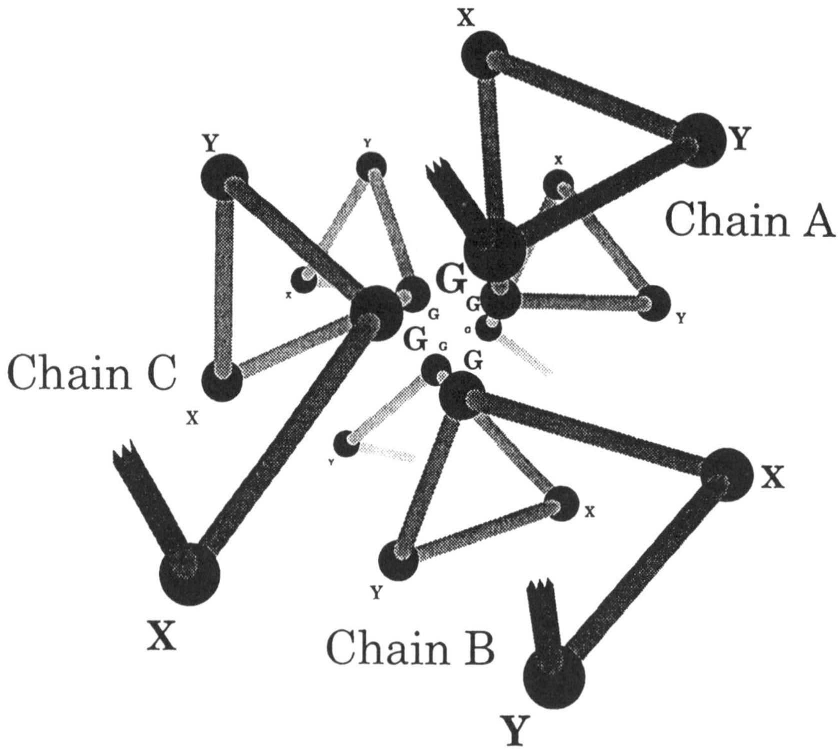

Figure 2. Cross-section of a collagen triple helix. The α carbons of the amino-acyl residues are represented by balls and the peptide bonds by sticks. The side chains of the residues are omitted. The sizes of the G, X and Y lettering, of the balls and of the sticks are varied according to the depth in the field. Note that the three chains are not equivalent since only one glycine can occupy the center of the helix at a given depth. It can be seen that all the X and Y residues are at the surface and will have their side chains pointing outward of the helix. Each polypeptide forms a left handed helix and assemble with the other two chains in a right handed superhelix. [Reprinted from van der Rest and Garrone (1990), with permission].

Imperfections (e.g., Gly-Xaa-Gly-Xaa-Yaa or Gly-Xaa-Yaa-Xaa-Yaa) are found frequently in the triple-helical domains of nonfibrillar collagens (Mayne and Burgeson, 1987). These imperfections are certainly responsible for subtle alterations of structure that may be functionally important but are yet poorly understood. On the other hand, single-based substitutions in glycine codons in type I collagen genes, changing a glycine for another amino acid, have been shown to be responsible for most cases of perinatal lethal Osteogenesis Imperfecta (Engel and Prockop, 1991).

The positions Xaa and Yaa can be occupied by any amino acid residue without disrupting this structure. Imino acids (proline and hydroxyproline) in these positions do favor the polyproline-like conformation, as their five-membered ring limits the rotation freedom around the bond between the imino group and the alpha-carbon (Ramachandran and Ramakrishnan, 1976). The restriction imposed to the orientation of a peptide bond involving the imino group in a *trans* conformation is just appropriate to stabilize the structure. In addition, hydroxyproline in the Yaa position allows the formation of additional water bridges between adjacent chains, therefore further stabilizing this structure. This increased stability is clearly demonstrated by the observation that nonhydroxylated collagen, obtained *in vitro* by inhibition of prolyl-hydroxylase, melts at a temperature 15 °C lower than the normally hydroxylated molecule (Berg and Prockop, 1973). In collagen, about one-third of the Xaa and Yaa positions are occupied by prolines and hydroxyprolines respectively (Fietzek and Kühn, 1976).

B. Properties

The presence of other amino acids in positions Xaa and Yaa is of major importance for the lateral interactions between a collagen molecule and its surrounding environment. As can be deduced from Figure 2, the side chains of residues Xaa and Yaa are pointing outward from the helix, and are therefore all available for lateral ionic or hydrophobic interactions. The triple-helix structure has the remarkable property of having two-thirds of the amino acids of the polypeptide chains at the surface of the protein. Lateral interactions are obviously most favored between adjacent triple helices, and thus play a fundamental role in fibril formation and in other forms of lateral aggregation of collagen molecules. These lateral interactions also impose some restrictions on the nature of the residues in positions Xaa and Yaa. For example, tryptophan residues are absent from most triple-helical domains while amino acids with short side chains, such as alanine, are overrepresented (Fietzek and Kühn, 1976).

The triple-helical domains can also be considered as molecular rods of ~1.5 nm in diameter and of ~30 nm in length/100 amino-acyl residues/chain. Thus they can be used in proteins to separate or to bridge functional domains. Type IX collagen nicely illustrates the main properties of this protein conformation (Figure 3). The two triple-helical domains nearest to the C-terminus (COL1 and COL2) are involved in lateral interactions with the fibril-containing type II collagen. A nontriple-helical domain (NC3), containing in the $\alpha 2$(IX) chain a glycosaminoglycan attachment site, serves as a hinge to the last triple-helical domain (COL3), which itself serves as a molecular rod to project the N-terminal globular domain of the $\alpha 1$(IX) chain (NC4) out of the fibril (van der Rest and Mayne, 1987; Vaughan et al., 1988).

The thermal stability of triple-helical domains are clearly related to their imino acid contents. While most triple-helical domains have denaturation temperatures

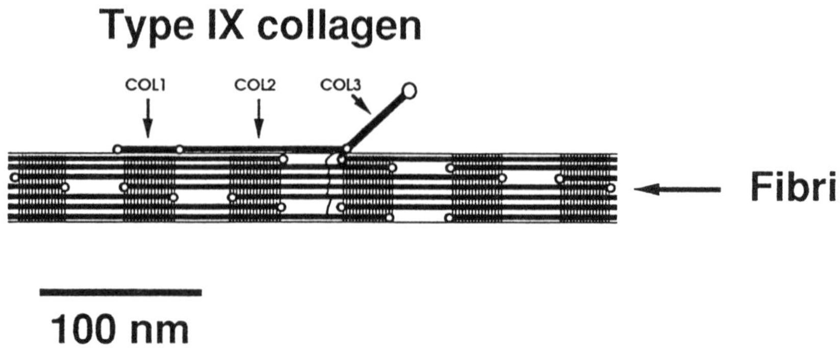

Figure 3. Schematic representation of the interaction between type IX collagen and a cartilage collagen fibril. The triple-helical domains COL1 and COL2 are thought to interact in an antiparallel fashion with type II collagen molecules while the triple-helical COL3 domain acts as an arm to project the globular NC4 domain out of the fibril. Modified from van der Rest and Mayne (1988).

slightly above the physiological temperatures, some triple helices have denaturation temperatures in the 45 to 50 °C range; for example, type X collagen and the COL3 domain of type IX collagen (Bruckner et al., 1983; Linsenmayer et al., 1986; van der Rest and Mayne, 1987). It should be remembered that the denaturation temperatures are usually measured on soluble molecules and that aggregated molecules may have substantially higher denaturation temperatures.

III. BIOSYNTHETIC STEPS IN TRIPLE-HELIX FORMATION

The collagen biosynthesis as usually described is that of type I collagen. However, several important posttranslational modification steps in the biosynthesis of type I collagen are not applicable to other collagen types, in particular to the nonfibrillar collagens. Here we will discuss only the biosynthetic steps that are necessary for the formation of the triple helices common to all the collagen types.

Like all secreted proteins, the collagen polypeptide chains are synthesized with a signal peptide that directs the polysomes to the surface of the endoplasmic reticulum in the lumen of which the nascent polypeptide is secreted. The specific posttranslational modifications start as the chains are still nascent.

A. Prolyl Hydroxylation

The folding of collagen triple helices is possible at physiological temperature only if enough prolyl residues in the Yaa positions of the (Gly-Xaa-Yaa)$_n$ repeats are hydroxylated into hydroxyprolyl residues (Prockop et al., 1976). The enzyme prolyl 4-hydroxylase is a tetrameric protein α2β2. cDNA clones for both subunits have been obtained and characterized (Parkkonen et al., 1988; Tasanen et al., 1988; Helaakoski et al., 1989). The individual subunits of the enzyme are catalytically inactive for prolyl hydroxylation, although it appears that the α subunit (64 kDa) contributes most of the catalytic site. Cooperativity of both subunits is required for activity (Kivirikko et al., 1989).

The β subunit (60 kDa) is a multifunctional polypeptide. It is the same gene product as the protein disulfide isomerase which is found in many different cell types and is active in the monomeric form. This same subunit may actually have three additional functions: (1) as thyroid hormone binding protein in the endoplasmic reticulum; (2) as cellular iodothyronine 5′-monodeiodinase; and (3) as the component binding to the glycosylation site (the sequence -Asn-Xaa-Ser/Thr-) in oligosaccharyl transferase (Kivirikko et al., 1989).

Prolyl 4-hydroxylase recognizes the sequence -Xaa-Pro-Gly- in collagen and in some other proteins. It requires Fe^{2+}, 2-oxoglutarate, molecular oxygen, and ascorbate for activity (Prockop et al., 1976). Oxygen and 2-oxoglutarate are used stoichiometrically in the hydroxylation reaction. The 2-oxoglutarate undergoes an oxidative decarboxylation into succinate, with the incorporation of one of the oxygen atoms. The second oxygen atom is incorporated into the hydroxyl group formed on the prolyl residue. Ascorbate is not used stoichiometrically. It appears to act as an alternative oxygen acceptor during the catalytic cycles in which decarboxylation of 2-oxoglutarate occurs uncoupled to prolyl hydroxylation, a side reaction that occurs even at saturating substrate concentration (Kivirikko et al., 1989).

Prolyl 4-hydroxylase is only active on unfolded collagen polypeptide chains and the level of prolyl hydroxylation is therefore limited by the rate of helix formation. In the studied mammalian collagens, most prolines in the Yaa position are however hydroxylated (Prockop et al., 1976).

A second enzyme, prolyl 3-hydroxylase, is involved in collagen posttranslational modifications of prolyl residues (Prockop et al., 1976). This enzyme hydroxylates very few prolyl residues that are in the Xaa position of the Gly-Xaa-Yaa triplet. In type I collagen, only one residue near the C-terminal end of the main triple helix is 3-hydroxylated. The conformational effect and the role of this modification are not known. 3-Hydroxyproline residues are usually observed near the C-terminal end of triple-helical domains (Fietzek and Kühn, 1976).

B. Lysyl Hydroxylation and Glycosylation

Three other enzymes are involved in collagen-specific posttranslational modifications: (1) peptidyl lysyl hydroxylase; (2) UDPGal-collagen galactosyl transferase; (3) and UDPGlc-collagen glucosyl transferase (see Prockop et al., 1976). The products of these reactions—peptidyl hydroxylysine, galactosyl-hydroxylysine, and glucosyl-galactosyl-hydroxylysine—are not necessary for the stability of the triple helix, but probably play a role in the interactions between collagen molecules and, in particular, in the formation of the cross-links (see below). Like prolyl-4-hydroxylase, these enzymes are only active on nontriple-helical chains. The extent of lysyl hydroxylation and of further glycosylation will thus depend on the rate of helix formation and on the concentrations of enzymes and substrates.

C. Chain Selection and Assembly

Collagen molecules can be assembled in homotrimers or heterotrimers with two or three different polypeptide chains (see Table 1). In addition, the same cell can synthesize simultaneously more than one collagen type (Gay et al., 1976). The accuracy of the chain selection in the assembly of the trimers is astonishing. For example, it is known that type I collagen can form heterotrimers $[\alpha1(I)]_2\alpha2(I)$ and homotrimers $[\alpha1(I)]_3$. In most tissues, however, no homotrimer can be detected. In a pathological condition in which a mutant proα2(I) chain is synthesized which cannot assemble with the normal proα1(I) chains, it is, however, the homotrimer which is formed (Nicholls et al., 1979; Nicholls et al., 1984).

In fibrillar collagens, it is clear that the C-propeptide domains are responsible for chain association and that the formation of the disulfide-bonded C-propeptide serves as the single nucleation site of triple-helix formation (Doege and Fessler, 1986). Several spontaneous mutations in this region of the molecule, including the one mentioned above, have been shown to prevent proper chain assembly (Pihlajaniemi et al., 1984; Deak et al., 1985; Bateman et al., 1989). It is, however, quite likely that chain selection involves more than the C-propeptide. Veis and Kirk (1989) have suggested that chain selection would already start at the level of the mRNAs, and that the translation of the type I collagen mRNAs are physically and spatially coordinated. No chaperone protein has yet been demonstrated for collagen chain selection, but the current data do not exclude the existence of such a molecule. Collagen-binding proteins from the endoplasmic reticulum have, however, been described that could actually be potential chaperone proteins (Nandan et al., 1990; Clarke et al., 1991).

In many other collagen types, there is no domain equivalent to the C-propeptide. For example, it has been known for some time that type IX collagen is a heterotrimer of three different chains, $\alpha1(IX)\alpha2(IX)\alpha3(IX)$. The nontriple-helical C-terminal domains (NC1) of the three chains are very short (15–25 residues) and show little sequence similarity between the three chains, except for a short region (6 residues) which is located at the junction of the last triple-helical domain (COL1) and contains two cysteines (van der Rest and Mayne, 1987). The mechanism of chain selection is not fully known yet.

In every case, the formation of disulfide bonds to stabilize the trimer appears important. Protein disulfide isomerase, which is regarded as the catalyst of disulfide bond formation *in vivo*, probably plays a role in this process, although this activity is not necessary for the catalytic mechanism of prolyl 4-hydroxylase (Myllylä et al., 1989).

D. Formation of the Helix

The formation of the helix has been extensively studied. It has been shown to proceed from the C-terminal end of the molecule to the N-terminus by a "zipper-like" process (Bruckner and Eikenberry, 1984). The *cis-trans* isomerization of the peptide bonds involving proline or hydroxyproline as nitrogen donors has been shown to be the rate limiting step of triple helix folding. This was demonstrated initially by kinetic studies (Bächinger et al., 1978; Bruckner and Eikenberry, 1984). More recently, it has been shown that this rate is increased in the presence of the enzyme prolyl *cis-trans* isomerase (Bächinger, 1987).

The prolyl and hydroxyprolyl residues are not evenly spaced in the triple-helical domains of collagen molecules and therefore all regions of these domains do not have the same stability and do not fold at the same rate. This has been particularly studied for collagen types V and XI (Morris et al., 1990). The regions containing the cross-linking sites or the cleavage site of tissue collagenase are remarkable for their low imino acid content. This may allow the triple helix to "breathe" so that these regions can perform their special functions (Eyre et al., 1984). On the other hand, the regions near the extremities of the triple-helical domains tend to be particularly rich in imino acids, thereby locking the conformation (Fietzek and Kühn, 1976). There are some exceptions that may be of critical biological importance. For example, the C-terminal ends of the triple-helical COL2 domains of the highly homologous type XII and type XIV collagens are made respectively of four and three Gly-Xaa-Yaa repeats that do not contain any imino acid, but a conserved RGD sequence that may be involved in cell interactions (Gordon et al., 1989; Dublet and van der Rest, 1991; Yamagata et al., 1991). It is quite likely that this short region does not fold as a perfect collagenous triple helix and this may be critical for potential interactions with cell receptors.

The triple helices of most nonfibrillar collagens contain a few imperfections of the repetitive sequence; the most common ones being Gly-Xaa-Yaa-Xaa-Yaa and Gly-Xaa-Gly-Xaa-Yaa. These imperfections are actually similar if one considers the second Gly of the latter imperfection as the first Yaa residue in the first motif. Exactly how these imperfections affect the conformation of the triple helix is not known, but their high conservation between collagens from the same subfamily [e.g., between type VIII and type X collagen chains (Ninomiya et al., 1990), or between the COL1 domains of the FACIT molecules (Gordon et al., 1987; Dublet and van der Rest, 1991)], suggests that they are essential for the functions of these molecules.

Most procollagen or collagen molecules contain more than one triple helical (COL) domain. The exceptions are collagen types VI, VIII, and X. The nontriple-helical (NC) domains separating the COL domains that have been described so far as always rather short (<50 amino acyl residues). How the triple helix propagates past these imperfections is not known. It is possible that these imperfections contain new nucleation sites for triple-helix formation.

IV. INTERACTIONS BETWEEN TRIPLE-HELICAL (COL) DOMAINS

The particular conformation of the collagen triple helices makes them remarkably fitted for lateral interactions between themselves since two-thirds of their aminoacyl residues have their side chains pointing outward of the triple helix, at the surface of the protein. The interactions between collagen triple-helical domains are governed by the same general principles of molecular association that have been described for other biological systems (Kadler et al., 1988). It is entropy driven, and the energy needed for association is provided primarily by the formation of hydrophobic interactions while specificity is provided by H-bonds and ionic bonds. These assemblies are often further stabilized by covalent bond formation (see section VII).

Collagen triple helices can be compared to cylinders whose lateral surfaces are the domains of binding available for intermolecular interactions (Hofmann et al., 1980). The positions of the interactive sites on these surfaces are determined by the amino acid sequence of the three constituent chains, their relative position in the helix (A, B, and C chains in Figure 2), and the geometrical parameters of the triple helices. In principle, it is possible to establish a detailed map of these interacting surfaces if these three sets of data are known. From there, one should be able to draw conclusions about the most favorable interactions between triple helices. Such an approach has been used with some success for type I and type III collagens (Hofmann et al., 1980), and can be envisaged for the other collagen types whose amino acid sequences and chain stoichiometries have been determined. It should be remembered, however, that the triple-helical domains are not perfectly regular objects since they often contain imperfections or regions of low imino acid content and, therefore, of low triple-helical stability. Such studies are therefore only valuable if they complete other biochemical, biophysical and morphological data.

A. Staggered Aggregates

The collagen fibril with a 67-nm striation, as observed in most connective tissues, is classically viewed as being formed by a quasi crystalline lattice of staggered molecules (see Figures 4 and 5; Hodge and Petruska, 1963; Olsen, 1963). The 67-nm periodicity of the fibril (D) corresponds to about 1/4.4 the length of an individual molecule. The analysis of the primary structure of the type I and type III collagen chains has revealed internal homologies between four related units of 234 residues, particularly within $\alpha 1(I)$ and $\alpha 1(III)$ chains (Hofmann et al., 1980). This length corresponds precisely to the D period. Charged amino acids are more variable than hydrophobic amino acids. These internal homologies can thus promote a staggered lateral aggregation of individual molecules as proposed in the classical model. Recently, a model has been proposed in which the D staggering would occur between microfibrils rather than between individual molecules (Ka-

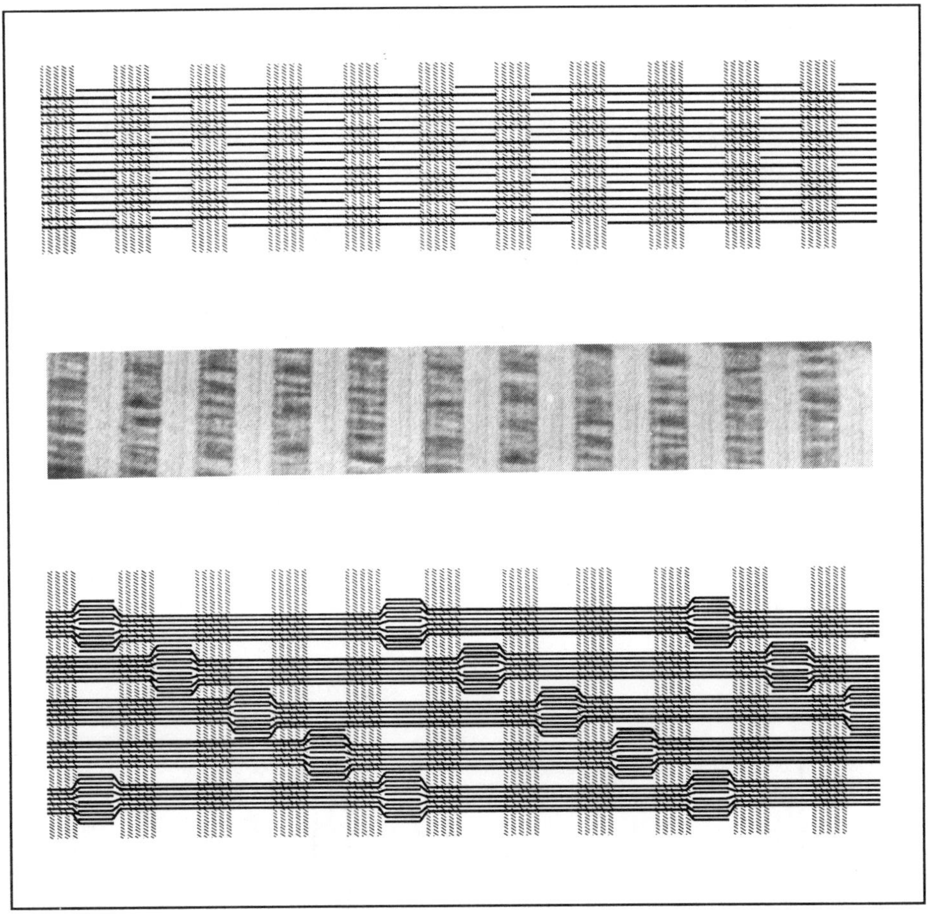

Figure 4. Schematic representations of aggregates formed by collagen molecules in striated fibrils. In the top part of the figure is the classical quarter-staggered fibril proposed by Hodge and Petruska (1962). Between the molecules of each row is a space called the gap that plays an important function in several processes such as the initial mineral deposition in mineralization. The gaps and overlaps are vertically aligned giving rise to the alternate clear (overlap) and dark (gap) bands seen in the negatively stained electron micrograph of a collagen fibril seen in the central part of the figure. The period of the fibril (D) is 67 nm. The fine transverse striated pattern is due to the regular alignment of charged residues between adjacent molecules. The lower diagram represents an alternate model that was recently proposed by Kajava (1991). In this model, a microfibril is formed by end-to-end association of nonstaggered units of six collagen molecules. Each of these units interacts with the adjacent unit by interdigitation of their extremities over a length of 0.4 D. The microfibrils themselves would be staggered in the fibril. This model can easily explain such observations as the pattern of intermolecular cross-linking described in Figure 12.

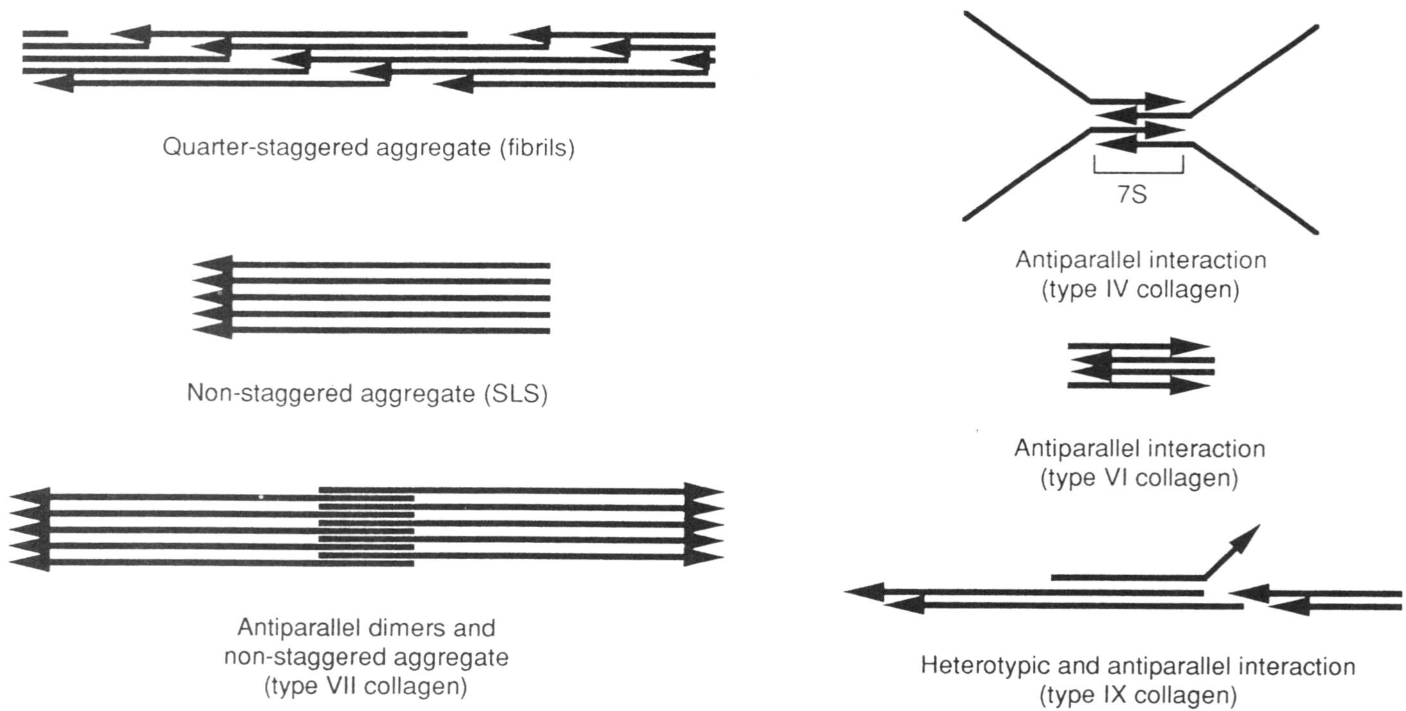

Figure 5. Various modes of interaction observed between collagen triple helices. The orientations of the triple helical domains, represented by solid bars, are indicated by the arrowheads which are pointing toward the N-terminal ends of the molecules.

java, 1991; see below). Such a model is also consistent with internal homologies corresponding to the D unit.

At least *in vitro*, all the collagen molecules from the fibrillar subgroup (types I, II, III, V, and XI) are able to form striated fibrils, although *in vivo*, all fibrils are probably heterotypic. It has also been proposed that the triple helical domains of type IV collagen molecules forming the legs of the "spider" interact laterally in a staggered fashion so that the network formed is denser than if it was formed by individual molecules (Yurchenco and Ruben, 1988; Yurchenco and Schittny, 1990).

B. Nonstaggered Molecules

Many collagen molecules are able to self-aggregate laterally in a nonstaggered parallel fashion *in vivo* and/or *in vitro*. This mode of aggregation is often referred to as 0D stagger, in reference to the D-staggering unit described above. Such a mode of aggregation has been demonstrated for a long time in the so-called segment-long spacing (SLS) aggregates of fibrillar collagens that can be obtained by precipitation of the molecules under acidic conditions in the presence of ATP (see Figure 7) (Bruns and Gross, 1973). Similar aggregates have been observed *in vivo* for the dimers of type VII collagen that form anchoring fibrils (Figure 5; see also Figure 9; Sakai et al., 1986; Morris et al., 1986).

For fibrillar collagens, the *in vivo* mode of aggregation is classically considered to occur by quarter-staggering of the individual molecules. However, loose non-staggered aggregates have been observed in secretory vesicles (Weinstock and Leblond, 1974). Recently, an interesting alternative model for striated fibrils has been proposed by Kajava (1991) based on the chemical and X-ray diffraction data from the literature. This model is based on a microfibril made of interdigitated units of six 0D-staggered molecules (see Figure 4). This model is particularly interesting since it easily explains observations such as the formation of intermolecular cross-links almost exclusively between telopeptides and lysyl residues, approximately 90 residues from the extremities of the triple helix (Eyre et al., 1984). The proposed mode of interdigitation between molecules also corresponds nicely to the initial intermolecular interaction that has been observed *in vitro* by light scattering (Silver and Trelstad, 1979).

Lateral self-aggregation between parallel and nonstaggered triple-helical domains of collagen molecules is obviously favored by hydrophobic interactions between the identically positioned hydrophobic amino acids. The ionic bonds between charged amino acids, which are classically viewed as conferring specificity in protein-protein interactions, are at first glance not favored in nonstaggered aggregates. However, a close examination of the primary structures of collagen molecules reveals a high proportion of pairs of charged amino acids, one acidic residue next or close to a basic one. For example, two-thirds of the basic residues in the sequence of the triple-helical region of the $\alpha 1(I)$ collagen chain are immediately adjacent or within three residues of a glutamate or aspartate residue (Fietzek and Kühn, 1976). This makes the formation of a large number of ionic bonds between nonstaggered molecules possible.

C. Antiparallel Aggregates

The interaction between identical collagen molecules has been shown to involve antiparallel aggregation between triple-helical domains in the 7S domain of type IV collagen (Timpl et al., 1981; Siebold et al., 1987), in type VI collagen (Engel et al., 1985) and at the C-terminal end of the P1 domain of type VII collagen (Morris et al., 1986).[2] These interactions involve two or four molecules and only part of their triple-helical domains. The resulting aggregates present an internal rotation symmetry, in contrast to the translation symmetry that can be attributed to the other homotypic collagen aggregates. In all cases, the extremity of the molecule opposite to the domain involved in the antiparallel aggregation consists of an adhesive nontriple-helical domain (see below). Thus, they are molecules that are able to interact as divalent or tetravalent linkers to form bridges (like type VII collagen), chains (like type VI collagen), or networks (like type IV collagen; see Figure 5). The interactions between the domains involved in antiparallel aggregations are all stabilized by symmetrical disulfide bonds. In addition, the 7S domain of type IV collagen has also been shown to contain sites for the formation of lysine-derived cross-links (Kühn et al., 1981).

D. Heterotypic Interactions

The aggregates discussed so far can all be described as homotypic. It should be remembered however that many structures, in particular the collagen fibrils, are actually heterotypic aggregates. Heterotypic fibrils containing collagens types I and III, types I and V, or types II, IX and XI have all been clearly demonstrated (Henkel and Glanville, 1982; Keene et al., 1987a; Birk et al., 1988; Mendler et al., 1989). It is actually very likely that in many tissues the striated fibrils contain simultaneously the collagen types I, III, and V. It is not clear however whether these heterotypic fibrils are made of homotypic subaggregates (microfibrils?) or are rather homogeneous mixtures of their constituents. The observation by Eyre et al. (1992) that most cross-links involving type XI collagen molecules are between molecules of the same type suggest the presence of some homotypic subaggregates.

Other collagen molecules, unable to form identifiable structures of their own, are found in association with other collagen types. This heterotypic mode of aggregation has been postulated for all the FACIT molecules (Shaw and Olsen, 1991), although it has only been firmly proven for type IX collagen. The characterization of covalent cross-links between type IX and type II collagens (Eyre et al., 1987; van der Rest and Mayne, 1988) and the direct visualization by rotary shadowing electron microscopy of type IX collagen molecules deposited onto cartilage collagen fibrils (Vaughan et al., 1988) have demonstrated that this collagen is involved in an heterotypic interaction. The original data did not permit orientation of type IX collagen in relation to type II collagen molecules. Very recent studies on the cross-linking of bovine type IX collagen (Eyre et al., 1992) and unpublished modelization of the interaction between type IX and type II collagen (Hofmann, H., Dublet, B., Kühn, K., and van der Rest, M., manuscript in preparation) suggest that the preferred orientation of the two molecules is antiparallel. The direct influence of copolymerization of type II collagen with type IX and XI collagens on the size of the resulting fibrils was recently demonstrated by Eikenberry et al. (1992). Only the proper mixture of these three molecules in their intact tissue form is able to result in the formation of fibrils similar to the ones observed in cartilage.

Type XII and type XIV collagens are closely, but not covalently, associated with fibrils containing type I collagen as their major constituent (Keene et al., 1991; Lunstrum et al., 1991; van der Rest et al., 1991). The homology between the COL1 domains of all FACIT molecules has been used as an argument to suggest that this domain is involved in the interaction of these molecules with the fibrils. There is at the moment no experimental proof that this is the case. The presence of subdomains homologous to von Willebrand factor A domains in the nontriple-helical NC3 domain of type XII indicates that, in these molecules, regions other that COL1 may be able to interact with collagens (Yamagata et al., 1991).

V. STRUCTURE AND FUNCTION OF NONTRIPLE-HELICAL (NC) DOMAINS

A. Propeptides of Fibril-Forming Collagens

The C-Propeptides

The C-propeptides of the fibril forming collagens have been shown to play a role as nucleation sites for the formation of the triple helix and in the process of chain selection (Prockop et al., 1979a,b). Some other functions, in particular in relation to fibrillogenesis, have also been postulated for the cleaved C-propeptides that sometimes accumulate in tissues (Fleischmajer et al., 1988), and in culture (Ruggiero et al., 1988). The accumulation of the C-propeptide of type II collagen in the mineralizing zone of growth plate cartilage led to the suggestion that it may play a role in the calcification process (Poole et al., 1984; van der Rest et al., 1986; Kujawa et al., 1989).

The structures of the C-propeptides of all the fibrillar collagen chains are closely homologous to each other (see Figure 1; Dion and Myers, 1987). They contain between 241 and 247 amino acid residues. The homology between C-propeptides extends to distant species, such as sponges and sea urchins (D'Alessio et al., 1989; Exposito and Garrone, 1990). It is characterized by the presence of conserved regions; in particular, the C-proteinase cleavage site, the cysteinyl residues, and an attachment site for a N-linked oligosaccharide.

The C-propeptides of the various fibrillar collagen α chains contain seven or eight cysteinyl residues. Six residues are involved in intrachain disulfide bonds (Dion and Myers, 1987). The presence of eight cysteines has been correlated with the ability of a given chain to participate in the formation of homotrimeric molecules, while chains participating in heterotrimers only would have C-propeptides containing only seven cysteines (Bernard et al., 1988). Between the highly conserved regions, several variable regions are noted which probably correspond to loops at the surface of the globular domain.

The same enzyme, called the C-proteinase, appears to be responsible for the cleavage of the C-propeptide of all the fibrillar collagens (Hojima et al., 1985).

The N-Propeptides

The N-propeptides of the fibrillar collagens are composed of a highly diverse N-terminal nontriple-helical domain (11 to ~ 435 residues). A triple helical domain (41–105 residues) connects the N-terminal nontriple-helical domain to a short nontriple-helical region (14–26 residues) containing the cleavage site for the N-protease in the collagens that are processed at this position. The triple-helical domain contained in the N-propeptide is interrupted in the known chains of type V and type XI collagens. The main structural features of the N-propeptides of fibrillar collagens are presented in Figure 1.

The N-propeptides are thus the most diverse domains in the fibrillar collagens, and it is therefore tempting to speculate that this structural diversity correlates to a functional diversity. The precise functions of the N-propeptides are not yet completely known. Two main functions are supported by experimental evidences. The first one is a role in the control of fibrillogenesis (for collagen types I and III; Fleischmajer et al., 1981), and the second is a feedback control on collagen synthesis (for collagen type I; Fouser et al., 1991).

The N-propeptides of type I and type II collagen are apparently cleaved by the same enzyme (Hojima et al., 1989). A distinct enzyme is, however, responsible for the cleavage of the N-propeptide of type III collagen (Halila and Peltonen, 1986).

The Propeptides and Fibril Formation

While it is easy to understand that the presence of N-propeptides still attached to collagen molecules (pN-collagen molecules) will affect fibril formation by steric hindrance, the precise effect is difficult to predict. *In vitro* fibrillogenesis of pN-collagen molecules has been shown to result in the formation of thin sheet-like structures (Holmes et al., 1991). *In vivo*, the presence of the N-propeptide of type III collagen has been demonstrated at the surface of thin fibrils, suggesting that it could act as a limiting element in fibril lateral growth (Fleischmajer et al., 1981). In disorders where the proteolytic cleavage of the N-propeptide of type I collagen is impaired, either by an enzymatic defect (dermatosparaxis; Lenaers et al., 1971) or by a mutation affecting the cleavage site on the procollagen molecule (Weil et al., 1989a,b; Vasan et al., 1991), abnormal fibrils are observed.

Obviously, the control of fibril formation is not limited to the presence or absence of N-propeptides and the processing of procollagen molecules both at the N- and C-terminal ends dramatically affects fibril formation. In addition, the heterotypic nature of the fibrils and the presence of noncollagenous fibril constituents such as the proteoglycan decorin (Pringle and Dodd, 1990; Fleischmajer et al., 1991) have to be taken into consideration.

B. NC1 Domain of Basement Membrane Collagens

The C-terminal nontriple-helical domain of type IV collagen (NC1) has been extensively studied (Weber et al., 1984; Siebold et al., 1988; Weber et al., 1988). It has been obtained in a sufficiently large quantity from human placenta to be crystallized, and its detailed structure derived from X-ray diffraction analysis can be expected soon (Stubbs et al., 1990). This domain is highly conserved between species. The mouse α1(IV) NC1 domain is 97% identical to the human one (Muthukumaran et al., 1989), and the *Drosophila* NC1 domain is 59% homologous to its mammalian counterparts (Blumberg et al., 1988). Between α1(IV) and α2(IV) NC1 domains, there is a 65% sequence homology (Hostikka and Tryggvason, 1988). The NC1 domains are all about 220 residues in length. The NC1 domain of each chain contains two subdomains with an internal homology of 35%.

The NC1 subdomains each contain a set of six cysteines. The pattern of disulfide bridge formation has been established and a model resembling a four-leaf clover has been derived (Weber et al., 1984). The main known function of the NC1 domain in matrix assembly is to provide the tail-to-tail interaction between two type IV collagen molecules (Timpl et al., 1981). The trimeric NC1 of one type IV molecule [two α1(IV) NC1 and one α2(IV) NC1] dimerizes into a highly ordered hexamer which is subsequently stabilized by intermolecular disulfide bond formation (Weber et al., 1988).

The existence of an additional basement membrane collagen chain was discovered during the characterization of the epitope of autoantibodies in Goodpasture syndrome, a very severe autoimmune disease characterized by the formation of anti-glomerular antibodies. The Goodpasture epitope was found to be in a NC1 domain homologous but distinct to those of α1(IV) and α2(IV) chains (Butkowski et al., 1990; Gunwar et al., 1991). The chain was called α3(IV) (Saus et al., 1988). Another novel NC1 domain was discovered in collagenase extracts of glomerular basement membranes and led to the characterization of the chain that was called α4(IV) (Gunwar et al., 1990). A fifth chain, α5(IV), was identified and shown to be defective in Alport's syndrome, an X-linked heritable nephritis (Hostikka et al., 1990; Myers et al., 1990). In all these three novel chains, the NC1 domains have the same overall structure as described above for the α1(IV) and α2(IV) chains. The detailed molecular organizations of these domains are not known.

C. Large Globular Domains of Type VI Collagen

Type VI collagen chains are characterized by the presence of large nontriple-helical regions at both extremities of the central triple helix. These regions are composed for the most part of repeats of domains with homologies to domains of other proteins (Bonaldo et al., 1989; Chu et al., 1989; Bonaldo and Colombatti, 1989; Chu et al., 1990b; Colombatti et al., 1990; Hayman et al., 1991). In the α1(VI) and α2(VI) chains, the N-terminal domains are 234 and 237 residues long, respectively, with one domain homologous to the type-A domain of von Willebrand factor. The C-terminal domains are 429 and 436 residues long, respectively, with two domains homologous to the same type-A domains. The sizes of the intact N-terminal and C-terminal regions of the human α3(VI) chain predicted from c-DNA sequence analysis, are 1804 and 803 residues long, respectively (Chu et al., 1990b). These regions are also characterized by the presence of about 200 residue-long repeats homologous to the A domains of von Willebrand factor. The N-terminal region contains nine such repeats. The C-terminal region contains two of these repeats, and, in addition, one domain similar to salivary proteins, one domain similar to fibronectin type III repeats and one domain similar to Kunitz type protease inhibitors. The functions of these domains remain to be elucidated.

Recently, alternative splicing has been demonstrated in the A domains of the α3(VI) chain, explaining in part the observations of multiple forms of type VI collagens (Little et al., 1989; Kielty et al., 1990; Saitta et al., 1990).

D. Three-Fingered N-Terminal Domains

Collagen types VII, XII, and XIV have large N-terminal nontriple-helical domains that have similar appearances on rotary shadowing electron micrographs (Burgeson, 1987; Dublet et al., 1989; Lunstrum et al., 1991; Aubert-Foucher et al., 1992).[2] Three fingers of sizes ranging from about 45 nm (in type VII collagen) to about 85 nm (in the long form of type XII collagen) extend from a central globule and are terminated by smaller globules. The apparent molecular weights of these domains are approximately 150 kDa/chain in type VII collagen (Burgeson, 1987), 190 kDa/chain in type XII (short form) and type XIV collagens (Dublet et al., 1989; Aubert-Foucher et al., 1992), and up to 310 kDa/chain in the longer form of type XII collagen (Yamagata et al., 1991). They all contain interchain disulfide bonds. These domains are called NC3 in collagen types XII and XIV and NC1 in type VII collagen.

The function of the NC1 domain in type VII collagen is to interact with basement membranes and anchoring plaques (Keene et al., 1987b). However, the nature and the detailed structure of this domain is not known. In contrast, the function of the NC3 domain of type XII collagen is not known but its complete primary structure has been derived from cDNA sequencing (Yamagata et al., 1991). The size of the NC3 domain of type XII collagen varies according to the tissue. In tendons of

17-day chick embryos, we have shown that its apparent molecular mass is approximately 190 kDa (Dublet et al., 1989). However, the reported cDNA sequence suggests that the NC3 domain contains 2727 amino acids. Immunoblotting performed with an antibody made to a fusion protein shows indeed that the molecular mass of the whole type XII collagen varies from 340 to 350 kDa in extracts from 4-day embryos down to about 220 kDa in tendons of 17-day chick embryos (Yamagata et al., 1991). The presence of these molecules of various sizes can only be explained by differences in the NC3 domain. We have shown that the tendon type XII collagen does not undergo proteolytic processing (Dublet et al., 1989). Partial sequencing at the protein level of the NC3 domain of the tendon molecule indicates that it contains only peptide from the C-terminal half of the published cDNA sequence (B. Dublet and M. van der Rest, unpublished data). This suggests that the primary transcript of the type XII collagen gene is subject to alternate splicing in the 5′end of the gene, or that this gene has an alternate promoter for a shorter transcript as has been described in the case of type IX collagen (see below).

The structure of the NC3 domain of type XII collagen is schematically represented on Figure 6. It should be viewed as a matrix glycoprotein and, like many such glycoproteins, it has a mosaic structure (Yamagata et al., 1991). The long form of the molecule contains 18 units homologous to fibronectin type III repeats that are interspersed by 4 units homologous to domain A motifs of von Willebrand factor. In addition, it contains, immediately upstream of the triple helical COL2 domain, a region that presents structural similarities to the N-terminal globular domain of other collagens (types IX and XI; Gordon et al., 1989; Yamagata et al., 1991).

While the full structure of the NC3 domain of type XIV collagen is not known yet, the current data demonstrate a very close homology to the organization of the corresponding domain in type XII collagen (Aubert-Foucher et al., 1992). Comparison of primary structure data obtained at the protein level on the NC3 domain of the chicken-type XIV collagen with those reported for the glycoprotein called "undulin" (Schuppan et al., 1990; Just et al., 1991) suggests that these two proteins may be products of the same gene (B. Dublet and M. van der Rest, unpublished data).

E. The NC Domains of Type IX Collagen

The α1(IX) chain of the cartilage-type IX collagen differs from the two other chains of this molecule by the presence of a large N-terminal globular domain (243 residues; van der Rest and Mayne, 1987; Vasios et al., 1988). In the other two chains, the NC4 domain is limited to a few amino acids only. The pI for this domain has been estimated from its amino acid composition to be 9.7 in the chicken (Vasios et al., 1988), and 10.55 in the human (Kimura et al., 1989). It is thought to interact with the acidic glycosaminoglycans of the cartilage matrix (see below).

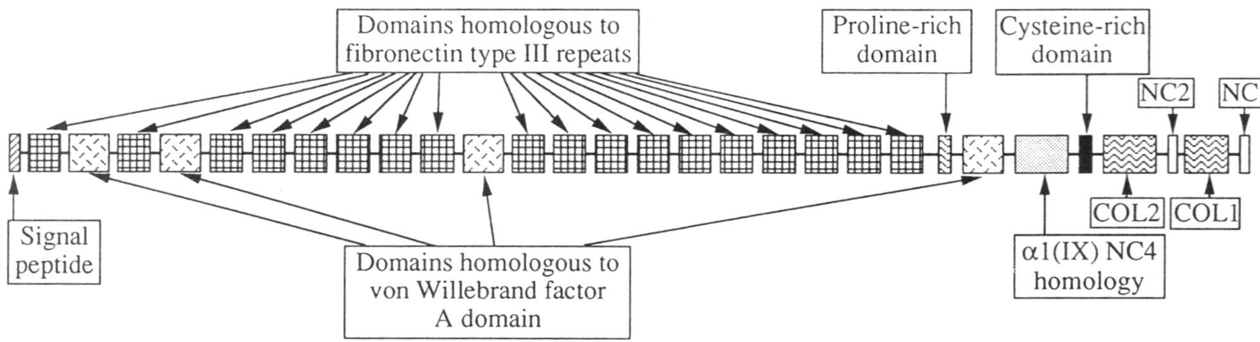

Figure 6. Domain structure of the long form of chicken type XII collagen. The triple-helical domains (COLI and COL2) represent only a small proportion (>10%) of the molecule. The N-terminal non-triple helical region, reported as NC3, is actually composed of 25 distinct domains, most of them having homologies with other known extracellular matrix proteins, as indicated in the figure.

Northern blot analysis of the α1(IX) mRNAs of the primary stroma of the cornea of the chick embryo has shown that the messages for this chain are shorter in this tissue than in cartilage (Svoboda et al., 1988). This shorter size is due to the usage of an alternate promoter which is located between exons 6 and 7 of the gene used for the cartilage transcript (Nishimura et al., 1989). The predicted translation product is almost completely missing the NC4 domain of the cartilage chain, retaining only its last amino acyl residue. An alternative first exon encodes for a different signal peptide. The remaining part of the molecule, including all the triple-helical domains, would be identical between the two gene products. The resulting short form of type IX collagen would thus be made of three highly similar chains and would only contain very short nontriple-helical domains.

The NC3 domain of the α2(IX) chain is 19 residue long and is characterized by the insertion of five amino acid residues as compared to the homologous domain in the other two chains (McCormick et al., 1987). These five residues (-Val-Glu-Gly-Ser-Ala-) define an atypical glycosaminoglycan attachment site since the "classical" attachment site is a -Ser-Gly-sequence. In cartilage, the glycosaminoglycan residue that is attached is about 40 kDa (Vaughan et al., 1985). In the avian vitreous humor, the chondroitin sulfate side chain is much larger (~300 kDa) and accounts for the largest part of the glycosaminoglycan content of this tissue (Yada et al., 1990; Brewton et al., 1991). Interestingly, the type IX collagen of this tissue appears to be the short form with no NC4 domain visible on rotary-shadowing electron micrographs.

F. The Globular Domains of Short-Chain (types VIII and X) Collagens

The primary structures of the α1(X), α1(VIII) and α2(VIII) were recently determined by cDNA and genomic DNA sequence analysis (Ninomiya et al., 1986; Yamaguchi et al., 1989; Thomas et al., 1990; Muragaki et al., 1991a,b). These three chains are very homologous. They consist of a central triple-helical domain (460, 454, and 457 residues respectively) bordered by two nontriple-helical domains. The C-terminal NC1 domains are 162, 173, and 167 residues respectively. This domain can be divided into two regions: (1) the N-terminal region (25% of the domain) has a sequence which shows no similarity between the three chains (<8%); (2) the C-terminal region (75% of this domain) is highly conserved, including one cysteine [61% sequence identity at the amino acid level between α1(VIII) and α1(X); 63% identity between α1(VIII) and α2(VIII)]. The N-terminal NC2 domains are not conserved between α1(VIII) and α1(X) (117 and 52 residues respectively). Only the last 20 residues of the NC2 domain of α2(VIII) have been determined and they show no clear similarity with the NC2 domains of the other two chains.

Although no function has been demonstrated for these domains, the participation of short-chain collagens in hexagonal lattices (Sawada et al., 1990; Kwan et al., 1991; see Figure 9), suggests that their NC domains may play a role as interactive elements in these structures.

VI. SUPRAMOLECULAR AGGREGATES MADE BY COLLAGENS

A striking property of collagen is its restrictive molecular "social" behavior. All collagen molecules are involved in supramolecular assemblies with other collagen molecules of either the same or of a different genetic type. These homogeneous interactions lead to the formation of large aggregates, *in vivo* as well as *in vitro*, which are usually easily recognizable by electron microscopy. In most cases, the number of molecules involved, the length and the regularity of their triple-helical domains result in the formation of highly ordered structures. Fibrils are formed by fibrillar collagens with a characteristic transverse banding pattern of 67-nm periodicity. Other assemblies are formed that involve fewer monomers, such as the long, beaded filaments (collagen type VI) and the short, anchoring fibers (collagen type VII). Other collagen molecules assemble differently, in sheets rather than in fibrils. They can then form structures appearing as hexagonal lattices (collagen types VIII and X), or as more complex meshworks in basement membranes (collagen type IV).

A. Fibrils and Filaments

Collagen-rich tissues, such as tendons, ligaments or dermis, are composed of bundles of filaments which were very early recognized as the main source of collagen. It has been known for more than a century that these filaments can be dissolved in acetic acid and reconstituted again when the acid is neutralized. It became clear that this property was the result of the polymeric nature of the collagenous filaments. In fact, as explained above, the interactive properties of the triple helix make each collagen type able to polymerize. Even fragments of the FACIT collagen type IX were shown to be able to aggregate *in vitro* in register, forming crystallites (Ayad et al., 1981; Shimokomaki et al., 1981; Ayad et al., 1982) and fibrils (Tiollier and Garrone, 1985).

Fibrillar Assemblies

Banded fibrils with a periodic 67-nm pattern are considered as the prototype of collagen assembly. Early electron microscope examination of mechanically separated, metal-shadowed collagen fibrils revealed a periodic succession of annelate structures, sometimes leading to the erroneous conclusion that they were made up of a series of discs or rings [for a review of these early observations, see Gross (1950)]. It is now well established that the use of negative staining (heavy metal salts used at a nonreactive pH) on native or reconstituted fibrils reveals a filamentous substructure and a period composed of alternatively clear and dark bands (Torri Tarelli and Petruccioli, 1971; Figure 4). The clear bands have retained few metallic compounds and are thus electron translucent. They correspond to tightly aggregated

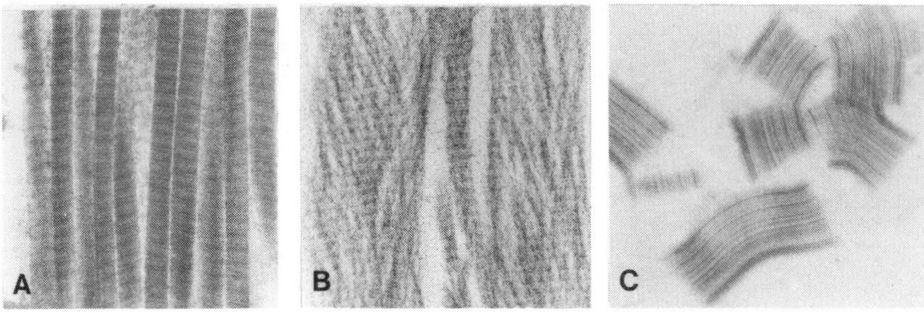

Figure 7. Transmission electron microscopy of aggregates of type I collagen (× 30,000, positive staining). **A.** Staggered aggregation forming the cross-banded fibrils with a 67 nm periodicity (rat skin). **B.** Collagen fibrils from rat tendons dehydrated with ethylene glycol and embedded in glycol methacrylate. The swelling of the fibrils reveals the filamentous units (courtesy of S. Franc). **C.** Ordered aggregates (SLS) obtained *in vitro* by dialysis of a purified bovine skin type I collagen solution against ATP. The length of the segments corresponds to the length of the collagen molecules. The width is variable.

domains of the fibrils (overlap region). The dark bands which contain more stain are then electron-dense and correspond to porous domains, and hence to loosely packed units (gap region). The 67-nm period comprises one clear plus one dark band. It is worth noting that noncollagenous components can accumulate at the level of gap regions and subsequently lower the contrast along the fibril axis (Eikenberry et al., 1984). Changing the pH of the staining solution, thereby modifying its reactivity, reveals a 67-nm period composed by a series of thin transverse bands (Figure 7). The pattern can change according to the nature of the solution used, but the period remains of 67 nm. Several detailed studies on native or reconstituted fibrils have clarified these aspects of collagen fibrils (Bairati et al., 1969), in particular by correlating the staining pattern with sequence data (Chapman, 1974; Meek et al., 1979; Chapman and Hulmes, 1984).

In addition to these general studies, several points have to be stressed:

1. The staining pattern is independent of the collagen type, at least for the three main fibrillar collagen types, I, II, and III. Reconstituted fibrils of type I and type III collagens are identical (Wiedemann et al., 1975).
2. Amazing variations can be observed between reconstituted collagen fibrils depending on the experimental conditions; for example, by using molecules with the telopeptides cleaved, or by changing the precipitation conditions, even with intact molecules [e.g., an oblique banding pattern has been observed in type I collagen fibrils reconstituted *in vitro* after trypsin treatment (Ghosh and Mitra, 1975); a staggered substructure has been obtained

for reconstituted fibrils obtained with cartilage collagen from lathyritic chicken and for reconstituted rat type I collagen (Bruns et al., 1973; Bruns, 1976); symmetrically banded fibrils (Fibrous Long Spacing or FLS) can be obtained when the collagen solution precipitates in the presence of glycosaminoglycans (Chapman and Armitage, 1972; Chapman and Hulmes, 1984)].

3. The chemical fixation used for processing samples in electron microscopy acts as a cross-linking step. In all cases, it modifies the staining pattern (Grant et al., 1967; Bairati et al., 1972; Meek and Chapman, 1985).
4. The axial periodicity is poorly visible on thin fibrils. Using small-angle X-ray diffraction, Veis et al. (1979) have determined that the minimum structure demonstrating axial periodicity is about 4 nm. However, an electron microscope demonstration of an axial periodicity requires fibrils of more than 10 nm.
5. Thin fibrils, possibly with a faintly visible periodicity, can assemble into thick fibrils displaying an obvious banding pattern. This has been observed in preparations of mechanically dispersed collagen fibrils; for example, from the corneal stroma (Meek and Holmes, 1983).
6. The surface of collagen fibrils can interact with other macromolecules such as fibronectin, proteoglycans, and the FACIT collagens. These latter two components participate in particularly well organized interactions. Proteoglycans have been revealed by cytochemical methods (Scott, 1988) forming projecting threads, regularly associated with the fibrils. More precisely, decorin, a proteodermatan sulfate, has been localized by immunoelectron microscopy regularly associated with definite transverse bands (Pringle and Dodd, 1990). Type IX collagen appears as thin threads with a globular end pointing out of type II collagen fibrils at regular intervals (Vaughan et al., 1988).

There is a discrepancy between the aspect of collagen fibrils after negative staining, where the filamentous substructure is obvious (Figure 4), and their appearance after positive staining, emphasizing the cross-striation rather than the longitudinal substriation (Figure 7). Special dehydration methods are able to reveal the constitutive filaments of native or reconstituted fibrils, even after positive staining (Bouteille and Pease, 1971; see Figure 7). Moreover, cytochemical methods suggested a carbohydrate-rich matrix inside native fibrils and clearly revealed an helical arrangement of the subfibrils (Pease and Bouteille, 1971). Attempts have been made to correlate this subfibrillar arrangement to the collagen type using freeze-fracturing of native collagen in tissues (Reale et al., 1981).

Collagen fibrils from primitive animals such as sponges (Porifera), sea-anemones, and jellyfish (Cnidaria) are thin (20 to 22 nm) and contain a simplified banding pattern. Most often the 67-nm period is bordered by sharp bands, sometimes forming protruding rings rather than striations, and contains two internal bands of lower intensity, separating the period into three similar segments (Garrone, 1984; Franc et al., 1985). In contrast, collagen fibers forming the cuticle of certain worms (Annelids) are coarse structures (up to 200 nm in diameter) but their surfaces appear absolutely smooth. Special cytochemical treatment can only reveal a transverse striation which is not of the conventional 67-nm type and which probably corresponds to visualization of polysaccharides associated with the fibers (Djaczenko and Calenda Cimmino, 1973). These fibers are composed of very long (up to 2400 nm) pepsin-resistant collagen monomers (Murray and Tanzer, 1983; Gaill et al., 1991b).

The ends of the fibrils are rarely seen when collagenous tissues are observed in the electron microscope. In certain cases, accumulations of collagen fibrils which taper to a sharp needle-like form have been described (Garrone, 1975; Connolly, 1981). This is particularly significant considering the demonstration that growing collagen fibrils have a pointed tip (Kadler et al., 1990).

Cross sections of collagen fibrils appear roughly circular in normal tissues. In pathological conditions, many variations can occur, the best known being the flower-shaped sections of dermatosparaxic collagen fibrils (O'Hara et al., 1970). However, a good preservation of fibrils during processing for electron microscopy revealed a quasi-hexagonal packing of molecules when electron micrographs of cross sections were submitted to optical diffraction (Hulmes et al., 1981). Real hexagonal cross sections of collagen fibrils were obtained with a quick-freezing (by liquid helium) and freeze-substitution procedure of tissues (Mallein-Gerin and Garrone, 1986). This method preserves blocks of internal crystalline domains and polygonal to hexagonal cross sections of the fibrils.

Once formed, collagen fibrils assemble differently according to tissues. A remarkable organization is the cross-ply arrangement found in the submucosa of the small intestine (Gabella, 1987) and in a large variety of tissues and organs. Analogies between *in vitro* self-assembled and *in vivo* cell-assembled collagen fibrils have been pointed out (Gaill et al., 1991a).

Filamentous Assemblies

Type VI collagen molecules assemble in an antiparallel fashion and form beaded filaments highly characteristic when viewed in the electron microscope after rotary shadowing (Figure 8). Special extraction procedures, under nonreducing and nondenaturing conditions, can be used to isolate intact filamentous assemblies of type VI collagen (Kielty et al., 1991). They appear to be strikingly similar to the reconstructed *in vitro* filaments (Furthmayr et al., 1983). In several tissues including elastic tissues and vessel walls, type VI collagen forms smooth, thin filaments, only distinguishable after labeling with specific antibodies (von der Mark et al., 1984). However, it has been suggested that long-spacing, cross-striated banded structures (Sun and White, 1975) occurring frequently in connective tissues (especially in pathological conditions) are actually filamentous aggregates of type VI collagen

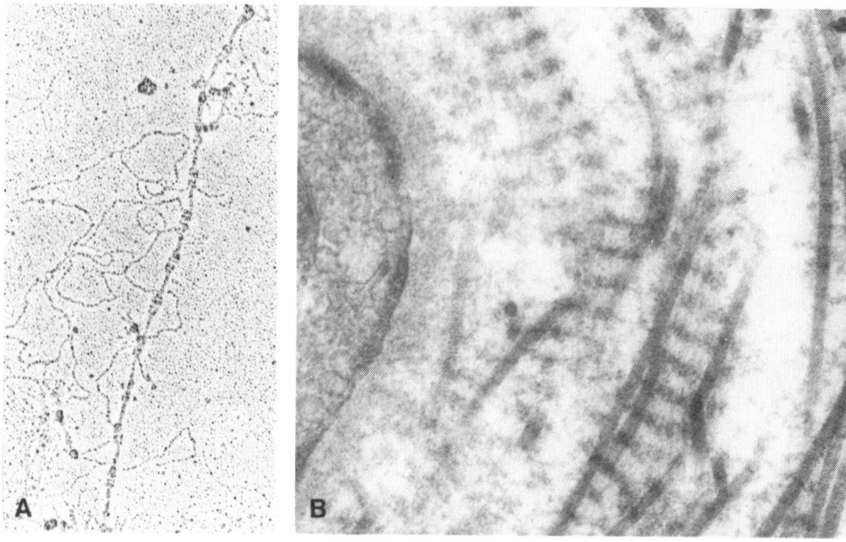

Figure 8. Transmission electron microscopy of type VI collagen filamentous assemblies (x 60,000). **A.** Low angle Pt/C rotary shadowed replica of unit filaments obtained from mechanical dispersion of bovine articular cartilage. The filaments in the background could be hyaluronic acid chains or DNA (courtesy from F. Ruggiero). **B.** Tissue aggregates of type VI collagen forming "Zebra collagen" structures in human gingiva.

(Bruns, 1984; Engel et al., 1985). Such fibrous forms, sometimes called "zebra collagen", have dense cross bands measuring about 30 to 40 nm in thickness, with a repeating periodicity of about 100 to 150 nm. The longitudinal filaments are about 8 nm in diameter and irregularly parallel. The width of the whole structure can reach several thousand nm. A specific interaction of type VI collagen with hyaluronic acid (McDevitt et al., 1991) has been recently demonstrated.

Segment Assemblies

Dialysis of a collagen solution against ATP led to precipitation of collagen molecules in register, forming bundles of a length equal to that of the molecule (Figure 7). These Segment Long Spacing (SLS) have been extensively characterized for the fibrillar collagen types (Kühn, 1982). Slightly changing the precipitation conditions resulted in the formation of narrow aggregates, or of extended ribbons, or even of fibrils made by the partial overlapping of SLS. SLS are commonly used to obtain an accurate measurement of the molecular length. They can be obtained also after digestion of the nonhelical telopeptides (by pepsin) or after collagenase splitting of the molecule. They have been largely considered as *in vitro* artefacts. However, as mentioned earlier, *in vivo* fibrillogenesis involving intracellular SLS-like aggregates and extracellular fibril assembly by bundles of nonstaggered molecules has been proposed (Trelstad and Hayashi, 1979; Kajava, 1991).

Type VII collagen constitutes the only *in vivo* example of normal, permanent segment assembly of collagen molecules. Due to the antiparallel overlapping of the C-terminal extremities of two molecules, the segments, which are in fact short fibrils, are centrosymetrical (Morris et al., 1986). This organization has been clearly demonstrated using the SLS technology (Bentz et al., 1983; Burgeson et al., 1985) which reveals monomer aggregates of 425 nm in length and overlapping dimer aggregates of 785 nm in length. These symmetrical segments, with fan-shaped ends, were known as anchoring fibrils, which are fibrous structures found between the dense layer of the basement membranes and the underlying stroma in various tissues (Sakai et al., 1986; Figure 9). Such structures were known long before the discovery of type VII collagen, and they were first described in the mid-1960s (Palade and Farquhar, 1965; Rowlatt, 1969; Susi, 1969; Briggaman et al., 1971). It now appears clearly that they form loops entrapping collagen fibrils, with their two ends anchored within the lamina densa of the basement membrane, or with one end attached to anchoring plaques formed by a condensation of material containing type IV collagen (Burgeson, 1987).

B. Sheets

In vertebrates, sheet organization of collagen molecules is found in all basement membranes for type IV collagen, in a special basement membrane, the Descemet's membrane for types IV and VIII collagen, and in hypertrophic cartilage, around cells for type X collagen (Figure 10). Another type of sheet organization can also be made by collagenous products often secreted by glandular cells or by epithelial layers. In vertebrates, they are only represented by the egg case of Selacians, but in lower animals they form cuticles, skeletal devices, and adhesive systems. Due to their mode of elaboration they have been called "secreted collagens" [for a review, see Bairati and Garrone (1985)]. In fact, they can be considered as collagens with an extracorporeal location, contrary to all the other known collagens which are intercellular products. In this respect, they can also be termed "outer collagens".

Basement Membranes

Basement membranes separate tissues such as epithelia and their glandular derivatives, endothelia, muscles of various types, and nerve and glial cells from connective tissues. Viewed in conventional transmission electron microscopy, a basement membrane (boundary membrane, basal lamina) appears as a network of fluffy linear elements forming a plane of more or less uniform thickness. However, this thickness can vary greatly (from 15 to 150 nm or even more) according to tissues and species. This layer is usually referred to as the lamina densa and it is

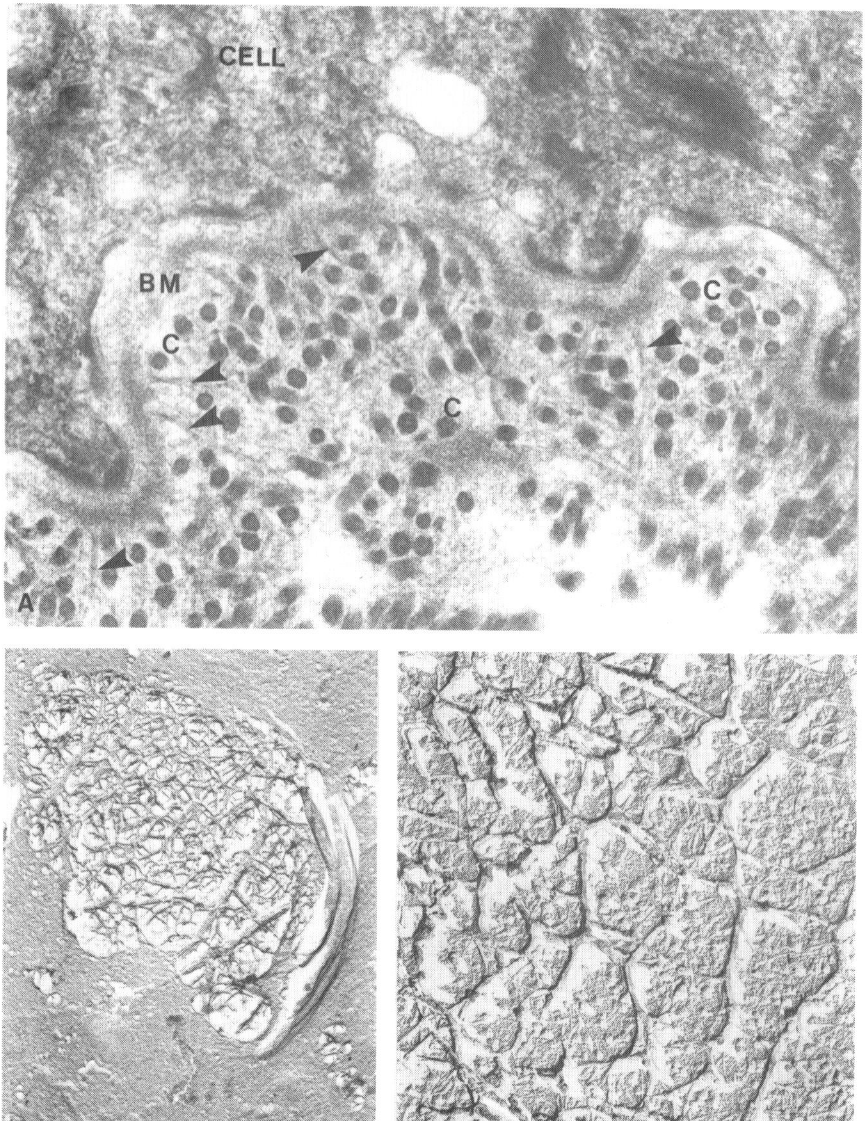

Figure 9. Transmission electron microscopy of anchoring fibrils and basement membranes. **A**. Conventional cross-section of the junctional area between the epithelial cells and the underlying connective tissue in human gingiva. The basement membrane (BM) underlines the basal side of the epithelial cells (CELL) and is connected to collagen fibrils (C), mostly seen in cross-sections, by anchoring fibrils (arrowheads) (× 65,000). **B** and **C**. Pt/C unidirectional replicas of basement membranes from chick embryo cornea after quick freezing and deep etching (courtesy of A. Barge). These "en face" views correspond to the lamina lucida side of the basement membrane. **B**. Partial extraction of the tissue by guanidinium chloride revealing the basement membrane network (× 45,000); **C**. Detailed view of the basement membrane network (× 67,500).

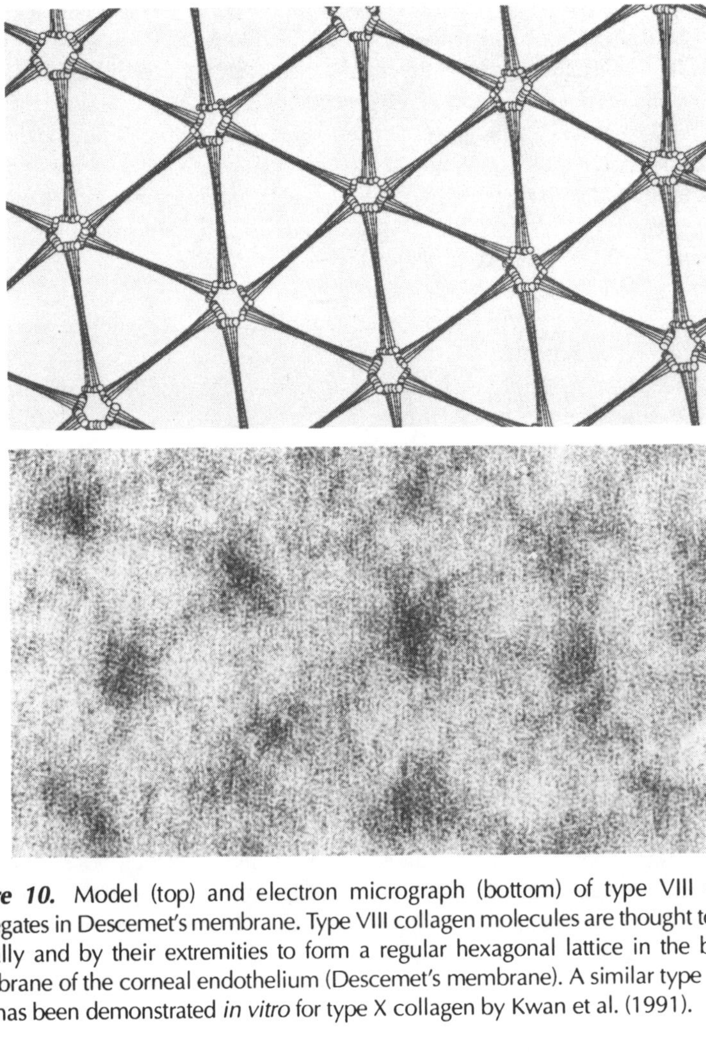

Figure 10. Model (top) and electron micrograph (bottom) of type VIII collagen aggregates in Descemet's membrane. Type VIII collagen molecules are thought to interact laterally and by their extremities to form a regular hexagonal lattice in the basement membrane of the corneal endothelium (Descemet's membrane). A similar type of aggregate has been demonstrated *in vitro* for type X collagen by Kwan et al. (1991).

one of the three layers composing the basement membranes. The lamina densa is bordered by the lamina lucida which lines the cell plasma membrane, and by the pars fibroreticularis constituting the transition zone between the lamina densa and the rest of the extracellular matrix (Inoue, 1989). The anchoring fibrils made of type VII collagen are components of this pars fibroreticularis. High-resolution ultrastructural studies of cross sectioned basement membranes reveal that the lamina densa is formed in fact by strands or "cords". These cords contain collagen type IV, laminin, nidogen, heparan sulfate proteoglycan, fibronectin, BM-40, and other glycoproteins, as demonstrated by immunostaining (Laurie et al., 1982; Schittny et al., 1988; see also the review of Leblond and Inoue, 1989). This composition is consistent with the biochemical characterization of basement membranes (see the reviews of Paulsson, 1987, and of Timpl, 1989). However, the best demonstration of type IV collagen organization in basement membranes came from ultrastructural studies of Yurchenco and Ruben (1987). These authors examined human amniotic basement membrane *en face* using stereoscopic high-resolution electron microscopy after freeze-drying and with the unidirectional platinum replication technique. Their observations were completed by antibody decoration and morphometry. They clearly demonstrated that the backbone of basement membrane is formed by type IV collagen organized into a crystalline-like lattice using lateral and end-domain interactions. This kind of organization was observed in the matrix of the Engelbreth-Holm-Swarm tumor of the mouse (Yurchenco and Ruben, 1988) as well as in *in vitro* networks (Yurchenco et al., 1986), and it confirms the originally postulated model (Timpl et al., 1981). Using quick-freezing by liquid helium and deep-etching, this fundamental network organization was observed in basement membranes deposited in culture and during development (Barge et al., 1991).

Hexagonal Lattices

The basement membrane (Descemet's membrane) deposited by the corneal endothelium contains, in certain species, stacks of hexagonal lattices extending in the plane of the membrane (Jakus, 1956). These lattices are composed of globular nodes and rod-like structures connecting the nodes. A collagen, forming long spacing fibrils, is present in this hexagonal lattice (Sawada et al., 1990). As type VIII collagen is a major component of Descemet's membrane, it is considered as a good candidate to be the main structural element of this lattice (Kapoor et al., 1988). Strong evidence has been provided by Sawada et al. (1990) who showed the decoration of the hexagonal lattice by monoclonal antibodies directed against type VIII collagen extracted from Descemet's membrane.

A regular hexagonal array, analogous to the hexagonal lattice of the Descemet's membrane, has been obtained recently *in vitro* using type X collagen (Kwan et al., 1991). Individual chicken type X collagen molecules incubated at 34 °C under neutral conditions form multimeric clusters via their carboxyterminal globular domain. The result is a large, central nodule with filaments (the triple helices) radiating outwards. The aggregation process continues with the lateral association of juxtaposed triple helices from adjacent clusters.

This kind of hexagonal arrangement of collagen molecules forming "geodesic domes" maintaining or reinforcing a curved architecture should have a biological significance for the shape of the cornea (for type VIII collagen), and for the environment of hypertrophic chondrocytes (for type X collagen). Finally, it is not surprising that the high degree of similarity in the primary structure of these two collagen types (see above) correlates with a remarkably close structural organization.

Outer Collagens

Most of these collagens were reviewed in detail by Bairati and Garrone (1985). The eggs of sharks are contained in a collagenous envelope which has been studied in detail in the dogfish (Knight and Hunt, 1976). It forms a structure in which the axial period is 37 nm. The electron microscope reveals crystalline regions with a square lattice of 11-nm sides. The proposed structural model implies that molecules are arranged in planes diagonally bisecting the square prisms (Galloway, 1986). This collagen is secreted by a special glandular apparatus. It is also a glandular system which releases large granules fusing to form the collagenous filaments of the byssal threads in mussels. The hypodermis (peripheral cell layer) of the nematode worms secretes a proteinaceous sheet containing several collagens clearly belonging to the nonfibrillar group (Cox, 1990). Generally the cuticle appears as an amorphous material, often multilaminated. In certain species, areas of striated material of about 20-nm periodicity are visible (Dick and Wright, 1973).

Cnidarians and sponges also contain large amount of a sheet-deposited collagenous material, appearing as tiny filaments (of about 10 nm in diameter) packed either with no regular organization, or forming striated arrangements and hexagonal lattices. In sponges, some of these collagens have been characterized at the cDNA and genomic level. They correspond to nonfibrillar, short-chain collagens, containing two short (66 and 171 amino acyl residues) triple-helical domains (Exposito et al., 1991).

VII. STABILIZATION OF AGGREGATES BY COVALENT CROSS-LINKING

Non-covalent interactions between charged and hydrophobic side chains of amino acids are responsible for the initial assembly of staggered collagen molecule into D-periodic fibrils. The further stabilization of collagen fibrils (particularly their tensile strength) depend largely upon the formation of intermolecular covalent cross-links.

Current data demonstrate at least three different pathways by which collagen molecules can be cross-linked: (1) lysyl oxydase-dependent, lysine-derived cross-links; (2) lysine-derived cross-links formed through nonenzymatic glycation; and (3) intermolecular disulfide bonds.

A. Lysyl Oxidase-Dependent, Lysine-Derived Cross-Links

At least three events leading to the stabilization of collagen fibrils in tissues can be defined during this cross-linking pathway: (1) the enzymatic oxidation of some specific lysine and/or hydroxylysine residues into the corresponding aldehydes; (2)

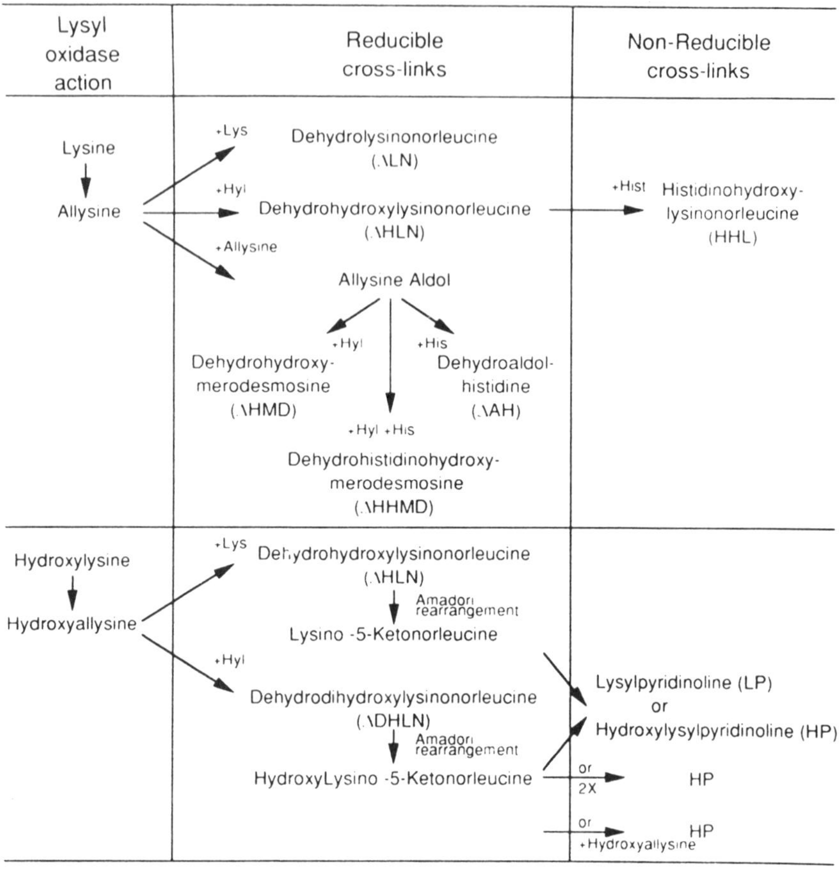

Figure 11. Allysine- and hydroxyallysine-based cross-links. Schematic representation of the two routes of lysyl oxidase-derived cross-linking in collagen. (The prefix Δ stands for dehydro).

the spontaneous formation of several different intra- and intermolecular reducible cross-links; and (3) the progressive replacement with aging of some reducible cross-links by stable nonreducible intermolecular compounds (Figure 11). [See comprehensive reviews by Light and Bailey (1982), Robins (1982), Eyre et al. (1984), Herbage et al. (1985), Yamauchi and Mechanic (1987), Ricard-Blum and Ville (1988), Robins (1988), Ricard-Blum and Ville (1989), Last et al. (1990)].

The Action of Lysyl Oxidase

The enzymatic cross-linking pathway is initiated by the oxidative deamination of specific lysine and hydroxylysine residues into the corresponding aldehydes at the nontriple helical ends (telopeptides) of collagen α chains. This posttranslational modification is catalyzed by lysyl oxidase (EC 1.4.3.13.). Most of the information available on the reaction pathway of lysyl oxidase was obtained by Kagan and his associates (Kagan, 1986). Lysyl oxidase has been purified from several sources in different forms. Recently it was demonstrated that rat aorta lysyl oxidase is translated from a 2.7-kb mRNA (of which 1.16 kb represent noncoding sequences) as an initial product of 46–48 kDa (Trackman et al., 1990). A 40-kDa precursor has been isolated from rat skin. It probably undergoes a proteolytic processing to give the 32-kDa protein usually isolated from elastin-rich or cartilaginous tissues (Romero-Chapman et al., 1991).

This peptidyl amine oxidase contains both a copper ion and an aromatic carbonyl as cofactors (Kagan, 1986). The organic cofactor is either pyrroloquinoline quinone (PQQ) or 6-hydroxydopa (Williamson et al., 1986). This latter cofactor is apparently derived from a tyrosine residue from the enzyme, as recently reported in bovine serum amine oxidase (Janes et al., 1990). Steady-state kinetic analyses and evidence for aldehyde production in the absence of oxygen have shown that lysyl oxidase follows a Ping Pong Bi Ter kinetic course (Williamson and Kagan, 1987). The oxidation of the amine to the corresponding aldehyde occurs in the first of two half-reactions. The enzyme is reoxidized by molecular oxygen in the second half-reaction.

Removal or chelation of copper from a preparation of lysyl oxidase results in the loss of enzyme activity and copper deficiencies reduce cross-linkage formation *in vivo*. Such evidence points to an essential role for copper in this enzyme, but very little information is available about the exact nature of its action. Recently, Gacheru et al. (1990) showed that 5 to 9 atoms of loosely bound copper and one atom of tightly bound copper are associated with the 32-kDa monomer of purified lysyl oxidase. The inability of the metal-free enzyme to catalyze the first half-reaction of amine oxidation suggests a role for copper in this catalytic phase. This does not fit however with the Ping Pong kinetic mechanism. Further studies are thus necessary to determine the exact role of the copper in the copper-dependent amine oxidases.

The native-staggered collagen fibril represents the optimal substrate form of collagen for lysyl oxidase. Collagen monomers appear not to be effective substrates (Siegel, 1974). Two binding sites for the enzyme have been proposed in highly cationic regions of collagen triple helices. They correspond to sequences conserved in the (α chains of type I, II, III, V, and XI collagens (Table 2) and, in 4D staggered molecules, are located opposite to the oxidizable lysines (9 N and 16 C) of the N- and C-terminal telopeptides. This hypothesis was supported by an *in vitro* demonstration of the binding of lysyl oxidase to the triple-helical portion of the collagen molecule in a fibrillar array (Cronlund et al., 1985) and with the immunolocalization of lysyl oxidase on collagen fibers in human placenta, skin, and aorta (Baccarani-Contri et al., 1989).

Table 2. Lysine and Histidine Residues Important in the Formation of Intermolecular Cross-Links in Fibrillar Collagens

	Lysine in Telopeptides (Greenspan et al. 1991)		*Position and Composition of the Sequences Containing Lys (K) and His (H) in the Triple Helical Domain*	
Collagen Chain	*N-Terminal*	*C-Terminal*	*N-Terminal*	*C-Terminal*
α_1(I)	+	+	87(KGHR)	930 (KGHR)
α_2(I)	+	–	87(KGIRGH)	933(KGHN)
α_1(II)	+	+	87(KGHR)	930(KGHR)
α(III)	+	+	96(KGHR)	939(KGHR)
α_2(V)	–	–	84(KGHR)	924(KGHP)
α_2(V)	+	–	87(KGHR)	933(KGHR)
α_1(XI)	–	+	84(KGHR)	924(KGHP)
α_2(XI)	ND	+	ND	927(KGHP)

Note: ND: not determined

Reducible Cross-Links

The crystal-like arrangement of collagen molecules in a fibril favors the spontaneous reaction of lysine-derived aldehydes with free amino groups on adjacent molecules. This results in the production of intermolecular Schiff-base cross-links. An Amadori rearrangement can occur for all cross-links derived from hydroxyallysine with the formation of chemically stable ketoamines. Intramolecular cross-links are also possible between two α chains within the same molecule by the formation of aldol cross-links (Figure 11).

The structure of these reducible cross-links has been extensively studied after stabilization and labeling by tritiated borohydride, particularly by the groups of Bailey and Robins in Langford and of the late Gerry Mechanic in Chapel Hill (Light and Bailey, 1982; Robins, 1982; Yamauchi and Mechanic, 1987; Robins, 1988).

Since these cross-links decrease in number during aging, it has been postulated that they are the chemical precursors of more complex nonreducible mature cross-links that would be responsible for an increase in the stability of the fibers.

Nonreducible Mature Cross-Links

Pyridinoline. A fluorescent 3-hydroxypyridinium compound was first identified by Fujimoto et al. (1977) in Achilles tendon. It is usually referred to as hydroxylysylpyridinoline (HP) (Figure 12). It has been shown to be the major cross-link in mature cartilage (Eyre and Oguchi, 1980; Robins, 1983). An analogue lacking the 3-hydroxyl group has been isolated by Barber et al. (1982) and called

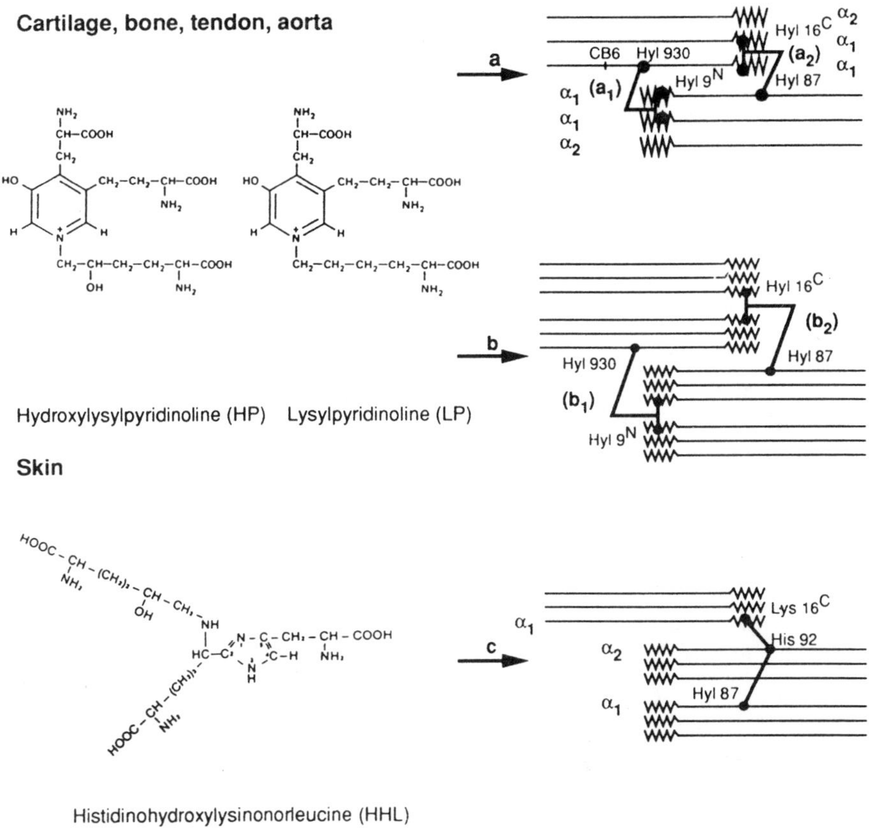

Figure 12. Structure of the non-reducible cross-links and their molecular locations. These cross-links have been determined (**a**) by Robins and Ducan (1987); (**b**) by Eyre et al. (1984); and (**c**) by Mechanic et al. (1987).

3-deoxypyridinoline or lysylpyridinoline (LP). The HP+LP content in a tissue depends on the extent of hydroxylation of the telopeptide lysine residues. The reported values (moles/mole of collagen) are 1.3 to 2.6 in cartilage and nucleus pulposus, and 0.3 to 0.7 in bone, dentin, tendon, ligament, and aorta. These cross-links are absent in skin and cornea (Eyre et al., 1984).

The content of these cross-links in human cartilage and bone increases during skeletal growth, reaching a maximum in the 15–25-year age group. Thereafter, it remains stable throughout life (Eyre et al., 1988). Similar values have been observed in human and bovine mature cartilage (Eyre et al., 1984). This particular cross-linking cannot, therefore, explain the difference in collagen solubility observed between these two tissues. After pepsin digestion and proteoglycan extraction, the human collagen remains insoluble while the bovine collagen is easily solubilized (Herbage et al., 1977; Ronzière et al., 1990).

A mechanism of formation of pyridinoline cross-links has been proposed by Eyre and Oguchi (1980). Two aligned divalent ketoamine cross-links in two adjacent pairs of 4D staggered collagen molecules would react to form a trivalent hydroxypyridinium cross-link with the reformation of one free peptidyl hydroxylysine residue. This is in good accordance with a stoichiometry of two reducible cross-links disappearing for every HP formed (Eyre et al., 1984). A different mechanism was proposed by Robins (1983) with the reaction of a ketoamine with a telopeptide hydroxyallysine residue.

The location of pyridinoline (Figure 12) has been investigated in cartilage, bone, dentin, tendon, and aorta collagen (Kuboki et al., 1981; Robins, 1983; Wu and Eyre, 1984; Light and Bailey, 1985; Henkel et al., 1987; Robins and Ducan, 1987). Two bonds of this trifunctional linkage involve the N- or C-telopeptides from a single molecule (Figure 12a) or from two different molecules (Figure 12b). The third bond is formed with the C- or N-terminal helical region of an adjacent molecule thereby stabilizing 4D staggered molecules. A pyridinoline cross-link involving three molecules (Figure 12b) instead of two (Figure 12a) would explain the correlation between the increase in the mechanical strength of the fibers and the content in nonreducible cross-links as well as the formation of heterotypic cross-linked peptides containing the N-telopeptides from both the $\alpha1(III)$ and $\alpha1(I)$ chains (Henkel and Glanville, 1982). The formation of such cross-links requires interaction between two 0D staggered molecules close enough for direct contact between amino acid side chains from the telopeptides (Figure 12b). This is an argument that has been used by Kajava (1991) to support the model of fibril formation that has been described above (see Figure 4).

Robins and his colleagues (Robins, 1982; Black et al., 1988a,b) have developed an enzyme-linked immunoassay (ELISA) for HP in urine and, more recently, a high performance liquid chromatography (HPLC) technique to measure both HP and LP simultaneously. This has allowed this group to use the values of the HP and LP cross-links as clinical markers of joint and bone degradation. The urinary concentration of LP which is derived mostly from bone collagen was increased significantly in rheumatoid arthritis (RA) and in osteoarthritis (OA), suggesting an accelerated bone degradation in both conditions. HP concentration was also increased, but to a larger extent in patients with RA. For these patients, this index correlated with clinical measures of joint involvement and biochemical parameters of inflammatory activity (Siebel et al., 1989). Furthermore, an increase in free HP and LP has been demonstrated in urine of rats with an increased bone resorption induced by estrogen deficiency (Black et al., 1989). All these data demonstrate the interest of measuring both HP and LP in urine to monitor degradative processes in arthritic and bone diseases.

Histidinohydroxylysinonorleucine (HHL). The exact structure and location of this nonreducible cross-link present in mature skin were precisely described by Mechanic et al. (1987) and Yamauchi et al. (1987). Described by Housley et al. (1975) as hydroxyaldolhistidine (HAH), histidinohydroxylysinonorleucine (HHL), a nonreducible cross-link, is made from one histidine, one hydroxylysine, and one lysine (Figure 12). The structure is consistent with a condensation of dehydrohydroxylysinonorleucine and histidine (Figure 11) (Yamauchi et al., 1987). The three amino acid residues participating in the formation of HHL in skin collagen were identified as α1(I)-chain Hyl 87, α1(I)-chain Lys 16C, and α2(I)-chain His 92 (Mechanic et al., 1987). It was proposed that HHL links three molecules as shown on Figure 12c. Such a cross-link is possible between 0D staggered adjacent molecules, as also postulated for the formation of HP or LP cross-links (Figure 12b). Furthermore, the HHL cross-linking pattern indicates a different arrangement of collagen molecules in skin fibrils compared to skeletal collagen fibrils (Mechanic et al., 1987). Such a difference had been already demonstrated by X-ray diffraction (Brodsky et al., 1980; Stinson and Sweeney, 1980).

Quantification of HHL in bovine and human skin at various ages indicates that it rapidly increases from birth through maturity, approaching 1 mole/mole of collagen. It is absent from other collagenous tissues such as dentin, bone, and tendon (Yamauchi et al., 1988). The quantity of HHL in adult human skin of the same individuals is significantly decreased in sun-exposed sites compared to unexposed sites (0.13 ± 0.07 and 0.69 ± 0.17 moles/mole of collagen, respectively) (Yamauchi et al., 1991). The decrease in the HHL concentration in sun-exposed skin is not due to photolysis of HHL or to prevention of HHL formation by UV irradiation. This demonstrates that the process of photoaging of dermal collagen is different from the chronologic aging of human skin.

Other Proposed Mature Nonreducible Cross-Links. Davison (1978) and Light and Bailey (1979, 1980) isolated from CNBr digest of insoluble tendon type I collagen a polymeric form of the C-terminal CNBr-peptide of the α1 chain (α1 CB6) that they called poly-α1 CB6. It was demonstrated that this cross-linking occurs through interactions involving the C-terminal nonhelical region of α1 CB6 in one molecule and the triple-helical peptide α1 CB5 in an adjacent 1D staggered molecule (Bailey et al., 1980). This cross-linked complex was extracted from skin, tendon, and bone and did not contain pyridinoline. A cross-link compound called M was partially characterized from an acid hydrolysate of poly-α1 CB6, but further proof is still needed to establish that this is a common mechanism for the maturation of all collagens (Barnard et al., 1987).

From proteolytic digests of collagen-rich tissues (bone, tendon, skin, cartilage, periosteum), Scott et al. (1981) isolated material that reacted rapidly at room temperature with p-dimethylaminobenzaldehyde (Ehrlich's reagent). This Ehrlich chromogen, tentatively identified as a pyrrole derived from the amino-ketone cross-links of collagen, was found in three-chain peptides isolated from human type III collagen and from the 7 S domain of type IV collagen (Scott et al., 1983). It was recently proposed (Hayase et al., 1989) that this chromophore could be formed through condensation of two adjacent lysyl-pyrraline residues, a glucose-derived pyrrole, and would thus represent a sugar-derived cross-link (see below).

Collagen Types and Enzyme-Dependent Cross-Links

The formation of reducible cross-links in fibrillar collagens and their maturation into nonreducible compounds depend on the presence of lysine (hydroxylysine) and histidine residues at specific sites, such as the N- and C-telopeptides and positions in the triple helix immediately adjacent to the N- and C-telopeptides in 4D staggered molecules.

Table 2 presents the positions of critical lysine and histidine residues in the sequences of the different α chains published so far. It is worth noting that type V collagen ([α1(V)]2 α2(V)) possesses only one oxidizable lysine per molecule, in the N-telopeptide of the α2(V) chain (Greenspan et al., 1991). A canonical sequence (KGHR or KGHP) thought to be important in cross-link formation is present in both N- and C-terminal triple-helical domains of every α chains at very conserved positions (Table 2). However, the participation of these amino acid residues in cross-links has only been established so far for type I, II, and III collagens. This canonical sequence may serve as an attachment site for lysyl oxidase as mentioned earlier or as an autocatalytic site (Eyre et al., 1984).

Among nonfibrillar collagens, the presence of enzyme-dependent cross-linking has been demonstrated only in type IV collagen and in type IX collagen which is cross-linked to type II collagen.

B. Lysine-Derived Cross-Links Formed through Glycation

Glycation is the name given to a nonenzymatic glycosylation of peptides. A sugar-derived cross-linking pathway in collagen fibrils was first suggested by Bensusan (1965). Glycation of collagen was reported by Robins and Bailey (1972) and by Tanzer et al. (1972). It has been shown to be increased in diabetes mellitus and in experimental diabetes (Cohen et al., 1980; Schnider and Kohn, 1980; Le Pape et al., 1981). More recently, the relationship between the browning products observed in aging collagen and the advanced Maillard products derived from glycated-lysine residues was demonstrated in several *in vitro* and *in vivo* studies. Finally, pentosidine was described as a specific cross-link of collagen and as an end product of the Maillard reaction. This demonstrated the importance of this cross-linking pathway in the age-related changes of proteins with a slow turnover such as collagen (Sell and Monnier, 1989).

If the lysyl oxidase-dependent cross-links can be considered as the normal and necessary way to stabilize collagen fibrils, the nonenzymatic glycation of collagen seems to induce alterations in the physicochemical properties and biological activities of collagen molecules and fibrils leading to pathophysiologic changes in connective tissues.

As excellent and recent reviews are available (Baynes and Monnier, 1989; Monnier et al., 1991; Reiser et al., 1991),we will limit our presentation to the recent important advances made in this field, mainly by Monnier and his collaborators.

The Maillard Reaction

The general scheme of the Maillard reaction is presented in Figure 13 (Monnier et al., 1991). Three major phases have been separated: initiation, propagation, and termination (in analogy to the early, intermediate, and advanced steps of Maillard reactions) (Baynes and Monnier, 1989).

The first step consists in the reaction of reducing sugars with free amino groups of either lysine or hydroxylysine. A Schiff base is formed that is stabilized by undergoing an Amadori rearrangement. Glucose, fructose, and pentose may react with collagen *in vitro* and *in vivo*. Glycation is limited to 2 to 4 moles of glucose/mole of protein when the proteins are incubated *in vitro* for extended period of time (Bitensky et al., 1989).

The second step involves the conversion of the initial Amadori products into more reactive compounds such as deoxyglucosones. In the advanced stage of the Maillard reaction, these latter compounds may react with free amino groups to form pyrrole-based yellow fluorescent pigments. An example of structure derived from the 3-deoxyglucosone is the pyrraline, a pyrrolic compound that has been detected *in vivo* by ELISA (Hayase et al., 1989). It was proposed, as noted above, that the Ehrlich chromogen of Scott et al. (1981) could have been formed from further condensation of two lysyl-pyrraline.

Fluorescent dimers (collagen b components) were observed in young rat tail tendons incubated in solutions containing D-ribose and in tail tendons from older rats (Tanaka et al., 1988b). Recently Hormel and Eyre (1991) isolated from human intervertebral disc a CNBr-peptide (CB12) of type II collagen containing a fluorescent chromophore that increases with age. They demonstrated that this chromophore was attached to the peptide at a lysine residue.

Pentosidine, the only fully characterized sugar-derived cross-link of collagen, involves a lysine and an arginine residue cross-linked via a pentose (Sell and Monnier, 1989). It was detected at high levels in several human adult tissues such as liver (330 pmol/mg collagen) tracheal cartilage (182 pmol/mg), cardiac muscle (139 pmol/mg), and dura mater (150 pmol/mg). In this latter tissue, a linear 10-fold increase throughout life was observed, reaching approximately 250 pmol/mg collagen in aged individuals (Sell and Monnier, 1989). In human skin, pentosidine was found to increase exponentially from 5 to 75 pmol/mg collagen over lifespan.

A 3- to 10-fold increase was noted in insulin-dependent diabetic and nondiabetic individuals with severe end-stage renal disease requiring hemodialysis (Sell and

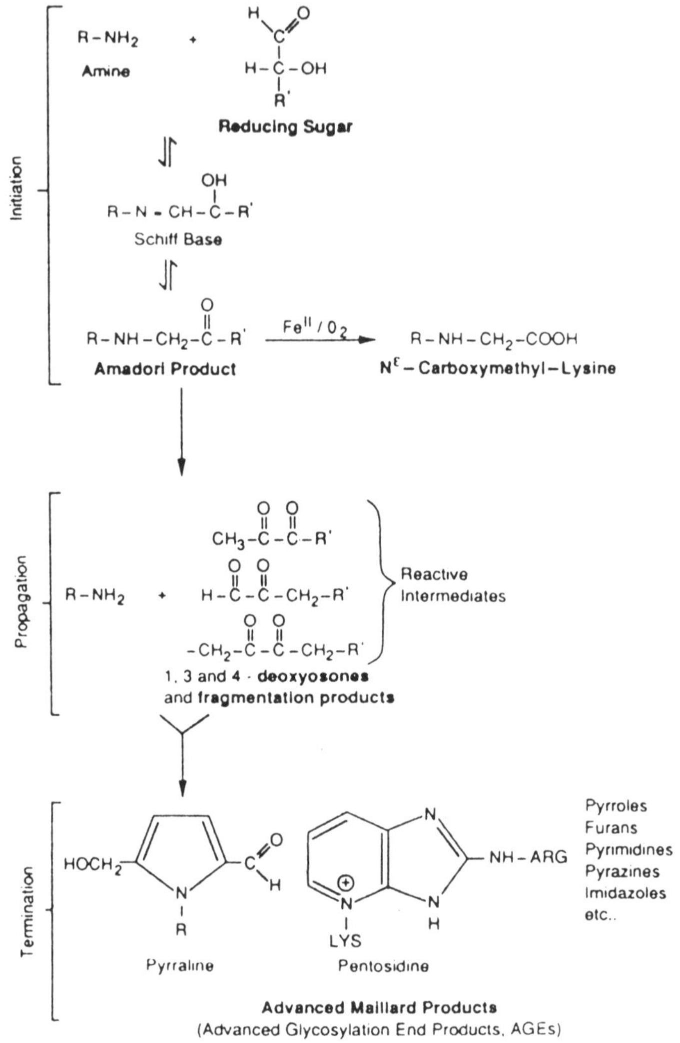

Figure 13. General scheme of the Maillard reaction. (From Monnier et al. (1991), with permission).

Monnier, 1990). In human articular cartilage, the amount of pentosidine per collagen increases linearly with age, while the amount of pyridinoline per collagen remained constant. On the other hand, the amount of pentosidine per pyridinoline increased exponentially during life (Uchiyama et al., 1991). Recently, it was shown (Grandhee and Monnier, 1991) that pentosidine can be formed not only from pentoses but also from hexoses and ascorbate.

There is evidence that formation of yellow-brown chromophores and fluorophores could involve pathways other than the Maillard reaction, such as the reaction of lipid peroxidation products on glycated residues (Hicks et al., 1989) or on constituent aromatic amino residues (Lunec et al., 1985).

Biological Implication of Sugar-Derived Cross-Links

Alterations in specific properties of collagen due to glycation have been reviewed in detail by Reiser (1991). Tanaka et al. (1988a) have shown that glycation induces expansion of the molecular packing of collagen with an increase in its intermolecular spacing. The mechanical properties of collagen fibers and their binding capacities to heparin, serum albumin, IgGs, and low density lipoproteins are progressively altered by this unregulated cross-linking. For example, Le Pape et al. (1983) observed an increased platelet aggregation by glycated collagen. Furthermore, Bassiouny et al. (1983) have shown that glucosylated collagen is capable of inducing the production of antibodies specifically directed against the modified sites in collagen. In addition, allogenic diabetic sera contain antibodies capable of binding *in vitro* glucosylated collagen. These authors postulated that altered vascular collagen may be the initiating event of an autoimmune response leading to the vascular complications of diabetes. The alterations in the binding properties of the type IV collagen molecules due to glycation are of particular importance since they may contribute to basement membrane structural abnormalities (Tsilibary et al., 1988).

As expected for potentially damaging reactions to proteins and DNA, the organism has developed powerful defense mechanisms. Several lines of defense against toxic carbonyl compounds have been considered by Monnier et al. (1991):

1. A tight control of the cellular and plasma concentrations of reactive sugars (pentoses) and aldehydes is maintained.
2. The Amadori products can be modified into unreactive carboxymethyllysine.
3. The Maillard reaction products can be enzymatically inactivated or degraded.
4. Protection can be obtained with scavenging agents (endogenous amines) of the reactive Maillard intermediates. In order to improve this line of defense, pharmacologic intervention is possible as demonstrated by treatments with aminoguanidine (Brownlee et al., 1986; Nicholls and Mandel, 1989), currently in Phase II of clinical trials.
5. Some cells can remove and degrade advanced Maillard molecules. As shown by Vlassara et al. (1987), macrophages have a specific receptor that mediates the degradation of proteins modified by advanced glycosylation end products. A possible negative effect of this reaction is the release by macrophages of tumor necrosis factor (TNFα) and interleukin 1, potent

initiators of tissue remodeling (Vlassara et al., 1988). For example, collagenase is increased during wound healing in diabetic rats (Hennessey et al., 1990).

C. Intermolecular Disulfide Bonds

Most of the cysteine residues present in fibrillar collagen are located in the N- and C-terminal nontriple helical domains and form inter- and intrachain disulfide bonds. For example, after aggregation of the carboxypropeptides of fibrillar collagens, the triple-helix formation can only occur after the stabilization of the trimer by disulfide bridges. The α1(XI) chain remains unique among fibrillar collagen chains in containing a cysteinyl residue located approximately in the middle of the triple-helical domain. However, this residue is not utilized in the formation of an interchain disulfide bond because reduction of the pepsin-treated molecule does not affect the migration of type XI chains on SDS-page (Bernard et al., 1988). In type III collagen, the very end of the helix is joined by interchain disulfide bridges.

Type IV, VI, and VII collagen are known to form aggregates stabilized by intermolecular disulfide bonds. Tetrameric aggregates of type IV collagen molecules are formed by self assembly of the 7S domain and are stabilized by intermolecular disulfide bonds and by lysine-derived cross-links. Furthermore, dimers can be generated by interactions between globular NC1 domains and stabilized by formation of disulfide bonds (Timpl, 1989).

A most unique feature of collagen VI is its assembly into dimers and tetramers stabilized by disulfide bonds and not by lysine-derived cross-links (Timpl and Engel, 1987). The human α1(VI), α2(VI), and α3(VI) chains contain 19, 20, and 29 cysteine residues respectively, but only one is present in the triple-helical domains. Two cysteines are required to form disulfide bridges between the C-terminal globule and the triple helix in the staggered collagen VI dimers. An unknown number of other disulfide bridges stabilizes the microfibrils formed by an overlapping end-to-end association of tetramers (Chu et al., 1990a).

Type VII collagen forms centro-symmetric dimers stabilized by disulfide bonds within an overlap of 60 nm of the carboxy-terminal ends of the triple helices. Further lateral association of these dimers, in a nonstaggered array, gives rise to the anchoring fibrils; important components of the junction between the basement membrane and the underlying matrix (Burgeson, 1987).

VIII. STRUCTURES OF COLLAGEN GENES

Excellent and comprehensive reviews on the structures and the organizations of the collagen genes have been recently published (Bornstein and Sage, 1989; Sandell and Boyd, 1990a; Vuorio and De Crombrugghe, 1990) and we will thus limit our presentation to a few general comments pertinent to the main issues raised in this chapter.

A. Exon Sizes and Ancestral Gene(s)

One of the objectives of this chapter was to underline the importance of the domain organization in collagen molecules. This organization is usually reflected at the gene level by the intron-exon structure, one of the best indicators for the determination of homologies between related proteins. As mentioned earlier, the collagen types can indeed be grouped in subfamilies according to their structural homologies (see Table 1).

Fibrillar Collagen Genes

The fibrillar collagen genes (types I, II, III, V, and XI) are closely related and clearly derived from a single ancestral gene. The large central triple-helical domain is encoded by a series of 41 or 42 exons whose sizes are related to a 54-bp unit (54 bp and multiples of 54 bp, 45 bp, 99 bp) (Sandell and Boyd, 1990b). All these exons are in phase with the repetitive triple helix protein sequence [(Gly-Xaa-Yaa)n]; that is, they all start by a complete Gly codon and end with a complete codon for an amino acid in the Yaa position. The intron-exon structure of all vertebrate fibrillar collagen genes is essentially identical for the region of the gene encoding the large triple helix. The C-propeptides are also very conserved and are encoded by four exons, the first one being a junction exon encoding also the C-terminal end of the triple helix. In evolutionary distant species, such as sea urchins and sponges, the intron-exon structures of the described fibrillar collagen genes are not identical to their vertebrate counterparts but are nevertheless clearly related (D'Alessio et al., 1989; Exposito and Garrone, 1990).

The greatest differences between the various fibrillar collagen chains are noted in the N-propeptide region (see Figure 1). This is also reflected at the gene level [see Figure 2a in Sandell and Boyd (1990b)]. The 5′ untranslated region, the signal peptide and its cleavage site are always contained in the first exon that ends by a split codon. For those chains that contain a cysteine-rich region, it is encoded in the second exon. In the case of the α2(I) chain, two very short exons encode for nine out of the 11 amino acids that separate the signal peptide cleavage site from the short triple helix. This short triple helix is encoded by exons that are not all related to the 54-bp unit or multiple of 9 bp. The intron-exon structure of this region is also quite variable since it contains between two and five exons. At the protein level, some of these domains contain interruptions of the triple helix [in particular in the pro-α1(II), pro-α1(V), and pro-α1(XI) chains] (Sandell and Boyd, 1990; Yoshioka and Ramirez, 1990; Greenspan et al., 1991). The nontriple-helical domain separating the main triple helix from the short one found in the N-propeptide is encoded in all vertebrate genes by a junction exon containing also the N-terminal end of the main triple helix (Sandell and Boyd, 1990).

Nonfibrillar Collagen Genes

In the nonfibrillar collagen genes, the intron-exon structures are much more diverse. The short-chain collagens [pro-α1(VIII), pro-α2(VIII), pro-α1(X)] have a much simpler gene structure (Ninomiya et al., 1990; Muragaki et al., 1991a). The pro-α1(X) gene contains only three exons. The first one contains an untranslated sequence. The second one encodes a short untranslated sequence, the signal peptide, and most of the N-terminal nontriple-helical domain. The last one is very large and encodes the end of this latter domain, the entire triple helix, and the C-terminal nontriple-helical domain. Similar organizations have been described for the genes of the two type VIII collagen chains.

The genes for type IV collagen chains are very fragmented [52 exons for the pro-α1(IV) gene and 47 exons for the pro-α2(IV) gene] (Sandell and Boyd, 1990). They contain a few exons that are related to the 54-bp unit found in fibrillar collagen (45, 54, 99, 108, and 162 bp) and many others that are not related to this basic unit. In the case of the genes encoding type IX collagen chains, the exons encoding the C-terminal COL1 domain, the C-terminal part of the COL2 domain, and the N-terminal part of the COL3 domain are not related to a 54-bp unit. However, the N-terminal part of the COL2 domain is encoded by a series of seven 54-bp exons and three other 54-bp exons are found in the region of the gene encoding the COL3 domain.

The nonfibrillar collagen genes that have been described in invertebrates also have a gene organization strikingly different from the fibrillar collagen genes. In sponges, Exposito et al. (1991) have described a subfamily of nonfibrillar collagen with a gene organization completely different from the fibrillar collagen gene from the same species. At least two distinct forms of collagen gene organizations thus also exist in the animal species that is most distant evolutionarily from vertebrates. The divergence between the fibrillar collagen genes and other collagen genes thus represents a very early event in the evolution of the animal kingdom or it may represent a case of convergent evolution.

B. Regulatory Elements

Several laboratories are very actively investigating the control of collagen gene expression and, in particular, regulatory elements that can bind stimulatory or inhibitory transcription factors (Bornstein and Sage, 1989; Sandell and Boyd, 1990a; Vuorio and de Crombrugghe, 1990; de Crombrugghe et al., 1991). Since many collagen molecules are heterotrimers, a very precise coordination of the production of the constituent chains is required. For type I collagen, the genes of the pro-α1(I) and pro-α2(I) chains are located on two different chromosomes and are thus expected to function independently, although their expressions are known to be tightly coordinated. For type IV collagen, the genes for the α1(IV) and α2(IV) chains are very close in the genome since they are adjacent and in opposite, "head to head" directions, separated only by approximately 120 bp (Pollner et al., 1990). Their transcription is thus regulated by a bidirectional promoter and direct interference between the two genes for their expression can thus be expected.

Several regulatory sequences have been described in the promoter element, and in the first intron of the studied collagen genes (Bornstein and Sage, 1989; de Crombrugghe et al., 1991). These are in particular the genes for the type I, II, III, and IV collagen chains. In the α2(I) gene, five regulatory elements binding stimulatory transcription factors and two elements binding inhibitory factors have been described in the first 400 bp upstream from the site of start of transcription. The functional activity of these sites have been studied by deletion and mutagenesis using a reporter gene such as that of bacterial chloramphenicol acetyltransferase (CAT). These elements include a CCAAT motif at around –80 that binds a dimeric CCAAT binding factor. An element at –300 binds nuclear factor I and was found to mediate the activation effect of transforming growth factor β (TGFβ) on the expression of this gene.

Important regulatory elements have also been described in the first intron. Insertion of a retrovirus in the first intron of the pro-α1(I) collagen gene results in the complete inactivation of the expression of this gene in most tissues of homozygous mice (called mov-13), as demonstrated by the group of Jaenisch (Harbers et al., 1984). This results in a lethal phenotype at about 12 days of gestation. It is remarkable that this gene is still active in some mov-13 cells, such as the odontoblast, when tooth rudiments from homozygous embryos are transplanted into normal mice (Kratochwil et al., 1989).

The detailed description of the recent important findings concerning the many regulatory elements of the various collagen genes would go much beyond the scope of this chapter and the reader is referred to the recent reviews of Bornstein and Sage (1989), Sandell and Boyd (1990a), Vuorio and de Crombrugghe (1990), and de Crombrugghe et al. (1991). Diversity among the various collagen chains is now also demonstrated in the regulatory elements of their genes, even for a given collagen type. The unraveling of these complex control mechanisms will permit a better understanding of the fine tuning of the expression of collagen genes that is observed in different tissues during development.

C. Transcription and Splicing Variants

Diversity in the collagen family of proteins is generated not only by the existence of a large number of different genes but also by the existence of different mRNAs that can be obtained from the same gene. These variations are produced either by the use of different transcription start sites or by alternative splicing of the primary transcript.

There are, at the moment, two well-documented examples of alternative transcription start sites. Two promoters have been described for the α1(IX) gene (Svoboda et al., 1988; Nishimura et al., 1989; Muragaki et al., 1990a,b). One promoter, used in cartilage, controls the expression of a long form of the protein that include a globular and basic N-terminal domain (NC4) that is encoded by the first seven exons. Between exons 6 and 7, another promoter is found which has been shown initially to direct the biosynthesis of a short form of α1(IX) in the primary stroma of the cornea in embryonic chick tendon. This alternative promoter is located upstream of an alternative first exon which is directly spliced to exon 8, which codes for the last amino acid of NC4 and the N-terminal extremity of the triple-helical COL3 domain. The resulting translation product must therefore have a very short NC4 domain since the total size of this domain, plus the signal peptide is only 25 amino acids. This short form is now known to be expressed in other tissues (Muragaki et al., 1990b). It has been detected in particular by rotary-shadowing electron microscopy in the avian vitreous humor (Brewton et al., 1991).

A cartilage-specific promoter has also been described in the second intron of the α2(I) collagen gene (Bennett and Adams, 1990). In this case, the short predicted translation product would not be in the same reading frame as the collagen chain and would therefore have a completely different amino acid sequence. In particular, it would not contain triple-helical domains. Whether a translation product is synthesized from this second reading frame and the possible functions of this protein remain to be established.

Alternative splicing has been demonstrated for transcription products from several collagen genes: in particular, those coding for the α1(II), α2(VI), α3(VI) and α1(XIII) chains. Except in this latter case, only regions coding for nontriple-helical sequences are alternatively spliced. In the case of the pro-α1(II) chain, two forms have been detected, one containing the cysteine rich region of the N-propeptide (see Figure 1), encoded by exon 2, and one missing it due to alternative splicing (Ryan and Sandell, 1990; Sandell et al., 1991). The alternatively spliced form is more abundant in fetal mRNA. Alternative splicing has been described for exons coding for the N-terminal nontriple-helical region of the α3(VI) chain. Three of the nine subdomains having homologies with type A domains of von Willebrand factor (N7 to N9) can be involved in alternative splicing (Stokes et al., 1991). Alternative splicing for the C-terminal nontriple-helical region of α2(VI) has also been reported (Saitta et al., 1990).

Finally, type XIII collagen gene has been shown to undergo a very complex process of alternative splicing, involving five different regions of the gene (Pihlajaniemi et al., 1990; Tikka et al., 1991). Triple helical sequences are involved in four cases, including the region corresponding to the junction between the C-terminal triple-helical COL3 domain and the last nontriple-helical NC4 domain. This raises interesting questions about the molecular assembly of the translation products since one would expect that all three chains of a given molecule should have triple-helical domains of approximately the same length.

IX. PERSPECTIVES

This chapter has underlined the diversity of structures found in the collagen family of proteins. In the last 20 years, the number of collagen types has increased very rapidly and there is no reason to believe that the list is closed. Actually several additional new collagens are being studied presently in different laboratories. However, the known collagen types account for the vast majority of the hydroxyproline (taken as an indicator of collagen triple helix) of vertebrate organisms. These additional collagen molecules can thus only represent minor components of large tissues or major components of small and very specialized structures.

Collagens, like many other macromolecules from the extracellular matrix, serve two main functions. The first one is structural. They provide a scaffold to the body, being responsible for most of the biomechanical properties of tissues and organs. The understanding of the multimolecular assemblies made by matrix macromolecules and the study of the biomechanical properties of these assemblies is far from being complete. The possible subtle variations to these properties by changes in molecular composition and in posttranslational modifications are still for a large part unclear and much experimental work is still required.

The extracellular matrix proteins are not passive architectural components. They are also involved in interactions with cells. Many cell receptors for collagens have been described and the diversity of collagen molecules offers to the cells a further opportunity for cells to refine their "feeling" for their environment. The fact that several of the recently described collagens contain large nontriple-helical domains with homologies to other noncollagenous matrix macromolecules may be of particular importance in this respect.

Finally it should be remembered that most tissues are constantly remodeled and, therefore, their constituent macromolecules, in particular collagen, have to be degraded. The mechanisms of degradation are different for the various collagen types. For example, although it is now known that type V and type I collagens form heterotypic fibrils, their susceptibilities to degradation by interstitial collagenase are different. Type I collagen is cleaved by this enzyme but type V is not. One can therefore hypothesize that a selective remodeling of these heterotypic fibrils is possible, with the type I collagen turning over and the type V collagen "core" being retained as a template for the fibril reconstruction. The detailed knowledge of the susceptibilities of the various collagen types to degradative enzymes is therefore of primary importance and this should be performed not only on isolated molecules but foremost on their aggregated tissue forms.

ACKNOWLEDGMENTS

The authors want to thank their colleagues, S. Shoshan, I. Pucci-Minafra, M. Moradi-Ameli and M. Aumailley, for their careful reading and criticism of this manuscript. They also acknowledge the electron micrographs contributed by A. Barge, S. Franc and F. Ruggiero, and the help of A. Bosch for the illustrations.

NOTES

1. Throughout this chapter we will use the terms triple helix and triple helical to designate the conformation found in collagen triple helices. The reader is reminded that triple α helices are found in some proteins, sometimes also containing collagen-like triple helices (Kodama et al., 1990).

2. The recent characterization and sequencing of a partial cDNA for type VII collagen strongly indicates that the orientation of the molecules is opposite to what was previously suggested (Parente et al., 1991). We are, nevertheless, keeping the original nomenclature of the domains of type VII collagen (NC1, NC2, P1, and P2), but indicate the correct orientation relative to the N- and C-termini.

REFERENCES

Aubert-Foucher, E., Eichenberger, D., Font, B., Goldschmidt, D., Léthias, C., & van der Rest, M. (1992). Purification and characterization of native type XIV collagen. J. Biol. Chem. 267: 15759–15764.

Ayad, S., Abedin, M.Z., Grundy, S.M., & Weiss, J.B. (1981). Isolation and characterization of an unusual collagen from hyaline cartilage and intervertebral disc. FEBS Lett. 123, 195–199.

Ayad, S., Abedin, M.Z., Weiss, J.B., & Grundy, S.M. (1982). Characterization of another short-chain disulphide-bonded collagen from cartilage, vitreous and intervertebral disc. FEBS Lett. 139, 300–304.

Baccarani-Contri, N., Vincenzi, D., Quaglino, D., Mori, G., & Pasquali-Ronchetti, I. (1989). Localization of human placenta lysyl oxidase on human placenta, skin and aorta by immunoelectronmicroscopy. Matrix 9, 428–436.

Bächinger, H.P., Bruckner, P., Timpl, R., & Engel, J. (1978). The role of *cis-trans* isomerization of peptide bonds in the coil-triple helix conversion of collagen. Eur. J. Biochem. 90, 605–613.

Bächinger, H.P. (1987). The influence of peptidyl-prolyl *cis-trans* isomerase on the *in vitro* folding of collagen. J. Biol. Chem. 262, 17144–17148.

Bailey, A.J., Light, N.D., & Atkins, E.D.T. (1980). Chemical cross-linking restrictions on model for the molecular organization of the collagen fiber. Nature 288, 408–410.

Bairati, A., Petruccioli, M.G., & Torri Tarelli, L. (1969). Studies on the ultrastructure of collagen fibrils. 1. Morphological evaluation of the periodic structure. J . Submicr. Cytol. 1, 113–141.

Bairati, A., Petruccioli, M.G., & Torri Tarelli, L. (1972). Submicroscopic structure of collagen fibrils. III. Periodic structure after fixation. J. Submicr. Cytol. 4, 171–197.

Bairati, A. & Garrone, R. (1985). Biology of Invertebrate and Low Vertebrate Collagens. Plenum, New York.

Barber, M., Bordoli, R.S., Elliott, G.J., Fujimoto, D., & Scott, J.E. (1982). The structure(s) of pyridinoline(s). Biochem. Biophys. Res. Commun. 109, 1041–1046.

Barge, A., Ruggiero, F., & Garrone, R. (1991). Structure of the basement membrane of corneal epithelium: Quick-freeze, deep-etch comparative study of networks deposited in culture and during development. Biol. Cell 72, 141–147.

Barnard, K., Light, N.D., Sims, T.J., & Bailey, A.J. (1987). Chemistry of the collagen cross-links. Origin and partial characterization of a putative mature cross-link of collagen. Biochem. J. 244, 303–309.

Bassiouny, A.R., Rosenberg, H., & McDonald, T.L. (1983). Glucosylated collagen is antigenic. Diabetes 32, 1182–1184.

Bateman, J.F., Lamande, S.R., Dahl, H.-H.M., Chan, D., Mascara, T., Cole, W.G. (1989). A frameshift mutation results in a truncated nonfunctional carboxyl-terminal proα1(I) propeptide of type I collagen in osteogenesis imperfecta. J. Biol. Chem. 264, 10960–10964.

Baynes, J.W., & Monnier, V.M. (1989). The Maillard Reaction in Aging, Diabetes and Nutrition. Alan R. Liss, New York.

Bennett, V.D., & Adams, S.L. (1990). Identification of a Cartilage-Specific Promoter within Intron 2 of the Chick α2(I) collagen gene. J. Biol. Chem. 265, 2223–2230.

Bensusan, H.B . (1965) A novel hypothesis for the mechanism of cross-linking in collagen In: Structure and Function of Connective and Skeletal Tissue (Fitton Jackson, S., Harkness, R.D., Partridge, S.M., & Tristram, G.R., eds.), pp. 42–46. Butterworths, London.

Bentz, H, Morris, N.P., Murray, L.W., Sakai, L.Y., Hollister, D.W., & Burgeson, R.E. (1983). Isolation and partial characterization of a new human collagen with an extended triple-helical structural domain. Proc. Natl. Acad. Sci. USA 80, 3168–3172.

Berg, R.A., & Prockop, D.J. (1973). The thermal transition of a non-hydroxylated form of collagen. Evidence for a role for hydroxyproline in stabilizing the triple helix of collagen. Biochem. Biophys. Res. Commun. 52, 15–120.

Bernard, M.P., Yoshioka, H., Rodriguez, E., van der Rest, M., Kimura, T., Ninomiya, Y., Olsen, B.R., & Ramirez, F. (1988). Cloning and sequencing of pro-α1(XI) collagen cDNA demonstrates that type XI belongs to the fibrillar class of collagens and reveals that the expression of the gene is not restricted to cartilagenous tissue. J. Biol. Chem. 263, 17159–17166.

Birk, D.E., Fitch, J.M., Babiarz, J.P., & Linsenmayer, T.F. (1988). Collagen type I and V are present in the same fibril in the avian corneal stroma. J. Cell Biol. 106, 999–1008.

Bitensky, M.W., Kowluru, A. and Kowluru, R.A. (1989). Nonenzymatic glycation and protein recognition In: The Maillard Reaction in Aging, Diabetes and Nutrition (Baynes, J.W. & Monnier, V.M., Eds.), pp. 185–204. Alan R. Liss, New York.

Black, D., Duncan, A.J., & Robins, S.P. (1988a). Quantitative analysis of the pyridinium cross-links of collagen in urine using ion-paired reverse-phase high-performance liquid chromatography. Anal. Biochem. 169, 197–203.

Black, D., Robins, S.P., Duncan, A., & Marabani, M. (1988b). Analysis by high performance liquid chromatography of two hydroxypyridinium cross-links of collagen in urine as indices of disease activity in rheumatoid arthritis. Biochem. Soc. Trans. 16, 817–818.

Black, D., Farquharson, C., & Robins, S.P. (1989). Excretion of pyridinium cross-links of collagen in ovariectomized rats as urinary markers for increased bone resorption. Calcif. Tissue Int. 44, 343–347.

Blumberg, B., MacKrell, A.J., & Fessler, J.H. (1988). *Drosophila* basement membrane procollagen α1(IV). II. Complete cDNA sequence, genomic structure, and general implications for supramolecular assemblies. J. Biol. Chem. 263, 18328–18337.

Bonaldo, P. & Colombatti, A. (1989). The carboxyl terminus of the chicken α3 chain of collagen VI is a unique mosaic structure with glycoprotein Ib-like, fibronectin type III, and Kunitz modules. J. Biol. Chem. 264, 20235–20239.

Bonaldo, P., Russo, V., Bucciotti, F., Bressan, G.M., & Colombatti, A. (1989). α1 Chain of chick type VI collagen. The complete cDNA sequence reveals a hybrid molecule made of one short collagen and three von Willebrand factor type A-like domains. J. Biol. Chem. 264, 5575–5580.

Bornstein, P. and Sage, H. (1980). Structurally distinct collagen types. Annu. Rev. Biochem. 49, 957–1003.

Bornstein, P. & Sage, H. (1989). Regulation of collagen gene expression. Prog. Nucleic Acid Res. Mol. Biol. 37, 67–106.

Bouteille, M. & Pease, D.C. (1971). The tridimensional structure of native collagenous fibrils, their proteinaceous filaments. J. Ultrastruct. Res. 35, 314–338.

Brewton, R.G., Wright, D.W., & Mayne, R. (1991). Structural and functional comparison of type IX collagen-proteoglycan from chicken cartilage and vitreous humor. J. Biol. Chem. 266, 4752–4757.

Briggam, R.A., Dalldorf, F.G., & Wheeler, C.E. (1971). Formation and origin of basal lamina and anchoring fibrils in adult human skin. J. Cell. Biol. 51, 384–395.

Brodsky, B., Eikenberry, E.F., & Cassidy, K. (1980). An unusual collagen periodicity in skin. Biochim. Biophys. Acta 621, 162–166.

Brown, K.E., Lawrence, R., & Sonenshein, G.E. (1991). Concerted modulation of α1(XI) and α2(V) collagen mRNAs in bovine vascular smooth muscle cells. J. Biol. Chem. 266, 23268–23273.

Brownlee, M., Vlassara, H., Kooney, A., Ulrich, P., & Cerami, A. (1986). Aminoguanidine prevents diabetes-induced arterial wall protein cross-linking. Science 232, 1629–1632.

Bruckner, P., Mayne, R., & Tuderman, L. (1983). p-HMW-Collagen, a minor collagen obtained from chick embryo cartilage without proteolytic treatment of the tissue. Eur. J. Biochem. 136, 333–339.

Bruckner, P. & Eikenberry, E.F. (1984). Formation of the triple helix of type I collagen in cellulo. Temperature-dependent kinetics support a model based on *cis-trans* isomerization of peptide bonds. Eur. J. Biochem. 140, 391–395.

Bruns, R.R. & Gross, J. (1973). Band pattern in the segment-long-spacing form of collagen. Its use in analysis of primary structure. Biochemistry 12, 808–815.

Bruns, R.R., Trelstad, R.L., & Gross, J. (1973). Cartilage collagen: a staggered substructure in reconstituted fibrils. Science 181, 269–271.

Bruns, R.R. (1976). Supramolecular structure of polymorphic collagen fibrils. J. Cell Biol. 68, 521–538.

Bruns, R.R. (1984). Beaded filaments and long-spacing fibrils: relation to type VI collagen. J. Ultrastruct. Res. 89, 136–145.

Burgeson, R.E., Morris, N.P., Murray, L.W., Duncan, K.G., Keene, D.R., & Sakai, L.Y. (1985). The structure of type VII collagen Ann. NY. Acad. Sci. 460, 47–57.

Burgeson, R.E. (1987). Type VII collagen In: Structure and Function of Collagen Types (Mayne, R. & Burgeson, R.E., Eds.), pp. 145–172. Academic Press, Orlando.

Burgeson, R.E., Lunstrum, G.P., Rokosova, B., Rimberg, C.S., Rosenbaum, L.M., & Keene, D.R. (1990). The structure and function of type VII collagen. Ann. NY Acad. Sci. 580, 32–43.

Butkowski, R.J., Shen, G.-Q., Wieslander, J., Michael, A.F., & Fish, A.J. (1990). Characterization of type IV collagen NC1 monomers and Goodpasture antigen in human renal basement membranes. J. Lab. Clin. Med. 115, 365–373.

Chapman, J.A. & Armitage, P.M. (1972). An analysis of fibrous long spacing forms of collagen. Connect. Tissue Res. 1, 31–37.

Chapman, J.A. (1974). The staining pattern of collagen fibrils. I. An analysis of electron micrographs. Connect. Tissue Res. 2, 137–150.

Chapman, J.A. & Hulmes, D.J.S. (1984). Electron microscopy of the collagen fibril In: Ultrastructure of the connective tissue matrix (Ruggeri, A. & Motta, P.M., Eds.), pp. 1–33. Martinus Nijhoff Publishers, Boston.

Chu, M.-L., Pan, T.-C., Conway, D., Kuo, H.-J., Glanville, R.W., Timpl, R., Mann, K., & Deutzmann, R. (1989). Sequence analysis of α1(VI) and α2(VI) chains of human type VI collagen reveals internal triplication of globular domains similar to the A domains of von Willebrand factor and two α2(VI) chain variants that differ in the carboxy terminus. EMBO J. 8, 1939–1946.

Chu, M.-L., Pan, T.-C., Conway, D., Saitta, B., Stokes, D., Kuo, H.-J., Glanville, R.W., Timpl, R., Mann, K., & Deutzmann, R. (1990a). The structure of type VI collagen. Ann. NY Acad. Sci. 580, 55–63.

Chu, M.-L., Zhang, R.-Z., Pan, T., Stokes, D., Conway, D., Kuo, H.-J., Glanville, R., Mayer, U., Mann, K., Deutzmann, R., & Timpl, R. (1990b). Mosaic structure of globular domains in the human type VI collagen α3 chain: Similarity to von Willebrand Factor, fibronectin, actin, salivary proteins and aprotinin type protease inhibitors. EMBO J. 9, 385–393.

Clarke, E.P., Cates, G.A., Ball, E.H., & Sanwal, B.D. (1991). A collagen-binding protein in the endoplasmic reticulum of myoblasts exhibits relationship with serine protease inhibitors. J. Biol. Chem. 266, 17230–17235.

Cohen, M.P., Urdanivia, E., Surma, M., & Wu, V. (1980). Increased glycosylation of glomerular basement membrane collagen in diabetes. Biochem. Biophys. Res. Commun. 95, 765–769.

Colombatti, A., Bonaldo, P., Russo, V, Bucciotti, F.J., & Mucignat, M.T. (1990). The α1 chain of chick type VI collagen is a hybrid molecule made of one short collagen and three von Willebrand factor type A-like domains. Ann. NY Acad. Sci. 580, 430–432.

Connolly, C.E. (1981). "Crystalline" collagen production by an unusual benign soft tissue tumor ("amianthioma"). Histopathology 5, 11–20.

Cox, G.N. (1990). Molecular biology of the cuticle collagen gene families of *Caenorhabditis elegans* and *Haemonchus contortus*. Acta Tropica 47, 269–281.

Cronlund, A.L., Smith, B.D., & Kagan, H.M. (1985). Binding of lysyl oxidase to fibrils of type I collagen. Connect. Tissue Res. 14, 109–119.

D'Alessio, M., Ramirez, F., Suzuki, H.R., Solursh, M., & Gambino, R. (1989). Structure and developmental expression of a sea urchin fibrillar collagen gene. Proc. Natl. Acad. Sci. USA 86, 9303–9307.

Davison, P.F. (1978). Bovine tendons, aging and collagen cross-linking. J. Biol. Chem. 253, 5635–5641.

Deak, S.B., van der Rest, M., & Prockop, D.J. (1985). Altered helical structure of a homotrimer of α1(I) chains synthesized by fibroblasts from a variant of Osteogenesis Imperfecta. Collagen Rel. Res. 5, 305–313.

de Crombrugghe, B., Vuorio, T., Karsenty, G., Maity, S., Rutheshouser, E.C., Goldberg, H. (1991). Transcriptional control mechanisms for the expression of type I collagen genes. Ann. Rheum. Dis. 50 Suppl. 4, 872–876.

Dick, T.A. & Wright, K.A. (1973). The ultrastructure of the cuticle of the nematode *Syphacia obvelata* (Rudolphi, 1802). I. The body cuticle of larvae, males, and females, and observations on its development. Can. J. Zool. 51, 187–196.

Dion, A.S. & Myers, J.C. (1987). COOH-terminal propetides of the major human procollagens. Structural, functional and genetic comparisons. J. Mol. Biol. 193, 127–143.

Djaczenko, W. & Calenda Cimmino, C. (1973). Visualization of polysaccharides in the cuticle of oligochaeta by the tris 1aziridinyl phosphine oxide method. J. Cell. Biol. 57, 859–867.

Doege, K.J. & Fessler, J.H. (1986). Folding of carboxyl domain and assembly of procollagen I. J. Biol. Chem. 261, 8924–8935.

Dublet, B., Oh, S., Sugrue, S.P., Gordon, M.K., Gerecke, D.R., Olsen, B.R., & van der Rest, M. (1989). The structure of avian type XII collagen. α1(XII) chains contain 190-kDa non-triple helical aminoterminal domains and form homotrimeric molecules. J. Biol. Chem. 264, 13150–13156.

Dublet, B. & van der Rest, M. (1991). Type XIV collagen, a new homotrimeric molecule extracted from bovine skin and tendon, with a triple helical disulfide-bonded domain homologous to type IX and type XII collagens. J. Biol. Chem. 266, 6853–6858.

Eikenberry, E.F., Childs, B., Sheren, S.B., Parry, D.A.D., Graig, A.S., & Brodsky, B. (1984). Crystalline fibril structure of type II collagen in lamprey notochord sheath. J. Mol. Biol. 176, 261–277.

Eikenberry, E.F., Mendler, M., Burgin, R., Winterhalter, K.H., & Bruckner, P. (1992). Fibrillar organization in cartilage In: Articular Cartilage and Osteoarthritis (Kuettner, K.E., Schleyerbach, R., Peyron, J.G. & Hascall, V.C., Eds.), pp. 133–149. Raven Press, New York.

Engel, J., Furthmayr, H., Odermatt, E., Von der Mark, H., Aumailley, M., Fleischmajer, R., & Timpl, R. (1985). Structure and macromolecular organization of type VI collagen. Ann. NY Acad. Sci. 460, 25–46.

Engel, J. & Prockop, D.J. (1991). The zipper-like folding of collagen triple helices and the effects of mutations that disrupt the zipper Annu. Rev. Biophys. Biophys. Chem. 20, 137–152.

Exposito, J.-Y. & Garrone, R. (1990). Characterization of a fibrillar collagen gene in sponges reveals the early evolutionary appearance of two collagen gene families. Proc. Natl. Acad. Sci. USA 87, 6669–6673.

Exposito, J.-Y., Le Guellec, D., Lu, Q., & Garrone, R. (1991). Short chain collagens in sponges are encoded by a family of closely related genes. J. Biol. Chem. 266, 21923–21928.

Eyre, D.R. & Oguchi, H. (1980). The hydroxypyridinium cross-links of skeletal collagens: their measurement, properties and a proposed pathway of formation. Biochem. Biophys. Res. Commun. 92, 403–410.

Eyre, D.R., Paz, M.A., & Gallop, P.M. (1984). Cross-linking in collagen and elastin. Ann. Rev. Biochem. 53, 717–748.

Eyre, D.R., Apone, S., Wu, J.-J., Ericsson, L.H., & Walsh, K.A. (1987). Collagen type IX: Evidence for covalent linkages to type II collagen in cartilage. FEBS Lett. 220, 337–341.

Eyre, D.R., Dickson, I.R., & Van Ness, K. (1988). Collagen cross-linking in human bone and articular cartilage. Biochem. J. 252, 495–500.

Eyre, D.R., Wu, J.-J., & Woods, P. (1992). Cartilage-specific collagens: Structural studies In: Articular Cartilage and Osteoarthritis (Kuettner. K.E., Schleyerbach, R., Peyron, J.G. & Hascall, V.C., Eds.), pp. 119–131. Raven Press, New York.

Fietzek, P.P. & Kühn, K. (1976). The primary structure of collagen. Int. Rev. Connect. Tissue Res. 7, 1–60.

Fleischmajer, R., Timpl, R., Tuderman, L., Raisher, L., Wiestner, M., Perlish, J.S., & Graves, P.N. (1981). Ultrastructural identification of extension aminopropeptides of type I and type III collagens in human skin. Proc. Natl. Acad. Sci. USA 78, 7360–7364.

Fleischmajer, R., Perlish, J.S., Timpl, R., & Olsen, B.R. (1988). Procollagen intermediates during tendon fibrillogenesis. J. Histochem. Cytochem. 36, 1425–1432.

Fleischmajer, R., Fisher, L.W., MacDonald, E.D., Jacobs, L., Jr., Perlish, J.S., & Termine, J.D. (1991). Decorin interacts with fibrillar collagen of embryonic and adult human skin. J. Struct. Biol. 106, 82–90.

Fouser, L., Sage, E.H., Clark, J., & Bornstein, P. (1991). Feedback regulation of collagen gene expression: A Trojan horse approach. Proc. Natl. Acad. Sci. USA 88, 10158–10162.

Franc, S., Ledger, P.W., & Garrone, R. (1985). Structural variability of collagen fibers in the calcareous axial rod of a sea pen. J. Morphol. 184, 75–84.

Fujimoto, D., Akiba, K., & Nakaruma, N. (1977). Isolation and characterization of a fluorescent material in bovine achilles tendon collagen. Biochem. Biophys. Res. Commun. 76, 1124–1129.

Furthmayr, H., Wiedemann, H., Timpl, R., Odermatt, E., & Engel, J. (1983). Electron-microscopical approach to a structural model of intima collagen. Biochem. J. 211, 303–311.

Gabella, G. (1987). The cross-ply arrangement of collagen fibers in the submucosa of the mammalian small intestine. Cell Tissue Res. 248, 491–497.

Gacheru, S.N., Trackman, P.C., Shah, M.A., O'Gara, C.Y., Spacciapoli, P., Greenaway, F.T., & Kagan, H.M. (1990). Structural and catalytic properties of copper in lysyl oxidase. J. Biol. Chem. 265, 19022–19027.

Gaill, F., Lechaire, J.-P., & Denèfle, J.-P. (1991a). Fibrillar pattern of self-assembled and cell-assembled collagen: Resemblance and analogy. Biol. Cell 72, 149–158.

Gaill, F., Wiedemann, H., Mann, K., Kühn, K., Timpl, R., & Engel, J. (1991b). Molecular characterization of cuticle and interstitial collagens from worms collected at deep sea hydrothermal vents. J. Mol. Biol. 221, 209–223.

Galloway, J. (1986). Kinky variations of collagen Nature 322, 497–498.

Garrone, R. (1975). Collagen resorption in sponges: involvement of bacteria and macrophages In: Protides of the Biological Fluids. 22nd colloquium (Peeters, H., Ed.), pp. 59–63. Pergamon Press, Oxford.

Garrone, R. (1984). Formation and involvement of extracellular matrix in the development of sponges, a primitive multicellular system In: The Role of Extracellular Matrix in Development (Trelstadt, R.L., Ed.), pp. 461–477. Alan R. Liss, Inc., New York.

Gay, S., Martin, G.R., Müller, P.K., Timpl, R., Kühn, K. (1976). Simultaneous synthesis of types I and III collagen by fibroblasts in culture. Proc. Natl. Acad. Sci. USA 73, 4037–4040.

Ghosh, S.K. & Mitra, H.P. (1975). Oblique banding pattern in collagen reconstituted in vitro after trypsin treatment. Biochim. Biophys. Acta 405, 340–346.

Gordon, M.K., Gerecke, D.R., & Olsen, B.R. (1987). Type XII collagen: Distinct extracellular matrix component discovered by cDNA cloning. Proc. Natl. Acad. Sci. USA 84, 6040–6044.

Gordon, M.K., Gerecke, D.R., Dublet, B., van der Rest, M., & Olsen, B.R. (1989). Type XII collagen. A large multidomain molecule with partial homology to type IX collagen. J. Biol. Chem. 264, 19772– 19778.

Grandhee, S.K. & Monnier, V.M. (1991). Mechanism of formation of the Maillard protein cross-link pentosidine. J. Biol. Chem. 266, 11649–11653.

Grant, R.A., Cox, R.W., & Horne, R.W. (1967). The structure and assembly of collagen fibrils. II. An electron-microscope study of cross-linked collagen. J. R. Micr. Soc. 87, 143–155.

Greenspan, D.S., Cheng, W., & Hoffman, G.G. (1991). The pro-α1(V) collagen chain. Complete primary structure, distribution of expression, and comparison with the pro-α1(XI) collagen chain. J. Biol. Chem. 266, 24727–24733.

Gross, J. (1950). Connective tissue fine structure and some methods for its analysis. J. Gerontol. 5, 343–360.

Gross, J. (1956). The behavior of collagen units as a model in morphogenesis. J. Biophys. Biochem. Cytol. Suppl. 2, 26.

Gunwar, S., Saus, J., Noelken, M.E., and Hudson, B.G. (1990). Glomerular basement membrane. Identification of a fourth chain, α4, of type IV collagen. J. Biol. Chem. 265, 5466–5469.

Gunwar, S., Ballester, F., Kalluri, R., Timoneda, J., Chonko, A.M., Edwards, S.J., Noelken, M.E., & Hudson, B.G. (1991). Glomerular basement membrane. Identification of dimeric subunits of the noncollagenous domain (hexamer) of collagen IV and the Goodpasture antigen. J. Biol. Chem. 266, 15318–15324.

Halila, R. & Peltonen, L. (1986). Purification of human procollagen type III N-proteinase from placenta and preparation of an antiserum. Biochem. J. 239, 47–52.

Harbers, K., Kuehn, M., Delius, H., & Jaenisch, R. (1984). Insertion of retrovirus into the first intron of α1(I) collagen gene leads to embryonic lethal mutation in mice. Proc. Natl. Acad. Sci. USA 81, 1504–1508.

Hayase, F., Nagaraj, R.H., Miyata, S., Njoroge, F.G., & Monnier, V.M. (1989). Aging of proteins: immunological detection of a glucose-derived pyrrole formed during Maillard reaction *in vivo.* J. Biol. Chem. 263, 3758–3764.

Hayman, A.R., Köppel, J., & Trueb, B. (1991). Complete structure of the chicken α2(VI) collagen gene. Eur. J. Biochem. 197, 177–184.

Helaakoski, T., Vuori, K., Myllylä, R., Kivirikko, K.I., & Pihlajaniemi, T. (1989). Molecular cloning of the α-subunit of human prolyl 4-hydroxylase: The complete cDNA-derived amino acid sequence and evidence for alternative splicing of RNA transcripts. Proc. Natl. Acad. Sci. USA 86, 4392–4396.

Henkel, W. & Glanville, R.W. (1982). Covalent crosslinking between molecules of type I and type III collagen. Eur. J. Biochem. 122, 205–213.

Henkel, W., Glanville, R.W., & Greifendorf, D. (1987). Characterization of a type I collagen trimeric cross-linked peptide from calf aorta and its cross-linked structure. Eur. J. Biochem. 165, 427–436.

Hennessey, P.J., Ford, E.G., Black, T., & Andrassy, R.J. (1990). Wound collagenase activity correlates direct with collagen glycosylation in diabetic rats. J. Pediat. Surg. 25, 75–78.

Herbage, D., Bouillet, J., & Bernengo, J.C. (1977). Biochemical and physico-chemical characterization of pepsin-solubilized type II collagen from bovine articular cartilage Biochem. J. 161, 303–312.

Herbage, D., Le Lous, M., Cohen-Solal, L., & Bazin, S. (1985). Study of collagen cross-links Front. Matrix Biol. 10, 59–91.

Hicks, M., Delbridge, L., Yue, D.K., & Reeve, T.S. (1989). Increase in crosslinking of nonenzymatically glycosylated collagen induced by products of lipid peroxidation. Arch. Biochem. Biophys. 268, 249–254.

Hodge, A.J. & Petruska, J.A. (1963). Recent studies with the electron microscope on the ordered aggregates of the tropocollagen molecule In: Aspects of Protein Structure (Ramachandran, G.N., Ed.), p. 289. Academic Press, New York.

Hofmann, H., Fietzek, P.P., & Kühn, K. (1980). Comparative analysis of the sequences of the three collagen chains $\alpha1$(I),$\alpha2$ and $\alpha1$(III). Functional and genetic aspects. J. Mol. Biol. 141, 293–314.

Hojima, Y., McKenzie, J., van der Rest, M., & Prockop, D.J. (1989). Type I procollagen N-proteinase from chick embryo tendons. Purification of a new 500-kDa form of the enzyme and identification of the catalytically active polypeptides. J. Biol. Chem. 264, 11336–11345.

Hojima, Y., van der Rest, M., & Prockop, D.J. (1985). Type I procollagen C-proteinase from chick embryo tendons. Purification and characterization. J. Biol. Chem. 260, 15996–16003.

Holmes, D.F., Mould, A.P., & Chapman, J.A. (1991). Morphology of sheet-like assemblies of pN-collagen, pC-collagen and procollagen studied by scanning transmission electron microscopy mass measurements. J. Mol. Biol. 220, 111–123.

Hormel, S.E. & Eyre, D.R. (1991). Collagen in the ageing human intervertebral disc: An increase in covalently bound fluorophores and chromophores. J. Biochim. Biophys. Acta Protein Struct. Mol. Enzymol. 1078, 243–250.

Hostikka, S.L. & Tryggvason, K. (1988). The complete primary structure of the $\alpha2$ chain of human type IV collagen and comparison with the $\alpha1$(IV) chain. J. Biol. Chem. 263, 19488–19493.

Hostikka, S.L., Eddy, R.L., Byers, M.G., Höyhtyä, M., Shows, T.B., & Tryggvason, K. (1990). Identification of a distinct type IV collagen α chain with restricted kidney distribution and assignment of its gene to the locus of X chromosome-linked Alport syndrome. Proc. Natl. Acad. Sci. USA 87, 1606–1610.

Housley, T., Tanzer, M.L., Hensen, E., & Gallop, P.M. (1975). Collagen cross-linking: isolation of hydroxyaldolhistidine, a naturally-occurring cross-link. Biochem. Biophys. Res. Commun. 67, 824–830.

Hulmes, D.J.S., Jesior, J.C., Miller, A., Berthet-Colominas, C., Wolff, C. (1981). Electron microscopy shows periodic structure in collagen fibril cross sections. Proc. Natl. Acad. Sci. USA 78, 3567–3571.

Inoue, S. (1989). Ultrastructure of basement membranes. Int. Rev. Cytol. 117, 57–98.

Jakus, M.A. (1956). Studies on the cornea. II. The structure of the Descemet's membrane. J. Biophys. Biochem. Cytol. 2, 243–252.

Janes, S.M., Wu, D., Wemmer, D., Smith, A.J., Kam, S., Maltby, D., Burlingame, A.L., & Klinman, J.P. (1990). A new redox cofactor in eukaryotic enzymes: 6-hydroxydopa at the active site of bovine serum amine oxidase. Science 248, 981–987.

Just, M., Herbst, H., Hummel, M., Dürkop, H., Tripier, D., Stein, H., & Schuppan, D. (1991). Undulin is a novel member of the fibronectin-tenascin family of extracellular matrix glycoproteins. J. Biol. Chem. 266, 7326–7332.

Kadler, K.E., Hojima, Y., & Prockop, D.J. (1988). Assembly of type I collagen fibrils *de novo*. Between 37 and 41 °C the process is limited by micro-unfolding of monomers. J. Biol. Chem. 263, 10517–10523.

Kadler, K.E., Hojima, Y., & Prockop, D.J. (1990). Collagen fibrils *in vitro* grow from pointed tips in the *C*- to *N*-terminal direction. Biochem. J. 268, 339–343.

Kagan, H.M. (1986). Characterization and regulation of lysyl oxidase. In: Biology of the Extracellular Matrix. (Mecham, R.P., Ed.), Vol. 1, pp. 321–398. Academic Press, Orlando.

Kajava, A.V. (1991). Molecular packing in type I collagen fibrils. A model with neighboring collagen molecules aligned in axial register. J. Mol. Biol. 218, 815–823.

Kapoor, R., Sakai, L.Y., Funk, S., Roux, E., Bornstein, P., & Sage, E.H. (1988). Type VIII collagen has a restricted distribution in specialized extracellular matrices. J. Cell Biol. 107, 721–730.

Keene, D.R., Sakai, L.Y., Bächinger, H.P., & Burgeson, R.E. (1987a). Type III collagen can be present on banded collagen fibrils regardless of fibril diameter. J. Cell Biol. 105, 2393–2402.

Keene, D.R., Sakai, L.Y., Lunstrum, G.P., Morris, N.P., & Burgeson, R.E. (1987b). Type VII collagen forms an extended network of anchoring fibrils. J. Cell Biol. 104, 611–621.

Keene, D.R., Lunstrum, G.P., Morris, N.P., Stoddard, D.W. & Burgeson, R.E. (1991). Two type XII-like collagen localize to the surface of banded collagen fibrils. J. Cell Biol. 113, 971–978.

Kielty, C.M., Boot-Handford, R.P., Ayad, S., Shuttleworth, C.A. & Grant, M.E. (1990). Molecular composition of type VI collagen. Evidence for chain heterogeneity in mammalian tissues and cultured cells. Biochem. J. 272, 787–795.

Kielty, C.M., Cummings, C., Whittaker, S.P., Shuttleworth, C.A., & Grant, M.E. (1991). Isolation and ultrastructural analysis of microfibrillar structures from foetal bovine elastic tissues. Relative abundance and supramolecular architecture of type VI collagen assemblies and fibrillin. J. Cell Sci. 99, 797–807.

Kimura, T., Mattei, M.-G., Stevens, J.W., Goldring, M.B., Ninomiya, Y., & Olsen, B.R. (1989). Molecular cloning of rat and human type IX collagen cDNA and localization of the α1 (IX) gene on the human chromosome 6. Eur. J. Biochem. 179, 71–78.

Kivirikko, K.I., Myllylä, R., & Pihlajaniemi, T. (1989). Protein hydroxylation: Prolyl 4-hydroxylase, an enzyme with four cosubstrates and a multifunctional subunit. FASEB J. 3, 1609–1617.

Knight, D.P. & Hunt, S. (1976). Fine structure of the dogfish egg case: a unique collagenous material. Tissue and Cell 8, 183–193.

Kodama, T., Freeman, M., Rohrer, L., Zabrecky, J., Matsudaira, P., & Krieger, M. (1990). Type I macrophage scavenger receptor contains a-helical and collagen-like coiled coils. Nature 343, 531–535.

Koller, E., Winterhalter, K.H., & Trueb, B. (1989). The globular domains of type VI collagen are related to the collagen-binding domains of cartilage matrix protein and von Willebrand factor. EMBO J. 8, 1073–1077.

Kratochwil, K., Von der Mark, K., Kollar, E.J., Jaenisch, R., Mooslehner, K., Schwarz, M., Haase, K., Gmachl, I., & Harbers, K. (1989). Retrovirus-induced insertional mutation in *Mov13* mice affects collagen I expression in a tissue-specific manner. Cell 57, 807–816.

Kuboki, Y., Tsuzaki, M., Sasaki, S., Liu, C.F., & Mechanic, G. (1981). Location of the intermolecular cross-links in bovine dentin collagen, solubilization with trypsin and isolation of cross-links peptides containing dihydroxylysinorleucine and pyridinoline. Biochem. Biophys. Res. Commun. 102, 119–126.

Kujawa, M.J., Weitzhandler, M., Poole, A.R., Rosenberg, L., & Caplan, A.I. (1989). Association of the C-propeptide of type II collagen with mineralization of embryonic chick long bone and sternal development. Connect. Tissue Res. 23, 179–199.

Kwan, A.P.L., Cummings, C.E., Chapman, J.A., & Grant, M.E. (1991). Macromolecular organization of chicken type X collagen *in vitro*. J. Cell Biol. 114, 597–604.

Kühn, K. (1982). Segment-long-spacing crystallites, a powerful tool in collagen research Collagen. Rel. Res. 2, 61–80.

Kühn, K., Wiedemann, H., Timpl, R., Risteli, J., Dieringer, H., Voss, T., & Glanville, R.W. (1981). Macromolecular structure of basement membrane collagens. Identification of 7 S collagen as a cross-linking domain of type IV collagen. FEBS Lett. 125, 123.

Last, J.A., Amstrong, L.G., & Reiser, K.M. (1990). Biosynthesis of collagen crosslinks. Int. J. Biochem. 6, 559–564.

Laurie, G.W., Leblond, C.P., & Martin, G.R. (1982). Localization of type IV collagen, laminin, heparan sulfate proteoglycan, and fibronectin to the basal lamina of basement membranes. J. Cell Biol. 95, 340–344.
Le Pape, A., Guitton, J., & Muh, J.P. (1981). Modifications of glomerular basement membrane cross-links in experimental diabetic rats. Biochem. Biophys. Res. Commun. 3, 1214–1221.
Le Pape, A., Gutman, N., Guitton, J., Legrand, Y., & Muh, J.P. (1983). Nonenzymatic glycosylation increases platelet aggregating potency of collagen from placenta of diabetic human beings. Biochem. Biophys. Res. Commun. 111, 602–610.
Leblond, C.P. & Inoue, S. (1989). Structure, composition, and assembly of basement membrane. Am. J. Anat. 185, 367–390.
Lenaers, A., Ansay, M., Nusgens, B., & Lapière, C. (1971). Collagen made of extended α-chains, procollagen in genetically defective dermatosparaxic calves. Eur. J. Biochem. 23, 533–543.
Light, N.D. & Bailey, A.J. (1979). Changes in cross-linking during aging in bovine tendon collagen. FEBS Lett. 97, 183–188.
Light, N.D. & Bailey, A.J. (1980). Polymeric C-terminal cross-linked material from type I collagen. Biochem. J. 189, 111–124.
Light, N.D. & Bailey, A.J. (1982). Covalent cross-links in collagen. Meth. Enzymol. 82, 360–372.
Light, N. & Bailey, A.J. (1985). Collagen cross-links: location of pyridinoline in type I collagen. FEBS Lett. 182, 503–508.
Linsenmayer, T.F., Gibney, & Schmid, T.M. (1986). Intracellular avian type X collagen *in situ* and determination of its thermal stability using a conformation-dependent monoclonal antibody. Exp. Cell Res. 166, 15–22.
Little, C.D., Piquet, D.M., Davis, L.A., Walters, L., & Drake, C.J. (1989). Distribution of laminin, collagen type IV, collagen type I, and fibronectin chicken cardiac jelly/basement membrane. Anat. Rec. 224, 417–425.
Lunec, J., Blake, D.R., McCleary, S.J., Brailsford, S., & Bacon, P.A. (1985). Self-perpetuating mechanism of immunoglobulin G aggregation in rheumatoid inflammation. J. Clin. Invest. 76, 2084–2090.
Lunstrum, G.P., Morris, N.P., McDonough, A.M., Keene, D.R., & Burgeson, R .E. (1991). Identification and partial characterization of two type XII-like collagen molecules. J. Cell Biol. 113, 963–969.
Mallein-Gerin, F. & Garrone, R. (1986). Preservation of polygonal sections and internal domains by quick-freezing of collagen fibrils. Int. J. Biol. Macromol. 8, 121–123.
Mayne, R. & Burgeson, R.E. (1987). Structure and Function of Collagen Types. Academic Press, Orlando, FL.
McCormick, D., van der Rest, M., Goodship, J., Lozano, G., Ninomiya, Y., & Olsen, B.R. (1987). Structure of the glycosaminoglycan domain in the type IX collagen-proteoglycan. Proc. Natl. Acad. Sci. USA 84, 4044–4048.
McDevitt, C.A., Marcelino, J., & Tucker, L. (1991). Interaction of intact type VI collagen with hyaluronan. FEBS Lett. 294, 167–170.
Mechanic, G.L., Katz, E.P., Henmi, M., Noyes, C., & Yamauchi, M. (1987). Locus of a histidine-based, stable trifuntional, helix to helix collagen cross-link: stereospecific collagen structure of type I skin fibrils. Biochemistry 20, 3500–3509.
Meek, K.M., Chapman, J.A., & Hardcastle, R.A. (1979). The staining pattern of collagen fibrils. Improved correlation with sequence data. J. Biol. Chem. 254, 10710–10714.
Meek, K.M. & Holmes, D.F. (1983). Interpretation of electron microscopical appearance of collagen fibrils from corneal stroma. Int. J. Biol. Macromol. 5, 17–25.
Meek, K.M. & Chapman, J.A. (1985). Glutaraldehyde-induced changes in the axially projected fine structure of collagen fibrils. J. Mol. Biol. 185, 359–370.
Mendler, M., Eich-Bender, S.G., Vaughan, L., Winterhalter, K.H. & Bruckner, P. (1989). Cartilage contains mixed fibrils of collagen types II, IX, and XI. J. Cell Biol. 108, 191–197.

Miller, E.J. (1976). Biochemical characterization and biological significance of the genetically distinct collagens. Mol. Cell. Biochem. 13, 165–192.

Miller, E.J., & Matukas, V.J. (1969). Chick cartilage collagen: a new type of α 1 chain not present in bone or skin of the species. Proc. Natl. Acad. Sci. USA 64, 1264–1268.

Monnier, V.M., Sell, D.R., Nagaraj, R.H., & Miyata, S. (1991). Mechanisms of protection against damage mediated by the Maillard reaction in aging. Gerontology 37, 152–165.

Morris, N.P., Keene, D.R., Glanville, R.W., Bentz, H., & Burgeson, R.E. (1986). The tissue form of type VII collagen is an antiparallel dimer. J. Biol. Chem. 261, 5638–5644.

Morris, N.P., Watt, S.L., Davis, J.M., Bachinger, H.P. (1990). Unfolding intermediates in the triple helix to coil transition of bovine type XI collagen and human type V collagens $\alpha 1_2 \alpha 2$ and $\alpha 1 \alpha 2 \alpha 3$. J. Biol. Chem. 265, 10081–10087.

Morris, N.P. & Bächinger, H.P. (1987). Type XI collagen is a heterotrimer with a composition (1α,2α,3α) retaining non-triple-helical domains. J. Biol. Chem. 262, 11345–11350.

Muragaki, Y., Kimura, T., Ninomiya, Y., & Olsen, B.R. (1990a). The complete primary structure of two distinct forms of human α1(IX) collagen chains. Eur. J. Biochem. 192, 703–708.

Muragaki, Y., Nishimura, I., Henney, A., Ninomiya, Y., & Olsen, B.R. (1990b). The α1(IX) collagen gene gives rise to two different transcripts in both mouse embryonic and human fetal RNA. Proc. Natl. Acad. Sci. USA 87, 2400–2404.

Muragaki, Y., Jacenko, O., Apte, S., Mattei, M.-G., Ninomiya, Y., & Reino Olsen, B. (1991a). The α2(VIII) collagen gene. A novel member of the short chain collagen family located on the human chromosome 1. J. Biol. Chem. 266, 7721–7727.

Muragaki, Y., Mattei, M.-G., Yamaguchi, N., Olsen, B.R., & Ninomiya, Y. (1991b). The complete primary structure of the human α1(VIII) chain and assignment of its gene (COL8A1) to chromosome 3 . Eur. J. Biochem. 197: 615–622.

Murray, L.W. & Tanzer, M.L. (1983). Characterization of a large fragment from annelid cuticle collagen and its relationship to the intact molecule Collagen. Rel. Res. 3, 445–458.

Muthukumaran, G., Blumberg, B. & Kurkinen, M. (1989). The complete primary structure for the α1-chain of mouse collagen IV. Differential evolution of collagen IV domains. J. Biol. Chem. 264, 6310–6317.

Myers, J.C., Jones, T.A., Pohjolainen, E.-R., Kadri, A.S., Goddard, A.D., Sheer, D., Solomon, E., & Pihlajaniemi, T. (1990). Molecular cloning of α5(IV) collagen and assignment of the gene to the region of the X chromosome containing the Alport syndrome locus. Am. J. Hum. Genet. 46, 1024–1033.

Myllylä, R., Kaska, D.D., & Kivirikko, K.I. (1989). The catalytic mechanism of the hydroxylation reaction of peptidyl proline and lysine does not require protein disulphide-isomerase activity. Biochem. J. 263, 609–611.

Nandan, D., Cates, G.A., Ball, E.H., & Sanwal, B.D. (1990). Partial characterization of a collagen-binding, differentiation-related glycoprotein from skeletal myoblasts. Arch. Biochem. Biophys. 278, 291–296.

Nicholls, A.C., Pope, F.M., & Schloon, H.G. (1979). Biochemical heterogeneity of Osteogenesis Imperfecta: a new variant. Lancet 1, 1193.

Nicholls, A.C., Osse, G., Schloon, H.G., Lenard, H.G., Deak, S.B., Myers, J.C., Prockop, D.J., Weigel, W.R.F., Fryer, P., & Pope, F.M. (1984). The clinical features of homozygous α2(I) collagen deficient Osteogenesis Imperfecta. J. Med. Genet. 21, 257–262.

Nicholls, K. & Mandel, T.E. (1989). Advanced glycosylation end-products in experimental murine diabetic nephropathy: effect of islet isografting and aminoguanidine. Lab. Invest. 60, 486–492.

Ninomiya, Y., Gordon, M.K., van der Rest, M., Schmid, T.M., Linsenmayer, T.F., & Olsen, B.R. (1986). The developmentally regulated type X collagen gene contains a long open reading frame without introns. J. Biol. Chem. 261, 5041--5050.

Ninomiya, Y., Castagnola, P., Gerecke, D.R., Gordon, M.K., Jacenko, P., McCarthy, M., Muragaki, Y., Nishimura, I., Oh, S., Rosenblum, N., Sato, N., Sugrue, S.P., Taylor, R., Vasios, G., Yamaguchi,

N., & Olsen, B.R. (1990) The molecular biology of collagens with short triple-helical domains In: Extracellular Matrix Genes (Sandell, L.J. & Boyd, C.D., Eds.), pp. 79–114. Academic Press, San Diego.

Nishimura, I., Muragaki, Y., & Olsen, B.R. (1989). Tissue-specific forms of type IX collagen-proteoglycan arise from the use of two widely separated promoters. J. Biol. Chem. 264, 20033–20041.

Niyibizi, C. & Eyre, D.R. (1989). Identification of the cartilage α1(XI) chain in type V collagen from bovine bone. FEBS Lett. 242, 314–318.

O'Hara, P.J., Read, W.K., Romane, W.M., & Bridges, C.H. (1970). A collagenous dysplasia in calves Lab. Invest. 23, 307–314.

Olsen, B.R. (1963). Electron microscope studies on collagen. II: Mechanism of linear polymerization of tropocollagen molecules Z. Zellforsch. 59, 184.

Palade, G.E. & Farquhar, M.G. (1965). A special fibril of the dermis. J. Cell Biol. 26, 263–291.

Parente, M.G., Chung, L.C., Ryynänen, J., Woodley, D.T., Wynn, K.C., Bauer, E.A., Mattei, M.-G., Chu, M.-L., & Uitto, J. (1991). Human type VII collagen: cDNA cloning and chromosomal mapping of the gene. Proc. Natl. Acad. Sci. USA 88, 6931–6935.

Parkkonen, T., Kivirikko, K.I., & Pihlajaniemi, T. (1988). Molecular cloning of a multifunctional chicken protein acting as the prolyl 4-hydroxylase, β-subunit, protein disulphide-isomerase and a cellular thyroid-hormone-binding protein. Comparison of cDNA-deduced amino acid sequences with those in other species. Biochem. J. 256, 1005–1011.

Paulsson, M. (1987). Noncollagenous proteins of basement membranes. Collagen. Rel. Res. 7, 443–461.

Pease, D.C. & Bouteille, M. (1971). The tridimensional ultrastructure of native collageneous fibrils, cytochemical evidence for a carbohydrate matrix. J. Ultrastruct. Res. 35, 339–358.

Pihlajaniemi, T., Dickson, L.A., Pope, F.M., Korhonen, V.R., Nicholls, A.C., Prockop, D.J., & Myers, J.C. (1984). Osteogenesis Imperfecta: Cloning of a pro α2(I) collagen gene with a frameshift mutation. J. Biol. Chem. 259, 12941–12944.

Pihlajaniemi, T. & Tamminen, M. (1990). The α1 chain of type XIII collagen consists of three collagenous and four noncollagenous domains, and its primary transcript undergoes complex alternative splicing. J. Biol. Chem. 265, 16922–16928.

Pollner, R., Fischer, G., Pöschl, E., & Kühn, K. (1990). Regulation of divergent transcription of the genes coding for basement membrane type IV collagen. Ann. NY Acad. Sci. 580, 44–54.

Poole, A.R., Pidoux, I., Reiner, A., Choi, H., & Rosenberg, L.C. (1984). Association of an extracellular protein (chondrocalcin) with the calcification of cartilage in endochondral bone formation. J. Cell Biol. 98, 54–65.

Pringle, G.A. & Dodd, C.M. (1990). Immunoelectron microscopic localization of the core protein of decorin near the d and e bands of tendon collagen fibrils by use of monoclonal antibodies. J. Histochem. Cytochem. 38, 1405–1411.

Prockop, D.J., Berg, R.A., Kivirikko, K.I., & Uitto, J. (1976). Intracellular steps in the biosynthesis of collagen In: Biochemistry of Collagen (Ramachandran, G.N. & Reddi, A.H., Eds.), pp. 163–273. Plenum Press, New York.

Prockop, D.J., Kivirikko, K.I., Tuderman, L., & Guzman, N.A. (1979a). The biosynthesis of collagen and its disorders. (First of two parts). N. Engl. J. Med. 301, 13–23.

Prockop, D.J., Kivirikko, K.I., Tuderman, L., & Guzman, N.A. (1979b). The biosynthesis of collagen and its disorders. (Second of two parts). N. Engl. J. Med. 301, 77–85.

Ramachandran, G.N. & Ramakrishnan, C. (1976). Molecular structure In: Biochemistry of Collagen (Ramachandran, G.N. & Reddi, A.H., Eds.), pp. 45–84. Plenum Press, New York.

Reale, E., Benazzo, F., & Ruggeri, A. (1981). Differences in the microfibrillar arrangement of collagen fibrils. Distribution and possible significance. J. Submicrosc. Cytol. 13, 135–143.

Reiser, K.M. (1991). Nonenzymatic glycation of collagen in aging and diabetes. Proc. Soc. Exp. Biol. Med. 196, 17–29.

Reiser, K.M., Crouch, E.C., Chang, K, & Williamson, J.R. (1991). Lysyl oxidase-mediated crosslinking in granulation tissue collagen in two models of hyperglycemia. Biochim. Biophys. Acta Mol. Basis Dis. 1097, 55–61.
Ricard-Blum, S. & Ville, G. (1988). Collagen cross-linking. Cell. Mol. Biol. 34, 581–590.
Ricard-Blum, S. & Ville, G. (1989). Collagen cross-linking. Int. J. Biochem. 21, 1185–1189.
Robins, S.P. & Bailey, A.J. (1972). Age-related changes in collagen: the identification of reducible lysine carbohydrate condensation products. Biochem. Biophys. Res. Commun. 48, 76–84.
Robins, S.P. (1982). Analysis of the crosslinking components in collagen and elastin. Meth. Biochem. Anal. 28, 329–379.
Robins, S.P. (1983). Cross-linking of collagen: isolation, structural characterization and glycosylation of pyridinoline. Biochem. J. 215, 167–173.
Robins, S.P. & Ducan, A. (1987). Pyridiniun cross-links of bone collagen and their location in peptides isolated from rat femur. Biochim. Biophys. Acta 914, 233–239.
Robins, S.P. (1988). Functional properties of collagen and elastin Bailliere's Clin. Rheum. 2, 1–36.
Romero-Chapman, N., Lee, J., Tinker, D., Uriu-Hare, J.Y., Keen, C.L., & Rucker, R.R (1991). Purification, properties and influence of dietary copper on accumulation and functional activity of lysyl oxidase in rat skin. Biochem. J. 275, 657–662.
Ronzière, M.-C., Ricard-Blum, S., Tiollier, J., Hartmann, D.J., Garrone, R., & Herbage, D. (1990). Comparative analysis of collagens solubilized from human foetal, and normal and osteoarthritic adult articular cartilage, with emphasis on type VI collagen. Biochim. Biophys. Acta Protein Struct. Mol. Enzymol. 1038, 222–230.
Rowlatt, C. (1969). Subepithelial fibrils associated with the basal lamina under simple epithelia in mouse uterus: possible tropocollagen aggregates. J. Ultrastruct. Res. 26, 44–51.
Ruggiero, F., Pfäffle, M., Von der Mark, K. & Garrone, R. (1988). Retention of carboxypropeptides in type-II collagen fibrils in chick embryo chondrocyte cultures. Cell Tissue Res. 252, 619–624.
Ryan, M.C. & Sandell, L.J. (1990). Differential expression of a cysteine-rich domain in the amino-terminal propeptide of type II (cartilage) procollagen by alternative splicing of mRNA. J. Biol. Chem. 265, 10334–10339.
Saitta, B., Stokes, D.G., Vissing, H., Timpl, R., & Chu, M.-L. (1990). Alternative splicing of the human α2(VI) collagen gene generates multiple mRNA transcripts which predict three protein variants with distinct carboxyl termini. J . Biol . Chem. 265, 6473–6480.
Sakai, L.Y., Keene, D.R., Morris, N.P., & Burgeson, R.E. (1986). Type VII collagen is a major structural component of anchoring fibrils. J. Cell. Biol. 103, 1577–1586.
Sandell, L.J. & Boyd, C.D. (1990a). Extracellular Matrix Genes. Academic Press, San Diego.
Sandell, L.J. & Boyd, C.D. (1990b). Conserved and divergent sequence and functional elements within collagen genes In: Extracellular Matrix Genes (Sandell, L.J. & Boyd. C.D., Eds.), pp. 1–56. Academic Press, San Diego.
Sandell, L.J., Morris, N., Robbins, J.R., & Goldring, M.B. (1991). Alternatively spliced type II procollagen mRNAs define distinct populations of cells during vertebral development: Differential expression of the amino-propeptide. J. Cell Biol. 114, 1307–1319.
Saus, J., Wieslander, J., Langeveld, J.P.M., Quinones, S., & Hudson, B.G. (1988). Identification of the Goodpasture antigen as the α3(IV) chain of collagen IV. J. Biol. Chem. 26, 13374–13380.
Sawada, H., Konomi, H., & Hirosawa, K. (1990). Characterization of the collagen in the hexagonal lattice of Descemet's membrane: Its relation to type VIII collagen. J. Cell Biol. 110, 219–227.
Schittny, J.C., Timpl, R., & Engel, J. (1988). High resolution immunoelectron microscopic localization of functional domains of laminin, nidogen, and heparan sulfate proteoglycan in epithelial basement membrane of mouse cornea reveals different topological orientations. J. Cell Biol. 107, 1599–1610.
Schnider, S.L. & Kohn, R.R. (1980). Glucosylation of human collagen in aging and diabetes mellitus. J. Clin. Inv. 66, 1179–1181.

Schuppan, D., Cantaluppi, M.C., Becker, J., Veit, A., Bunte, T., Troyer, D., Schuppan, F., Schmid, M., Ackermann, R., & Hahn, E.G. (1990). Undulin, an extracellular matrix glycoprotein associated with collagen fibrils. J. Biol. Chem. 265, 8823–8832.

Scott, J.E., Hughes, E.W., & Shuttleworth, A. (1981). A collagen-associated Ehrlich chromogen: a pyrrolic cross-link? Biosci. Rep. 1, 611–618.

Scott, J.E., Qian, R.G., Henkel, W., & Glanville, R.W. (1983). An Ehrlich chromogen in collagen cross-links. Biochem. J. 209, 263–264.

Scott, J.E. (1988). Proteoglycan-fibrillar collagen interactions. Biochem. J. 252, 313–323.

Sell, D.R. & Monnier, V.M. (1989). Structure elucidation of a senescence cross-link from human extracellular matrix. J. Biol. Chem. 264, 21597–21602.

Sell, D.R. & Monnier, V.M. (1990). End-stage renal disease and diabetes catalyze the formation of a pentose-derived crosslink from aging human collagen. J. Clin. Invest. 85, 380–384.

Shaw, L.M. & Olsen, B.R. (1991). FACIT collagens: Diverse molecular bridges in extracellular matrices. Trends. Biochem. Sci. 16, 191–194.

Shimokomaki, M., Duance, V.C., & Bailey, A.J. (1981). Identification of two further collagenous fractions from articular cartilage. Bioscience Rep. 1, 561–570.

Siebel, M.J., Duncan, A., & Robins, S.P. (1989). Urinary hydroxypyridinium cross-links provide indices of cartilage and bone involvement in arthritic diseases. J. Rheumatol. 16, 964–970.

Siebold, B., Qian, R.-Q., Glanville, R.W., Hofmann, H., Deutzmann, R., & Kühn, K. (1987). Construction of a model for the aggregation and cross-linking region (7S domain) of type IV collagen based upon an evaluation of the primary structure of the α1 and α2 chains in this region. Eur. J. Biochem. 168, 569–575.

Siebold, B., Deutzmann, R., & Kühn, K. (1988). The arrangement of intra- and intermolecular disulfide bonds in the carboxyterminal, non-collagenous aggregation and cross-linking domain, of basement-membrane type IV collagen. Eur. J. Biochem. 176, 617–624.

Siegel, R.C. (1974). Biosynthesis of collagen cross-links: increased activity of purified lysyl oxidase with reconstituted collagen fibrils. Proc. Natl. Acad. Sci. USA 71, 4826–4830.

Silver, F.H. & Trelstad, R.L. (1979). Linear aggregation and the turbidimetric lag phase: type I collagen fibrillogenesis *in vitro* J. Theor. Biol. 81, 515–526.

Stinson, R.H. & Sweeny, P.R. (1980). Skin collagen has an unusual d-spacing. Biochim. Biophys. Acta 621, 158–161.

Stokes, D.G., Saitta, B., Timpl, R., Chu, M.-L. (1991). Human α3(VI) collagen gene. Characterization of exons coding for the amino-terminal globular domain and alternative splicing in normal and tumor cells. J. Biol. Chem. 266, 8626–8633.

Stubbs, M., Summers, L., Mayr, I., Schneider, M., Bode, W., Huber, R., Ries, A., & Kühn, K. (1990). Crystals of the NC1 domain of human type IV collagen. J. Mol. Biol. 211, 683–684.

Sun, C.N. & White, H.J. (1975). Extracellular cross-striated banded structures in human connective tissue. Tissue and Cell 7, 419–432.

Susi, F.R. (1969). Anchoring fibrils in the attachment of epithelium to connective tissue in oral mucous membranes. J. Dent. Res. 48, 144–148.

Svoboda, K.K., Nishimura, I., Sugrue, S.P., Ninomiya, Y., & Olsen, B.R. (1988). Embryonic chicken cornea and cartilage synthesize type IX collagen molecules with different amino-terminal domains. Proc. Natl. Acad. Sci. USA 85, 7496–7500.

Tanaka, S., Avigad, G., Brodsky, B., & Eikenberry, E.F. (1988a). Glycation induces expansion of the molecular packing of collagen. J. Mol. Biol. 203, 495–505.

Tanaka, S., Avigad, G., Eikenberry, E.F., & Brodsky, B. (1988b). Isolation and partial characterization of collagen, chains dimerized by sugar-derived cross-links. J. Biol. Chem. 263, 17650–17657.

Tanzer, M.L., Fairweather, R., & Gallop, P.M. (1972). Collagen cross-links. Isolation of reduced N-hexosyl hydroxylysine from borohydride reduced calf skin insoluble collagen. Arch. Biochem. Biophys. 151, 137–141 .

Tasanen, K., Parkkonen, T., Chow, L.T., Kivirikko, K.I., & Pihlajaniemi, T. (1988). Characterization of the human gene for a polypeptide that acts both as the β subunit of prolyl 4-hydroxylase and as protein disulfide isomerase. J. Biol. Chem. 263, 16218–16224.

Thomas, J.T., Boot-Handford, R.P., Marriott, A., Kwan, A.P.L., Ayad, S., & Grant, M.E. (1990). Isolation of cDNAs encoding bovine type X collagen. Ann. NY Acad. Sci. 580, 477–479.

Tikka, L., Pihlajaniemi, T., Henttu, P., Prockop, D.J., & Tryggvason, K. (1988). Gene structure for the α1 chain of a human short-chain collagen (type XIII) with alternatively spliced transcripts and translation termination codon at the 5′ end of the last exon. Proc. Natl. Acad. Sci. USA 85, 7491–7495.

Tikka, L., Elomaa, O., Pihlajaniemi, T., & Tryggvason, K. (1991). Human α1(XIII) collagen gene. Multiple forms of the gene transcripts are generated through complex alternative splicing of several short exons. J. Biol. Chem. 266, 17713–17719.

Timpl, R., Wiedemann, H., van Delden, V., Furthmayr, H., & Kühn, K. (1981). A network model for the organization of type IV collagen molecules in basement membranes. Eur. J. Biochem. 120, 203–211.

Timpl, R. & Engel, J. (1987). Type VI collagen. In: Structure and Function of Collagen Types (Mayne, R. & Burgeson, R.E., Eds.), pp. 105–143. Academic Press, Orlando, FL.

Timpl, R. (1989). Structure and biological activity of basement membrane proteins. Eur. J. Biochem. 180, 487–502.

Tiollier, J. & Garrone, R. (1985). Ultrastructure des collagènes mineurs du cartilage In: Cartilage 85 (Harmand, M.-F. & Ducassou, D., Eds.), pp. 23–30. Biomat, Bordeaux.

Torri Tarelli, L., & Petruccioli, M.G. (1971). Studies on the ultrastructure of collagen fibrils. II Filamentous structure with negative staining. J. Submicr. Cytol. 3, 153–170.

Trackman, P.C., Pratt, A.M., Wolanski, A., Tang, S.S., Offner, G.D., Troxler, R.F., & Kagan, H. (1990). Cloning of rat aorta lysyl oxidase cDNA: Complete codons and predicted amino acid sequence. Biochemistry 29, 4863–4870.

Trelstad, R.L. & Hayashi, K. (1979). Tendon collagen fibrillogenesis: Intracellular subassemblies and cell surface changes associated with fibril growth. Dev. Biol 71, 228–242.

Tsilibary, E.C., Charonis, A.S., Reger, L.A., Wohlhueter, R.M., & Furcht, L.T. (1988). The effect of nonenzymatic glycosylation on the binding of the main noncollagenous NC1 domain to type IV collagen. J. Biol. Chem. 263, 4302–4308.

Uchiyama, A., Ohishi, T., Takahashi, M., Kushida, K., Jnone, T., Fujie, M., & Horiuchi, K. (1991). Flourophores from aging human articular cartilage. Biochem. J. 110, 714–718.

van der Rest, M., Rosenberg, L.C., Olsen, B.R., & Poole, A.R. (1986). Chondrocalcin is identical to the C-propeptide of type II procollagen. Biochem. J. 237, 923–925.

van der Rest, M. & Mayne, R. (1987). Type IX collagen In: Structure and Function of the Collagen Types (Mayne, R. & Burgeson, R.E., Eds.), pp. 195–221. Academic Press, Orlando, FL.

van der Rest, M. & Mayne, R. (1988). Type IX collagen from cartilage is covalently crosslinked to type II collagen. J. Biol. Chem. 263, 1615–1618.

van der Rest, M., Aubert-Foucher, E., Dublet, B., Eichenberger, D., Font, B., & Goldschmidt, D. (1991). Structure and function of the fibril-associated collagens. Biochem. Soc. Trans. 19, 820–824.

van der Rest, M. & Garrone, R. (1990). Collagens as multidomain proteins. Biochimie 72, 473–484.

van der Rest, M. & Garrone, R. (1991). The collagen family of proteins. FASEB J. 5, 2814–2823.

Vasan, N.S., Kuivaniemi, H., Vogel, B.E., Minor, R.R., Wootton, J.A.M., Tromp, G., Weksberg, R., & Prockop, D.J. (1991). A mutation in the proα2(I) gene (COLIA2) for type I procollagen in Ehlers-Danlos syndrome type VII: Evidence suggesting that skipping of exon 6 in RNA splicing may be a common cause of the phenotype. Am. J. Hum. Genet. 48, 305–317.

Vasios, G., Nishimura, I., Konomi, H., van der Rest, M., Ninomiya, Y., & Olsen, B.R. (1988). Type IX collagen-proteoglycan contains a large amino-terminal globular domain encoded by multiple exons. J. Biol. Chem. 263, 2324–2329.

Vaughan, L., Winterhalter, K.H., & Bruckner, P. (1985). Proteoglycan Lt from chicken embryo sternum identified as type IX collagen. J. Biol. Chem. 260, 4758–4763.

Vaughan, L., Mendler, M., Huber, S., Bruckner, P., Winterhalter, K.H., Irwin, M.H., & Mayne, R. (1988). D-periodic distribution of type IX collagen along cartilage fibrils. J. Cell Biol. 106, 991–997.

Veis, A., Miller, A., Leibovich, S.J., & Traub, W. (1979). The limiting collagen microfibril. The minimum structure demonstrating native axial periodicity. Biochim. Biophys. Acta 576, 88–98.

Veis, A. & Kirk, T.Z. (1989). The coordinate synthesis and cotranslational assembly of type I procollagen. J. Biol. Chem. 264, 3884–3889.

Vlassara, H., Valinsky, J., Brownlee, M., Cerami, C., Nishimoto, S., & Cerami, A. (1987). Advanced glycosylation endproducts on the erythrocyte cell surface induce receptor-mediated phagocytosis by macrophages: a model for turnover of aging cells. J. Exp. Med. 166, 539–530.

Vlassara, H., Brownlee, M., Manogue, K., Dinarello, C., & Pasagia, A. (1988). Cachectin/TNF and IL-1 induced by glucose-modified proteins: role in normal tissue remodeling. Science 240, 1546–1549.

von der Mark, H., Aumailley, M., Wick, G., Fleischmajer, R., & Timpl, R. (1984). Immunochemistry, genuine size and tissue localization of collagen VI. Eur. J. Biochem. 142, 493–502.

Vuorio, E. & de Crombrugghe, B. (1990). The family of collagen genes. Annu. Rev. Biochem. 59, 837–872.

Weber, S., Engel, J., Wiedemann, H., Glanville, R.W., & Timpl, R. (1984). Subunit structure and assembly of the globular domain of basement-membrane collagen type IV. Eur. J. Biochem. 139, 401–410.

Weber, S., Dölz, R., Timpl, R., Fessler, J.H., & Engel, J. (1988). Reductive cleavage and reformation of the interchain and intrachain disulfide bonds in the globular hexameric domain NC1 involved in network assembly of basement membrane collagen (type IV). Eur. J. Biochem. 175, 229–236.

Weil, D., D'Alessio, M., Ramirez, F., De Wet, W.J., Cole, W.G., Chan, D., & Bateman, J.F. (1989a). A base substitution in the exon of a collagen gene causes alternative splicing and generates a structurally abnormal polypeptide in a patient with Ehlers-Danlos syndrome type VII. EMBO J. 8, 1705–1710.

Weil, D., D'Alessio, M., Ramirez, F., Steinmann, B., Wirtz, M.K., Glanville, R.W., & Hollister, D.W. (1989b). Temperature-dependent expression of a collagen splicing defect in the fibroblasts of a patient with Ehlers-Danlos syndrome Type VII. J. Biol. Chem. 264, 16804–16809.

Weinstock, M. & Leblond, C.P. (1974). Synthesis, migration, and release of precursor collagen by odontoblasts as visualized by radioautography after [3H] proline administration. J. Cell Biol. 60, 92–127.

Wiedemann, H., Chung, E., Fuji, T., Miller, E.J., & Kühn, K. (1975). Comparative electron microscope studies on type III and type I collagens. Eur. J. Biochem. 51, 363–368.

Williamson, P.R., Moog, R.S., Dooley, D.M., & Kagan, H.M. (1986). Evidence for pyrroloquinolinequinone as the carbonyl cofactor in lysyl oxidase by absorption and resonance Raman spectroscopy. J. Biol. Chem. 261, 16302–16305.

Williamson, P.R. & Kagan, H.M. (1987). α-Proton abstraction and carbanion formation in the mechanism of action of lysyl oxidase. J. Biol. Chem. 262, 8196–8201.

Wu, J.-J. & Eyre, D.R. (1989) Covalent interactions of type IX collagen in cartilage. Connect. Tissue Res. 20, 241–246.

Wu, J.J. & Eyre, D.R. (1984). Identification of hydroxypyridinium cross-linking sites in type II collagen of bovine articular cartilage. Biochemistry 23, 1850–1857.

Yada, T., Suzuki, S., Kobayashi, K., Kobayashi, M., Hoshino, T., Horie, K., & Kimata, K. (1990). Occurrence in chick embryo vitreous humor of a type IX collagen proteoglycan with an extraordinarily large chondroitin sulfate chain and short α1 polypeptide. J. Biol. Chem. 265, 6992–6999.

Yamagata, M., Yamada, K.M., Yamada, S.S., Shinomura, T., Tanaka, H., Nishida, Y., Obara, M., & Kimata, K. (1991). The complete primary structure of type XII collagen shows a chimeric molecule with reiterated fibronectin type III motifs, von Willebrand factor A motifs, a domain homologous to a noncollagenous region of type IX collagen and short collagenous domains with an Arg-Gly-Asp site. J. Cell Biol. 115, 209–221.

Yamaguchi, N., Benya, P.D., van der Rest, M., & Ninomiya, Y. (1989). The cloning and sequencing of α1(VIII) collagen cDNAs demonstrate that type VIII collagen is a short chain collagen and contains triple-helical and carboxyl-terminal non-triple-helical domains similar to those of type X collagen. J. Biol. Chem. 264, 16022–16029.

Yamauchi, M., London, R.E., Guenat, C., Hashimoto, F., & Mechanic, G.L. (1987). Structure and formation of a stable histidine-based trifunctional cross-link in skin collagen. J. Biol. Chem. 262, 11428–11434.

Yamauchi, M. & Mechanic, G.L. (1987). Cross-linking of collagen. In: Collagen (Nimni, M.E., Ed.), pp. 157–170. CRC Press, Boca Raton.

Yamauchi, M., Woodley, D.T., & Mechanic, G.L. (1988). Aging and cross-linking of skin collagen. Biochem. Biophys. Res. Commun. 152, 898–903.

Yamauchi, M., Prisayanh, P., Haque, Z., & Woodley, D.T. (1991). Collagen cross-linking in sun-exposed and unexposed sites of aged human skin. J. Invest. Dermatol. 97, 938–941.

Yoshioka, H. & Ramirez, F. (1990). Pro-α1(XI) collagen. Structure of the amino-terminal propeptide and expression of the gene in tumor cell lines. J. Biol. Chem. 265, 6423–6426.

Yurchenco, P.D., Tsilibary, E.C., Charonis, A.S., & Furthmayr, H. (1986). Models of self-assembly of basement membrane. J. Histochem. Cytochem. 34, 93–102.

Yurchenco, P.D. & Ruben, G.C. (1987). Basement membrane structure *in situ*: Evidence for lateral association in the type IV collagen network. J. Cell Biol. 105, 2559–2568.

Yurchenco, P.D. & Ruben, G.C. (1988). Type IV collagen lateral associations in the EHS tumor matrix: Comparison with amniotic and *in vitro* networks. Am. J. Pathol. 132, 278–291.

Yurchenco, P.D. & Schittny, J.C. (1990). Molecular architecture of basement membranes. FASEB J. 4, 1577–1590.

PROTEOGLYCAN GENE FAMILIES

John R. Hassell, Thomas C. Blochberger, Jody A. Rada, Shukti Chakravarti, and Douglas Noonan

Advances in Molecular and Cell Biology
Volume 6, pages 69–113.

ISBN: 1-55938-515-4

I. STRUCTURE

Proteoglycans consist of a core protein containing covalently bound glycosaminoglycan (GAG) side chains (see Hascall et al., 1991; Kjellén and Lindahl, 1991; Wight et al., 1991 for recent reviews). The number of GAG side chains can vary from 1 to over 100, depending upon the nature of the core protein and the cell type in which the core protein is expressed. GAGs are repeating disaccharides with covalently bound sulfate esters. The structure of the various GAGs and their linkage regions have been covered extensively in previous reviews (Fransson, 1985; Kjellén and Lindhal, 1991). There are basically three different kinds of GAG side chains: (1) chondroitin/dermatan sulfate, which consists of a repeating disaccharide of galactosamine and either glucuronic acid or iduronic acid; (2) heparin and heparan sulfate, which consists of a repeating disaccharide of glucosamine and either glucuronic acid or iduronic acid; and (3) keratan sulfate, which consists of a repeating disaccharide of glucosamine and galactose. All of these repeating disaccharides are unbranched and are sulfated in various positions. Hyaluronate is also a repeating disaccharide but it is not sulfated and is not known to be bound to any protein. It exists in a free form and not as a proteoglycan. Keratan sulfate lacks uronic acid and is, therefore, somewhat distinct from chondroitin/dermatan sulfate, heparin, heparan sulfate, and hyaluronate.

The GAG side chains are attached to the core protein through a sequence of sugars known as the linkage region. The linkage region for chondroitin/dermatan sulfate, heparin, and heparan sulfate is glucuronic acid-galactose-galactose-xylose-serine. Keratan sulfate can be linked to either serine/threonine or asparagine. In aggrecan, the major proteoglycan of cartilage, the linkage region for keratan sulfate is galactose-N-acetylglucosamine-N-acetylglucosamine-serine/threonine. In lumican, the major proteoglycan of the corneal stroma, the linkage region for keratan sulfate is a complex type N-linked oligosaccharide. Proteoglycans are also often substituted with O-linked and N-linked oligosaccharides. The N-linked oligosaccharides can be either the complex or high mannose type.

The core proteins of proteoglycans exhibit diverse structure. They contain most of the structural motifs (i.e., transmembrane segments, cysteine rich regions, leucine rich repeats, globular domains) seen in many glycoproteins. A hypothetical proteoglycan, fantastican, is shown with diagrammatic representations of the structural features contained in the core proteins and the side chains of proteoglycans is shown in Figure 1. Fantastican performs many of the same functions as imagican (Kjellén and Lindahl, 1991). The structure and function of core proteins is the subject of this review.

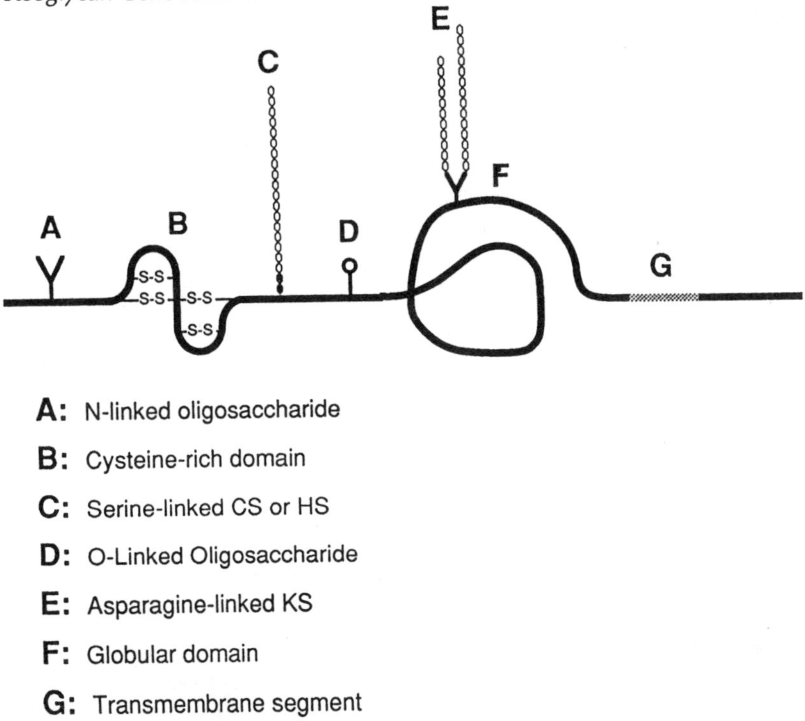

Figure 1. Fantastican: A hypothetical composite proteoglycan that illustrates the structural features found on proteoglycans.

II. BIOSYNTHESIS

The general steps involved in the synthesis of proteoglycans is shown in Figure 2. The initial translation product of the proteoglycan, the precursor protein, takes place from mRNA in the rough endoplasmic reticulum. High mannose-containing N-linked oligosaccharides are also transferred to the precursor protein at certain asparagine residues that are followed by any amino acid (except proline) and either serine or threonine. The partially glycosylated precursor protein then moves to the Golgi where the majority of posttranslational modifications take place. These modifications include: (1) The conversion of certain high mannose N-linked oligosaccharides to complex type oligosaccharides; (2) addition of O-linked oligosaccharides into appropriate serine and threonine residues; (3) addition of linkage regions to certain serine residues that are followed by a glycine residue; (4) addition of GAG to linkage regions; and (5) sulfation of GAGs. The completed proteoglycan may then enter a storage granule for secretion via the regulated pathway or be constitutively secreted as either an extracellular matrix proteoglycan or externalized as a cell surface proteoglycan with an intercalated transmembrane segment.

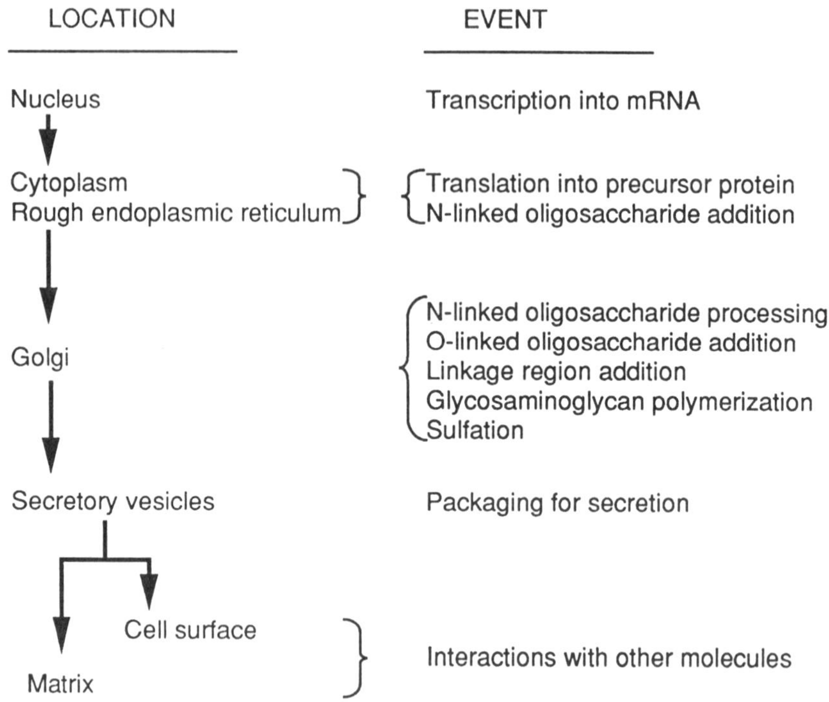

Figure 2. Major steps involved in the biosynthesis and secretion of proteoglycans.

III. CLASSIFICATION AND NOMENCLATURE

Historically, GAGs were isolated and characterized first. When it was later discovered that GAGs were attached to protein, the term proteoglycan was invented and proteoglycans were named after the type of GAG side chain present. Attention also changed to investigating the structure and properties of the intact proteoglycan. One of the first properties discovered was the ability of some proteoglycans to aggregate. This activity was shown to be due to a region of the core protein that could bind to hyaluronic acid and lead to a nomenclature that, in addition to naming the GAG type, also defined a proteoglycan as either aggregating or nonaggregating. Proteoglycan nomenclature was also expanded to include its size (i.e., large vs. small), its buoyant density in CsCl (i.e., high vs. low), and the name of the tissues from which the proteoglycan was isolated. As a result, many proteoglycans have long, multiword names that reflect all of these characteristics.

The information gleaned from core protein characterization, however, did not necessarily relate to this multiword nomenclature. Antibodies (both polyclonal and monoclonal) prepared against proteoglycans showed that the core proteins of some

Table 1. Chromosome Localizations of Proteoglycan Core Protein Genes

Name	*Chromosome*	*Reference*
Syndecan	mouse chromosome 12	Oettinger et al., 1991
Biglycan	human chromosome X27-ter	Fisher et al., 1991
Decorin	human chromosome 129	McBride et al., 1990
Perlecan	human chromosome 1p36.1	Wintle et al., 1990 Kallunki et al., 1991
	mouse chromosome 4 linked to Ak-2	Chakravarti et al., 1991
Fibroglycan	human chromosome 8q23	Marynen et al., 1991

proteoglycans had related, if not identical, epitopes, while the core proteins of other proteoglycans were immunologically distinct. Antibodies were also used to identify precursor proteins to some proteoglycans. Furthermore, the use of purified chondroitinase ABC, heparatinase, and keratanase to remove GAG side chains on proteoglycans without proteolysis allowed estimates of core protein sizes to be obtained by SDS-PAGE. As a result of these characterizations, distinct families of proteoglycans began to emerge (for review see Hassell et al., 1986).

Knowledge of the structure of the core protein of proteoglycans increased dramatically when recombinant DNA technology was used to isolate cDNA clones to proteoglycans. The first clone to a proteoglycan, serglycin, was obtained in 1985 (Bourdon et al.) and complete cDNA clones for at least 16 different proteoglycans have been obtained, sequenced, and published as of April 1992. Chromosome localizations have been done for a few of these proteoglycans (Table 1). The amino acid sequence deduced from the cDNA clones has provided the complete core protein structure for many proteoglycans and established their identity. Analysis of these core protein sequences showed that there is no relationship between the type of GAG side chain and any structural domain on the core protein, except, of course, for the 2–3 amino acids at the GAG attachment site. Furthermore, there are no structural domains common to all proteoglycans, like the G-X-Y repeat in collagen, that can be used to tie them together as a group. Consequently, there is a growing trend away from the multiword nomenclature and to refer to each proteoglycan by a single simple name (i.e., aggrecan, syndecan, decorin, serglycin, etc.) that is based on some structural and/or function properties of the core protein. This nomenclature is more consistent with that of the other components of the extracellular matrix, where a Roman numeral, in the case of collagen, or a simple name, in the case of glycoproteins, is used to identify the gene or the gene product.

Despite the lack of a single unifying structural domain common to the core protein of all proteoglycans, there are distinguishing characteristics that allow them to be grouped into three broad categories. The first group consists of those whose core proteins show only little or no homology (i.e., similarity) with each other or with known proteins or glycoproteins. These may be, or may have been derived from, the first proteoglycans to evolve. The second group of proteoglycans consists of those whose core proteins show considerable homology to known glycoproteins and, in some cases, to each other as well. These proteoglycans likely evolved from the ancestorial genes for extracellular matrix glycoproteins by the simple recruitment and addition of a GAG binding region to a structural domain(s) from existing glycoproteins. The third group of proteoglycans has previously been identified and designated as "part-time proteoglycans" (for reviews see Kjellén and Lindahl, 1991; Wight et al., 1991). These are proteoglycans whose complete core protein also exists as a glycoprotein. In these cases, the same gene is used for the production of both a glycoprotein and a proteoglycan, and what differs is whether or not the linkage regions and/or GAG side chains are put on the translation product. These proteoglycans are probably at a transition stage during evolution. This review will focus on the structural features and functional properties of the structurally distinct proteoglycans and the proteoglycans that are homologous to known glycoproteins, the proteoglycans found primarily in the first two groups.

IV. STRUCTURALLY DISTINCT PROTEOGLYCANS

This group of proteoglycans is typified by their notable lack of similarity of their core proteins. The majority of proteoglycans in this group have a transmembrane segment and because of this are cell surface proteoglycans. Since cell surfaces developed very early in evolution and the amino acid sequence in their core proteins are unique, showing little or no similarity to known glycoproteins, it may be that these proteoglycans represent the first proteoglycans to evolve. This group can be divided into two major families: proteoglycans associated with the secretory pathway, and the cell surface proteoglycans. The secretory pathway associated class has but a single known member, serglycin.

A. Serglycin: A Secretory Granule Proteoglycan

Serglycin was the first proteoglycan whose complete sequence was determined (Bourdon et al., 1986). It received several names but serglycin was finally adopted since it aptly describes the major feature of this molecule, a series of serine-glycine repeats. Serglycin was originally isolated from rat yolk sac tumor cells as the predominant chondroitin sulfate proteoglycan produced by these cells (Bourdon et al., 1985; Bourdon et al., 1986). The deduced amino acid sequence for the core protein shows it to be very small (19 kDa). It has since been cloned from rat basophilic leukemia-1 cells (Avraham et al., 1988), human promyelocytic leukemia cells (Stevens et al., 1988), human platelets (Alliel et al., 1988), mouse mast cells (Avraham et al., 1989b), mouse mastocytoma cells (Kjellén et al., 1989), hemopoietic cells (Stellrecht and Saunders, 1989), and rat natural killer cells (Giorda et al.,

```
Hum  MMQKLLKCSRLVLALALILVLESSVQGYPTQRARYQWVRCNPDSNSANCLEEKG
Mus  ..MQVPVGSRLVLALAFVLVWGSSVQGYPARRARYQWVRCKPNGFFANCIEEKG
Rat  .MRQVPVGTRLVLALAFVLVWGSSVQGYPARRARYQWVRCKPDGIFANCIEEKG

     PMFELLPGESNKIPRLRTDLFPKTRIQDLNRIFPLSEDYSGSGFGSGSGSGS
     PQFDLIDESNNIGPPMNNPVLMEGPSKDF...ISNYDDYGSGSGSGSGSGSG
     PRFDLIAEESNVGPPMTDPVLMRGFPNDF...FPISDDYSGSGSGSGSGSGS

     GSG............................FLTEMEQDYQLVDESDAFHDN
     SGSGSG.........................FLGDMEWEYQPTDESNIVYFN
     GSGSGSGSGSGSGSGSGSGSGSGSGSGSGSGSLADMEWEYQPTDENNIVYFN

     LRSLDRNLPSDSQDLGQHGLEEDFML
     YKPFDRIL.TE.QNQDQ..PEDDFII
     YGPFDRML.TE.QNQEQ..P.GDFII
```

Figure 3. Alignment of human (hum), murine (mus), and rat (rat) serglycin core protein sequence.

1990). The major species differences lie in the number of serine-glycine (S-G) repeats characteristic of the molecule; the rat has 24 repeats, the mouse 10.5, and the human 8 with a single interruption (see Figure 3). The human sequence also shows some sequence divergence in the C-terminal portion of the molecule.

The S-G repeat of serglycin was originally speculated to be the standard sequence for glycosaminoglycan side chain addition. However, this motif now appears to be relatively rare among proteoglycans. It is found only as tandem pairs in syndecan, fibroglycan, and ryudocan, and as a triplet in glypican. The S-G repeat sequence does serve as the attachment site for glycosaminoglycans to the serglycin core protein. This region of the core protein is densely substituted with large numbers of either chondroitin sulfate or heparin side chains.

The gene structure for serglycin is now known for both the mouse (Avraham et al., 1889a; Angerth et al., 1990) and human genes (Nicodemus et al., 1990). It consists of three closely spaced exons, the S-G repeat is contained in the third exon. Substantial sequence similarity between the two species is also found in the 5′ flanking region of the gene (Nicodemus et al., 1990), indicating a conserved regulatory mechanism may be present.

Generally isolated from hemopoietic cells, where this proteoglycan enjoys predominant expression, serglycin is contained within the abundant secretory granules found in these cells. Serglycin is differentially expressed during hematopoieses, showing higher levels in promyelocytes, eosinophils, and basophils (Stellrecht et al., 1991). The major function of this molecule may well be to provide the substrate for the synthesis of glycosaminoglycan chains which have biological activity. The completed molecule may also serve in the packaging of enzymes and other proteins into the secretory granules of many cell types (Huang et al., 1982; Stevens, 1987). Unlike all other proteoglycans, a role in the assembly or function of the extracellular or pericellular matrix is not known for serglycin. There is also no known pathology associated with this molecule.

B. Cell Surface Proteoglycans

Proteoglycans which occur on the cell surface are associated with the cell membrane either as an integral membrane protein or anchored via a phosphoinositol linkage. All of these proteins also show the unique characteristic of being unrelated to other extracellular proteins. These molecules can be further divided into four distinct subfamilies based on the molecular sequences of the known members. The first subfamily consists of related cell surface proteoglycans: syndecan, fibroglycan, and ryudocan. The other three subfamilies consist (at this time) of one member each: the integral membrane proteoglycans betaglycan and NG2, and glypican, a proteoglycan anchored to the cell surface by a phosphoinositol linkage. In spite of their sequence differences, these molecules may share some common functions, most notably functions attributable to the heparan sulfate side chains, the binding of growth factors, matrix components, and antithrombin III.

The Syndecated Proteoglycans: Syndecan, Fibroglycan, and Ryudocan

This subfamily of proteoglycans consists of unique molecules characterized by highly similar transmembrane and cytoplasmic domains as shown in Figure 4 (Marynen et al., 1989; Saunders et al., 1989; Kiefer et al., 1990; Mali et al., 1990; Kojima et al., 1992b, Pierce et al., 1992). There is also a "pre" segment of nine amino acids prior to the transmembrane segment which is also conserved. In syndecan and fibroglycan, the "pre" segment contains a dibasic sequence which has been proposed as the cleavage site for proteoglycan release; such a site is not present in ryudocan. Other notable features shared by these molecules are a lack of cysteine in the extracellular or intracellular domains, and the presence of a S-G-S-G potential glycosaminoglycan attachment site. Although the primary sequences of the extracellular domains differ substantially in these three molecules, they all appear to take on an "extended" conformation; the core proteins characteristically appear larger on SDS-PAGE electrophoresis than the sequence predicts.

The C-terminal cytoplasmic domain shared by these molecules appears to be able to bind cytoskeletal elements. This evidence comes largely from studies on syndecan. The molecule cosediments with F actin (Rapraeger et al., 1986), a subset colocalizes with actin filament bundles, and when the molecule is cross-linked at the cell surface, they cluster and are aggregated by the cytoskeleton (Rapraeger et al., 1986). The C-terminous contains three conserved tyrosine residues, the possibility of phosphorylation of which has been proposed, though these residues do not conform to typical phosphorylation sites.

```
Fibroglycan { Hum  .............MRRAWILLTLGLVACVSAESRAELTSDKDMYLDNSSIEEASGVYPIDDDDYASASGSGA.................................................
              Rat  ...MRVRATSPGNMQRAWILLTLGLMACVSAETRAELTSDKDMYLDSSSIEEASGLYPIDDDDYSSASGSGA.................................................
Ryudocan  —   Rat  MAPVCLFAPLLLLLLGGFPVAPGESIRETEVIDPQDLLEGRYFSGALPDDEDAGGLEQ..DSDF.ELSGSGD..................................................
Syndecan    { Hum  MRRAALWLWLCALALSLQLALP.......................QIVATNLPPEDQDGSGDDSDNFSGSGAGALQDITLSQQTPSTWKDTQLLTAIPTSPEPTGLEATAASTSTL
              Mus  MRRAALWLWLCALALRLQPALP.......................QIVAVNVPPEDQDGSGDDSDNFSGSGTGALPD.TLSRQTPSTWKDVWLLTATPTAPEPTSSNTETAFTSVL
              Ham  MRRAALWLWLCALALRLQPVLP.......................QIVTVNVPPEDQDGSGDDSDNFSGSGTGALPDITLSRQASPTLKDVWLLTATPTAPEPTSRDAQATTTSIL
              Rat  MRRAALWLWLCALALRLQPALP.......................QIVTANVPPEDQDGSGDDSDNFSGSGTGALPDMTLSRQTPSTWKDVWLLTATPTAPEPTSRDTEATLTSIL

Fibroglycan { Hum  ..................................................................................DEDVESPELTTTRPLPKILLTSAAPKVETTTLNIQN
              Rat  ..................................................................................YEDKGSPDLTTSQLIPRISLTSAAPEVETMTLKTQS
Ryudocan  —   Rat  ..................................................................................LDDTEEPRTFPEVISPLVPLDNHIPENAQPGIRVPS
Syndecan    { Hum  PAGEGPKEGEAVVLPEVEPGLTAR..EQEATPRPRETTQLPTTHQASTT.TATTAQEPATSHPHRDMQPGHHETSTPAGPSQADLHTPHTEDGGPS..ATERAAEDGASSQLPAAE
              Mus  PAGEKPEEGEPVLHVEAEPGFTARDKEKEVTTRPRETVQLPITQRASTV.RVTTAQAAVTSHPHGGMQPGLHETSAPTAPGQPDHQPPRVEGGGTS..VIKEVVEDGTANQLPAGE
              Ham  PAAEKPGEGEPVLTAEVDPGFTARDKESEVTTRPRETTQLLITHWVSTA.RATTAQAPVTSHPHRDVQPGLHETSAPTAPGQPDQQPP...SGGTS..VIKEVAEDGATNQLPTGE
              Rat  PAGEKPEEGEPVAHVEAEPDFTARDKEKEATTRPRETTQLPVTQQASTAARATTAQASVTSHPHGDVQPGLHETLAPTAPGQPDHQPPSVEDGGTS..VIKEVVEDETTNQLPAGE

                                                                        Pre       Transmembrane Domain              Cytoplasmic Domain
Fibroglycan { Hum  KIPAQTKSPEETDKEKVHLSDSERKMDPAEEDTNVYTEKHSDSLFKRTEVLAAVIAGGVIGFLFAIFLILLLVYRMRKKDEGSYDLGERK.PSSAAYQKAPTK..EFYA
              Rat  ITPTQTESPEETDKKEFEISEAEEKQDPAVKSTDVYTEKHSDNLFKRTEVLAAVIAGGVIGFLFAIFLILLLYYRMRKKDEGSYDLGERK.PSSAAYQKAPTK..EFYA
Ryudocan  —   Rat  .EPKELEENEVIPKR.VP.SDVGD.DDVSNK.VSMSSTSQGSNIFERTEVLAALIVGGVVGILFAVFLILLLVYRMKKKDEGSYDLG.KK.P...IYKKAPTN..EFYA
Syndecan    { Hum  GSGEQDFTFETSGENTAVVAVEPDRRNQSPV.DQGAT.GASQGLLDRKEVLGGVIAGGLVGLIFAVCLVGFMLYRMKKKDEGSYSLEEPKQANGGAYQK.PTKQEEFYA
              Mus  GSGEQDFTFETSGENTAVAAVEPGLRNQPPV.DEGAT.GASQSLLDRKEVLGGVIAGGLVGLIFAVCLVAFMLYRMKKKDEGSYSLEEPKQANGGAYQK.PTKQEEFYA
              Ham  GSGEQDFTFETSGENTAVAAVEPDQRNQPPV.DEGAT.GASQGLLDRKEVLGGVIAGGLVGLIFAVCLVGFMLYRMKKKDEGSYSLEEPKQANGGAYQK.PTKQEEFYA
              Rat  GSGEQDFTFETSGENTAVAGVEPDLRNQSPV.DEGAT.GASQGLLDRKEVLGGVIAGGLVGLIFAVCLVAFMLYRMKKKDEGSYSLEEPKQANGGAYQK.PTKQEEFYA
```

Figure 4. Alignment of fibroglycans, ryudocan, and syndecan core protein sequences; human (hum), rat (rat), murine (mus), and hamster (ham).

Syndecan. Syndecan, cloned in mouse (Saunders et al., 1989), human (Mali et al., 1990), hamster (Kiefer et al., 1990), and rat (Kojima et al., 1992b), is highly conserved in these species with a 32-kDa core protein. Other than the S-G-S-G potential glycosaminoglycan attachment site, there are three S-G-D/E potential attachment sites conserved. The hamster sequence contains an additional S-G; however, this occurs in a region of relative interspecies variability and is not likely used for glycosaminoglycan attachment. A single cysteine residue is present in the transmembrane domain (another cysteine residue is found in the signal peptide sequence). Syndecan is notably larger than fibroglycan or ryudocan, and has two more potential glycosaminoglycan attachment sites than these latter two molecules. The single gene for syndecan has been mapped to human chromosome 2 (Ala-Kapee et al., 1990).

Syndecan contains both heparan sulfate and chondroitin sulfate side chains. These chains appear to be responsible for much of the ectodomain activities. Syndecan binds to collagen types I, II, and V, and fibronectin, tenascin, and thrombospondin (via the heparan sulfate chains). These interactions indicate that the molecule can serve as an extracellular matrix receptor, linking certain extracellular matrix molecules to the cell cytoskeleton. Further support for this function is the restoration of an epithelial morphology by expression of syndecan in testosterone-treated mammary cells (Leppà et al., 1992). These cells, when treated with testosterone, lose their epithelial morphology and have reduced syndecan expression; transfection of the syndecan gene eliminates this response to testosterone.

However, matrix attachment does not appear to be the only function of syndecan. The heparan sulfate chains can also bind various growth factors, in particular bFGF. The importance of this binding is underlined by the fact that the hamster syndecan sequence was identified by ligand-affinity cloning using bFGF (Kiefer et al., 1990). While the exact function this molecule plays in the bFGF recognition is not known, it does appear to be of importance. The affinity of bFGF for syndecan is lower than that of the high-affinity receptors. However, these high-affinity receptors are incapable of binding to bFGF in the absence of heparan sulfate proteoglycans (Rapraeger et al., 1991; Yayon et al., 1991). One possibility is that the binding of bFGF to heparan sulfate "activates" the FGF, allowing it to then bind to the high-affinity receptors; another is that the heparan sulfate activates the high-affinity receptors themselves (see Klagsbrun and Baird, 1991, for review). The cytoplasmic cytoskeleton-binding domain of syndecan may also be involved in delivering the FGF-heparan sulfate complex to the high-affinity receptors. It should also be noted that the growth factor receptor activity is probably not restricted to syndecan. All the cell surface proteoglycans, with the exception of NG2, appear as heparan sulfate proteoglycans and could have similar or even identical functions. It is possible that they could also serve as competitive inhibitors, particularly in the case of glypican, which could remain "free" in the membrane.

In the adult animal, syndecan is predominantly expressed on epithelial cells. It is also present on the surfaces of other cell types, and is transiently expressed by mesenchymal cells during development in a number of systems (Thesleff et al., 1988; Brauker et al., 1991; Trautman et al., 1991). Polymorphisms in the glycosaminoglycan chain composition and length are also evident for syndecan. The number of chondroitin sulfate chains, and the length of both chondroitin sulfate and heparan sulfate chains is different in a number of cell types and during lung development (Sanderson and Bernfield, 1988; Boutin et al., 1991; Brauker et al., 1991). These changes may reflect altered synthesis patterns or the changes in tissue composition (Brauker et al., 1991). However, such changes could also be the result of changes in the quantities and types of growth factors present in these systems. TGFβ affects the glycosaminoglycan chain composition in NMuMG mammary cells (Rasmussen and Rapraeger, 1988; Rapraeger, 1989), and also appears to alter the glycosaminoglycan composition of betaglycan (Andres et al., 1992). Other growth factors may also have effects on glycosaminoglycan composition. It has been reported that IL-3, IL-5, and GM-CSF all increase the size of the chains attached to serglycin in human eosinophils (Rothenberg et al., 1988).

TGFβ does not appear to alter the synthetic rates of syndecan in NMuMG cells, while both TGFβ and bFGF substantially increases the synthesis of syndecan in 3T3 cells (Elenius et al., 1992). These two growth factors acted synergistically on syndecan synthesis in 3T3 cells, also increasing the shedding of syndecan into the media. In view of the studies which show that the FGF-like *int 2* gene product and TGFβ expression correlate with syndecan expression in the developing tooth (Thesleff et al., 1988; Wilkinson et al., 1989) and in developing lung (Heine et al., 1987; Brauker et al., 1991), it has been proposed that these growth factors may be the signals mediating the transient expression of syndecan in mesenchymal cells during development (Brauker et al., 1991; Elenius et al., 1992).

Fibroglycan. Cloned from human fibroblasts (Marynan et al., 1989) and rat hepatocytes (Pierce et al., 1992), the core protein of fibroglycan (23 kDa) is substantially smaller than syndecan. Fibroglycan has only three occurrences of S-G repeats; two are in a tandem repeat like that in syndecan. The gene appears to be unique and has been mapped to human chromosome 8q23 (Marynen et al., 1989). Fibroglycan is highly expressed in liver and kidney, though unlike syndecan, it appears to be widely distributed among different adult tissues (Pierce et al., 1992).

Ryudocan. Rat ryudocan was isolated and cloned from endothelial cells (Kojima et al., 1992b) as a proteoglycan which, along with syndecan, has a subfraction containing heparan sulfate chains that are active in anticoagulation and antithrombin III binding (Kojima et al., 1992a). The name ryudocan was derived from ryùdousel, meaning fluidity in Japanese. The core protein of ryudocan (22 kDa) is similar to fibroglycan in size, and, to a lesser extent, in sequence. In addition to the common transmembrane region and cytoplasmic sequence, it has a S-G-S-G sequence as well as one other S-G site. Unlike syndecan and fibroglycan, ryudocan has no potential N-linked oligosaccharide sequences. A major difference between ryudocan and syndecan or fibroglycan is that it lacks the dibasic sequence which appears to serve as the proteolytic cleavage site that releases syndecan and fibroglycan from the cell surface. However, this molecule was found in medium preparations (Kojima et al., 1992a); the mechanism for removal from the cell surface may be substantially different from syndecan and fibroglycan, indicating a potentially different function.

The specific biological functions of fibroglycan and ryudocan remain to be established, as well as their tissue distribution and roles, if any, in development. Since these molecules contain relatively short extracellular segments, their major functions may be mediated through their glycosaminoglycan side chains. Both of these molecules have been identified as heparan sulfate proteoglycans, thus all the functions ascribed to cell surface heparan sulfate proteoglycans could also be ascribed to these molecules; namely, anticoagulant activity, growth factor binding and presentation to the high-affinity receptors, and interactions with matrix components. The conserved cytoplasmic domain that appears to be associated with the cytoskeleton, as in syndecan, may be involved in these latter two functions.

Betaglycan

Betaglycan (Lòpez-Casillas et al., 1991), like the members of the syndecated family, has a transmembrane sequence and relatively short cytoplasmic tail (43 amino acids) and glycosaminoglycan chains, both heparan and chondroitin sulfate. Apart from these structural features, these molecules share no other similarities. The core protein of betaglycan is larger (92 kDa) and both the extracellular and cytoplasmic sequences of betaglycan are completely different. The extracellular domain of betaglycan is much larger than those of the syndecated family, and contains 16 cysteines, which are absent in the extracellular domain of the syndecan family. One of these cysteines occurs in the probable signal peptide sequence, raising the possibility of an unpaired cysteine in betaglycan, though disulfide bonding has not been noted for this molecule. There are six S-G repeats, none of which are in tandem. Two of the sites are proposed as the most likely glycosaminoglycan attachment sites. The cytoplasmic sequence of betaglycan is also unusual, rich in threonine and serine, including one threonine which may be a substrate for protein kinase C. There is also a single cysteine in the betaglycan cytoplasmic domain.

The betaglycan transmembrane and cytoplasmic domains bear a strong similarity to another transmembrane protein, endoglin. Endoglin is a major membrane protein isolated from human vascular endothelium (Gougos and Letarte, 1990). A role in cell-cell adhesion has been proposed for endoglin; it contains an RGD sequence which may be recognized by integrins. Such a function would also imply an interaction between the endoglin (and by analogy betaglycan) and the cytoskeleton, as is thought for members of the syndecan family.

A major function of betaglycan is TGFβ binding. Betaglycan appears to be an intermediate affinity receptor for TGFβ (Cheifetz et al., 1988; Segarini and Seyedin, 1988). This binding is mediated by the core protein; the glycosaminoglycan chains of betaglycan are not necessary (Cheifetz and Massaguè, 1989). The role of betaglycan in the TGFβ signaling system is still unknown as yet. Three possibilities have been proposed (Lòpez-Casillas et al., 1991): (1) to activate TGFβ action by capturing TGFβ from the pericellular environment for presentation to the high-affinity receptors; (2) to suppress TGFβ action by removing it from the pericellular environment; and (3) to signal TGFβ responses directly, though such a mechanism is unknown as yet. Betaglycan also can bind bFGF through its heparan sulfate side chains and, in a manner similar to that proposed for the other cell surface proteoglycans, may be involved in bFGF responses (Andres et al., 1992).

Like syndecan and fibroglycan, betaglycan is released from the cell surface as a soluble form (Andres et al., 1989). This could occur through the dibasic sequences present near the transmembrane domain as in syndecan and fibroglycan, or through a separate sequence, L-A-V-V, which is identical to a cleavage sequence for TGFα (Lòpez-Casillas et al., 1991). The functional roles for soluble betaglycan—or for that matter, soluble syndecan, fibroglycan, and ryudocan—are not yet clear.

NG2

While NG2 has a transmembrane segment (Nishiyama et al., 1991), it differs from the other proteoglycans discussed in this section by criteria other than sequence. NG2 has: a very large (251 kDa) core protein; chondroitin sulfate side chains rather than heparan sulfate; and a large cytoplasmic sequence (76 amino acids). Structurally, NG2 is divided into five domains. The extracellular segment contains three of these domains: a cysteine containing domain 1, a cysteine-lacking domain 2 which appears to contain the glycosaminoglycan side chains, and the cysteine-containing domain 3. The cysteines in domain 1 appear to be interchain disulfide-bonded, since this segment can be released intact by proteolysis (Nishiyama et al., 1991). The transmembrane segment and cytoplasmic domain make up the remaining two domains of this molecule. Like most large molecules, NG2 does show an internally repetitive sequence, with four repeats of 200 amino acids and two other short repeats. The large repeats are of significance since they show some degree of similarity to repeats found in chick N-cadherin, which have been proposed to be calcium-binding domains. A single gene appears to exist for the molecule, but the chromosomal localization is not yet known.

Isolated from developing neural tissue and cells, this molecule generally appears to be absent or expressed at very low levels in adult tissues. NG2 has been reported to bind collagen VI, and may act as a receptor for this molecule, but other functions are not yet known. It is notable that NG2 appears to be the rat homologue of the human melanoma specific antigen studied extensively by Harper and Reisfeld (1987). N-terminal sequence from this molecule is comparable with that of NG2

(Nishiyama et al., 1991). The expression of this molecule on malignant melanoma cells, which are of neural crest origin, may indicate a loss of differentiation of these cells, reverting to a more primitive neural precursor phenotype that, along with increased NG2 expression, also have increased migratory and invasive potential.

Glypican

Glypican differs from the other cell surface proteoglycans not only in amino acid sequence but in mechanism of cell surface attachment. It maintains its association with the cell surface via a phosphoinositol linkage rather than a transmembrane segment (David et al., 1990). This molecule has a short hydrophobic sequence near the C-terminus, and a S-A-A sequence 10 residues before the hydrophobic sequence that agrees with this site being used for the phosphoinositol linkage. The overall sequence is quite different from that of the syndecan family, being more similar to betaglycan in the size of its core protein (62 kDa) and the presence of numerous cysteines. A triplet of tandem S-G residues which appears near the C-terminus, which along with a S-G-E sequence near the N-terminus are proposed to contain the glycosaminoglycan linkage sites. Several dibasic residue pairs make this proteoglycan a candidate for cleavage by the same protease that acts on syndecan, fibroglycan, and betaglycan. In fact this molecule appears to be rapidly shed from the cell surface.

Phosphoinositol-linked cell surface proteoglycans have been detected in several cell types including rat liver (Ishihara et al., 1987), ovarian granulosa cells (Yanagashita and McQuillan, 1989), Schwann cells (Carey and Evans, 1989), and embryonic heart cells (Chajek-Shaul et al., 1989). Interestingly, it is this molecule, or a related one, that has been proposed to contribute the heparan sulfate chains which localize to the nucleus in confluent liver cells and are associated with the cell cycle (Ishihara et al., 1987). Found as a heparan sulfate proteoglycan, this molecule may also have all the functions ascribed to the heparan sulfate chains of the above proteoglycans. However, unlike these molecules, glypican does not appear to have a direct relationship with the cytoskeleton, and may therefore actually have a competitive role for these functions.

V. PROTEOGLYCANS THAT ARE STRUCTURALLY RELATED TO GLYCOPROTEINS

This group of proteoglycans have core proteins that show considerable homology to known glycoproteins and in some cases considerably homology with each other as well. These proteoglycans are primarily present in extracellular matrices and are often referred to as interstitial proteoglycans. Since extracellular matrices developed later in evolution than cell surfaces, and the core proteins of these interstitial proteoglycans show homology to glycoproteins (many of which are also present in extracellular matrices), it is likely that these proteoglycans evolved, in part, from domains for ancestorial genes for glycoproteins. These proteoglycans, although homologous to some glycoproteins, are clearly a distinct and separate gene product from the related glycoprotein. Members of this group can be divided into three major families based on the structure and interaction of their core protein.

A. Proteoglycans with Leucine-Rich Repeats: Decorin, Lumican, Fibromodulin, Biglycan, and PG-Lb

This family of proteoglycans has the largest number of members and has often been referred to as the "small interstitial proteoglycans". Decorin was originally cloned from a human fibroblast cell line (Krusius et al., 1986), biglycan from human bone cells (Fisher et al., 1989), fibromodulin from bovine tracheal chondrocytes (Oldberg et al., 1989), lumican from chick corneas (Blochberger et al., 1992), and PG-Lb from chick epiphyseal cartilage (Shinomura et al., 1992). Decorin and biglycan have been cloned from more than one species (Table 2). These five proteoglycans are present in a variety of connective tissues with many tissues containing more than one of these proteoglycans. Their core proteins are remarkably similar in size (Table 2) and they all consist of a centrally located leucine-rich repeat domain flanked by N- and C-terminal domains containing cysteines. The genes for biglycan and decorin have been mapped in humans to Xql3-qter and 12pl2.1-qter, respectively (McBride et al., 1990). Based solely on similarities in amino acid sequence, these proteoglycans can be further divided into three subfamilies. The first subfamily consists of fibromodulin and lumican, and the second subfamily consists of biglycan and decorin. The members of these two subfamilies are approximately 50% identical with each other and approximately 30% identical with members of the other subfamily. PG-Lb is the lone member of the third subfamily.

Structure and Homology

The most striking feature of this family is the presence of a domain containing a repetitive leucine-rich sequence with the structure L-X-X-L-X-L-X-X-N-X-L/I, where X- is any amino acid. This leucine-rich domain comprises over 60% of the core protein of lumican, decorin, biglycan, and fibromodulin. Alignment of the leucine repeats in chick decorin (Li et al., 1992) and in chick lumican (Blochberger et al., 1992) reveals that there are three variations, or motifs, of this leucine-rich repeat. The three motifs are tandemly linked to form a unit and three units are tandemly linked to form the nine leucine-rich repeats in this domain. There is a distinct spacing pattern between the motifs in each unit and between the different units that is consistent in all four proteoglycans. Using the asparagine in each motif as a reference point, there are 21 amino acids between motifs in adjacent units, and 24 to 26 amino acids between motifs within a unit (Blochberger et al., 1992; Li et al., 1992). PG-Lb does not contain as many leucine repeats as the other four

Table 2.

Proteoglycan	*Species*	*Core Protein*[a]	*Amino Acid #*	*Tissue*	*Reference*
Decorin (PG-II, PG-40, PG-S2)	Human	36,319 Da	329	Embryonic fibroblast Cell line IMR 90	(Krusius et al., 1986)
	Bovine	36,383 Da	330	Bone	(Day et al., 1987)
	Chick	35,970 Da	327	Cornea	(Li et al., 1992)
Biglycan (PG-I, PG-S1)	Human	37,983 Da	343	Adult Bone Cell	(Fisher et al., 1989)
	Bovine	37,280 Da	331	Articular Cartilage	(Neame et al., 1989)
	Rat	37,993 Da	332	Vascular Smooth Muscle Cells	(Dreher et al., 1990
Fibromodulin	Bovine	39,270 Da	357	Trachael chondrocytes	(Oldberg et al., 1989)
Lumican (Corneal Keratan Sulfate Proteoglycan)	Chick	35,750 Da	325	Cornea	(Blochberger et al., 1992)
Proteoglycan-Lb (PG-Lb)	Chick	32,230 Da	293	Embryonic epiphyseal Cartilage	(Shinomura et al., 199:

Note: [a]The size of each proteoglycan's core protein was determined from the deduced amino acid sequence of cDNA clones, with one exception. The amino acid sequence of bovine biglycan was determined directly from the protein. Size is defined here as the molecular weight of the protein after its signal peptide and propeptide (if present) have been removed, but before the protein has received any glycosylation.

proteoglycans and has a different spacing of leucine-rich repeats. Leucine-rich repeats have been reported in at least 19 other proteins from a diverse range of species. The spacing in many of these proteins occurs at approximately 24 amino acid intervals or at multiples of 24. This spacing may be required for proper functioning of the leucine-rich repeat domain. No obvious correlation exists between the units and to the exons in the human gene for biglycan (Fisher, 1991).

Another feature shared by the five members of this proteoglycan family is the number and positioning of cysteine residues. The amino terminus of each protein contains four cysteines and the carboxy terminus contains two. Amino acid alignment shows that the cysteines align to the same position. Interestingly, bovine osteoinductive factor (Bentz et al., 1990; Madison et al., 1990), drosophila toll protein (Hashimoto et al., 1988), and drosophila chaoptin (Reinke et al., 1988), which all contain leucine-rich repeats, also have four cysteine residues at their N-terminus that align with the proteoglycans' cysteines. In addition, bovine osteoinductive factor has two cysteines at the C-terminus that align to the proteoglycans' C-terminal cysteines (Fisher et al., 1991).

Although the primary structure of the small interstitial proteoglycans is defined, one can only speculate about their three-dimensional shape. Chick decorin (Li et al., 1992) contains the following sequence at both the amino and carboxy terminal ends of each of its units: P-X-β-X-P-X-β, where X = any amino acid, and β = aspartic acid, asparagine, glycine, or serine (amino acids commonly found in β bends). Only half of this sequence, however, is found at the carboxy terminus of unit 3. Decorin cloned from other species also displays this pattern with little variation (Krusius et al., 1986; Day et al., 1987; Dreher et al., 1990). Biglycan (Fisher et al., 1989; Neame et al., 1989), fibromodulin (Oldberg et al., 1989), and lumican (Blochberger et al., 1992) deviate further from this sequence, but in all cases two prolines flank both ends of all three units except, like decorin, only one proline is found at the carboxy-terminus of unit 3. PG-Lb (Shinomura et al., 1992) diverges even further from this pattern, but still contains proline residues between some of its leucine repeats. The leucine motif has been shown to contain amphipathic β-sheet structure (Krantz et al., 1991). Thus, one could envision a macromolecule where the β bend sequences cause the units to stack up together and align with one another by their β-sheet structure. Biglycan from bovine articular cartilage has been shown to contain disulfide bonds between the first and fourth cysteines and between the fifth and sixth cysteines (Neame et al., 1989). This means each end of the molecule would be curled upon itself due to disulfide bonds. This would agree with rotary shadowed images viewed by electron microscopy which shows decorin and lumican to be "tadpole" shaped, with a globular protein head (core protein) and GAG chains which resemble tails (Scott et al., 1990).

There are also differences between the five proteoglycans. Most of these differences are at the N- and C-terminal ends. Decorin has been shown to possess a propeptide (Sawhney et al., 1991) at the N-terminal end and biglycan may as well (Fisher et al., 1989). No evidence for a propeptide exists for fibromodulin, lumican, and PG-Lb. Fibromodulin possesses a tyrosine-rich region near its amino terminus (Oldberg et al., 1989). Tyrosylprotein sulfotransferase will sulfate tyrosine residues adjacent to acidic residues (Niehrs et al., 1990). Five such tyrosines exist in fibromodulin's N-terminus. Fibromodulin was shown to possess sulfated tyrosine on its amino terminus (Antonsson et al., 1991). How sulfated tyrosine affects a protein's biological activity is unclear. The authors (Antonsson et al., 1991) noted that tyrosine sulfation was shown to increase the rate of protein secretion from cells (Friederich et al., 1988) and to have an effect on the half-life of proteins released into the circulation (Pauwels et al., 1987). Interestingly, lumican (Blochberger et al., 1992) possesses four N-terminal tyrosine residues and PG-Lb (Shinomura et al., 1992) two tyrosines which could potentially be sulfated. Biglycan and decorin lack potential tyrosine sulfation sites.

Biglycan is thought to contain two chondroitin sulfate/dermatansulfate (CS/DS) chains attached at amino acids 5 and 10, and two N-linked oligosaccharides (Fisher et al., 1987; Choi et al., 1989; Neame et al., 1989), while decorin contains one CS/DS chain at amino acid 4 (Chopra et al., 1985; Fisher et al., 1987; Choi et al., 1989; Mann et al., 1990), one O-linked and three putative N-linked oligosaccharides (Midura et al., 1989; Scott et al., 1990). Whether biglycan or decorin receive CS or DS chains is partly affected by the type of tissue they are found in (Neame et al., 1989). PG-Lb has been shown to contain dermatan sulfate (Shinomura et al., 1992). Its two potential dermatan sulfate addition sites are in alignment with biglycan's two dermatan sulfate sites.

Lumican from cornea and fibromodulin from cartilage contain no CS/DS, but do contain N-linked oligosaccharides with keratan sulfate (KS) chains attached onto their ends. Four out of the five possible sites for KS addition (i.e., those sites which receive N-linked oligosaccharides) on fibromodulin do receive KS (Plaas et al., 1990). Lumican has been estimated to contain five N-linked oligosaccharides, of which two to three of them are substituted with KS (Midura et al., 1989). Interestingly, the first four sites (from the N-terminal end) of fibromodulin which receive KS are located in the leucine-rich region and align with the first four potential KS addition sites found on lumican. Their fifth potential KS sites lie outside of the leucine-rich region and are not in alignment (Blochberger et al., 1992).

Tissue Distribution

The localization of both mRNAs and core proteins of biglycan and decorin was examined in both human skeletal and nonskeletal tissues using [^{35}S]-labeled RNA probes and antibodies directed against dissimilar regions on the two protein cores (Bianco et al., 1990). Their tissue distributions were shown to differ significantly, sometimes to the point of being mutually exclusive. In developing bones, biglycan expression was localized to articular areas and territorial matrices, whereas decorin expression was found in nonarticular areas and in interterritorial matrices. In nonskeletal tissues, decorin was found in all connective tissues rich in collagen

Type I or II such as dermis, tendon, and cornea. Biglycan was absent from these matrices, but found in specialized connective tissue cells (endothelial cells, skeletal myofibers), or other cell types in which it was located on the cell surface (keratinocytes, renal tubular epithelia). Immunofluorescent staining of chick limb embryo cartilage with antibodies directed to the core protein of PG-Lb showed the protein to be present only in the zone of flattened chondrocytes (Shinomura et al., 1984).

Bovine fibromodulin was detected via radioimmunoassay in different types of fibrillar connective tissues such as tendon, sclera, aorta, and possibly skin. Little or no core protein was found in bone, lens, kidney, vitreous, skeletal muscle, liver, and cornea. Indirect immunofluorescence showed fibromodulin was present in the extracellular matrix of cartilage, primarily in the territorial matrix (Heinegård et al., 1986).

The presence of chick lumican was examined in various tissues via northern blot analysis and immunoprecipitation of lumican precursor protein (Blochberger et al., 1992). Cornea, corneal fibroblasts, muscle, and intestine contained the highest levels of lumican mRNA with lesser amounts in heart, liver, and sclera, respectively. Precursor protein synthesis levels for lumican correlated reasonably well with the mRNA levels. However, antibodies to the keratan sulfate epitopes showed a more restricted range in tissues (Funderburgh et al., 1987). This suggests that KS addition to lumican or fibromodulin may be regulated by tissue specific posttranslational modifications.

Functions

The interstitial proteoglycans can also be divided into groups based on tissue distribution and function. Decorin, fibromodulin, and lumican constitute one group, and both biglycan and PG-Lb would be placed into their own separate categories. The proteoglycans in the first group are found in fibrillar matrices, bind fibrillar collagen, and affect collagen fibrillogenesis (Vogel et al., 1984; Brown et al., 1989; Hedbom et al., 1989; Rada et al., 1993). Biglycan is found in the pericellular space of specialized cell types or associated with the surface of cells (Bianco et al., 1990) and does not bind fibrillar collagen or affect its synthesis (Brown et al., 1989). No interaction of PG-Lb with collagen has been reported.

The transparency of the corneal stroma depends, in part, on the small size of the collagen fibrils and the small size of the interfibrillar spaces (Benedek, 1971). Both decorin and lumican (Midura and Hascall, 1989; Schrecengost et al., 1992) are present in the corneal stroma and they are suspected to play a role in maintaining the small size of the collagen fibrils and their spacing. Both decorin and lumican have also been shown to bind to type I collagen and limit the size of the collagen fibrils forms *in vitro*. Their GAG chains, have no effect on fibrillogenesis (Brown et al., 1989; Rada et al., 1991a). The highly charged glycosaminoglycan chains on these proteoglycans impart to the stroma its hydrophilic character, and provide the water retention necessary for preserving the proper interfibrillar distances necessary for corneal transparency. In addition, the GAG chains on the corneal proteoglycans may be providing a "yardstick" function (Scott, 1991) to maintain fibril spacing. Electron micrographs of Cupromeronic Blue-stained rabbit corneas show glycan filament formation. These filaments are composed of two or more GAG chains aligned side by side and end to end. Scott (1991) has proposed that the proteoglycans bind to collagen fibrils and extend their GAG chains out through the interfibrillar space until they reach an adjacent collagen fibril. The core proteins would bind to the surface of the collagen fibril and the GAG chains in the interfibrillar space would hybridize to each other. This arrangement would help to maintain the proper distance between the fibrils. Thus, both the core proteins and GAG components of lumican and decorin are playing separate roles in maintaining collagen fibril diameters and spacing.

Lumican is not always expressed as a proteoglycan and therefore be considered a "part time proteoglycan". Bovine lumican was shown to be present in aorta almost exclusively as a glycoprotein and not a proteoglycan (Funderburgh et al., 1991). This indicates that the lumican exerts its biological activity in aorta through its core protein, and thus may be involved in some aspect of collagen fibril regulation and not water retention.

Another putative function of some of the members of this gene family may involve regulation of cell growth and development. Osteoinductive factor (OIF) (Bentz et al., 1990; Madisen et al., 1990) is a protein which is quite homologous to the small interstitial proteoglycans. It contains the leucine-rich repeat domain and the six cysteine residues which align to the same position as in all five proteoglycans. OIF purified from bovine bone matrix (Bentz et al., 1989) has been shown to stimulate osteoblast and bone marrow cell proliferation, promote alkaline phosphatase activity, and prevent osteoclast cell development (Kukita et al., 1990). Transforming growth factor β (TGF-β) type 1 or type 2 is needed for OIF's activity. One report (Fisher et al., 1991) speculated that because the mature form of OIF contains only the last three leucine repeats and the two C-terminal cysteines, the TGF-β binding domain may be located in the carboxylterminal segment of the protein. This same region of decorin is suspected of being involved in inhibiting collagen fibrillogenesis (Rada et al., 1991). Biglycan and decorin have been shown to bind TGF-β1 (Yamaguchi et al., 1990). In one case, proliferation of Chinese hamster ovary cells was inhibited upon increased decorin expression (Yamaguchi et al., 1988). This inhibition is due, in part, by decorin binding TGF-β, which can then no longer stimulate cell growth. Thus, some of these proteoglycans may be involved in another aspect of matrix organization aside from collagen fibril regulation.

Fisher et al. (1991) and Bionco et al. (1990) suggested several possible functions for biglycan based on its localization. Since biglycan is structurally homologous to the other three proteoglycans, it may be binding to collagen type VI or VIII which are present in the pericellular space, but not in type I or II collagen-containing fibrillar matrices. Biglycan expression was also seen in all the tissues examined where there was ongoing morphogenesis and differentiation. This suggests that biglycan may play a role in developmental aspects.

The most fascinating aspect of the small interstitial proteoglycans is the presence of the leucine repeat and the question of what role it plays in their functions. The leucine-rich segment may act as a binding domain, allowing the proteoglycans to interact with other proteins and cell membranes. Alternatively, it may serve as scaffolding, providing a backbone structure which would allow other areas of the protein, such as the N- or C-termini, to interact in a specific fashion with other cellular components. Insertion and deletion mutagenesis experiments have shown that the leucine motif is required for human placental ribonuclease inhibitor to bind RNase A (Lee et al., 1990), for yeast adenylate cyclase to interact with RAS proteins (Suzuki et al., 1990), and for human U2 snRNP A′ protein to reconstitute with the other U2 snRNP proteins (Fresco et al., 1991). The leucine-rich repeats are not likely, however, to be directly involved in collagen binding or regulate collagen fibrillogenesis. Reduction and alkylation of lumican and decorin destroys its fibrillogenesis-regulating activity (Brown et al., 1989; Rada et al., 1991a). Furthermore, biglycan, which has nine leucine motifs that align with the nine motifs in lumican, decorin, and fibromodulin (Figure 1), does not bind collagen and has no fibrillogenesis-regulating activity (Brown et al., 1989). Decorin shows normal activity in the fibrillogenesis assay after removal of the N-terminal 17 amino acids. This indicates that the N-terminal region of the protein does not contain the collagen fibrillogenesis-regulating region. Most of the activity was lost, however, when the C-terminal half of decorin was removed (Rada et al., 1991a). This suggests that the C-terminus, where biglycan differs significantly from the other three proteoglycans, contains the fibrillogenesis-regulating activity.

Tissue Pathology

In the corneal stroma, lumican exists as a highly sulfated proteoglycan and decorin as an undersulfated proteoglycan. Lumican can bind more water than decorin, but can also release water more readily than decorin (Bettelheim et al., 1975). These properties are thought to be important in maintaining the proper water level in the stroma, and thus promoting transparency. There are several examples of altered proteoglycans production during corneal opacities. During healing of corneal wounds which penetrated the stroma, scar tissue develops and the cornea becomes opaque. A low-sulfated lumican and a highly sulfated decorin are found in the healing stroma. After the wound healed and the corneal stroma became transparent again, lumican was found to be highly sulfated and decorin undersulfated (Funderburgh et al., 1989). In the disease keratoconus, the central part of the cornea protrudes leading to scarring and thus corneal opacity. Again, lumican was found to be undersulfated and decorin oversulfated (Funderburgh et al., 1990; Sawaguchi et al., 1991). Macular corneal dystrophy is an inherited disease in which a haze develops in the center of the corneal stroma. Deposits form in the extracellular matrix. Blindness develops within 10 to 20 years and vision must be restored by keratoplasty (Hassell et al., 1989). In Type I macular corneal dystrophy, lumican receives no sulfate on its poly-N-acetyl lactosamine chains, whereas decorin is somewhat oversulfated (Hassell et al., 1980). Thus, it appears that the keratocytes which synthesize these proteoglycans in the stroma respond to different situations in a "wound healing" manner (Funderburgh et al., 1991). Synthesis of highly sulfated lumican and low-sulfated decorin in the transparent state switches to low-sulfated lumican and highly sulfated decorin in the opaque state. The sulfation state of proteoglycans in the opaque cornea is that which is normally found for proteoglycans in connective tissues other than cornea.

B. Hyaluronate-Binding Proteoglycans

Hyaluronate is thought to play an important role in embryonic development and repair processes by providing a highly hydrated, relatively unresisting matrix through which cells can migrate. Furthermore, hyaluronate forms a coat around many types of cells, profoundly affecting their adhesive and migratory processes (Toole, 1991). Several proteoglycans have been identified which exhibit related structural homologies and specifically interact with hyaluronate (Norling et al., 1978; Coster et al., 1979; Radhakrishnamurthy et al., 1980; Kiang et al., 1981; Anderson, 1982; Iozzo and Wight, 1982; Chang et al., 1983; Bumol et al., 1984; Heinegard et al., 1985; Johansson et al., 1985; Kimata et al., 1986; Goldstein et al., 1989; Morgelin et al., 1989; Stamenkovic et al., 1989; Zimmerman and Ruoslahti, 1989; Heinegard et al., 1990; Rauch et al., 1991). cDNA clones to three proteoglycans—aggrecan, versican, and CD44—have revealed the presence of hyaluronate binding domains. Although these proteoglycans have domains with other functions, the hyaluronate binding domain makes them members of this family.

Aggrecan

Aggrecan, the major proteoglycan of cartilage, is composed of a large core protein to which are attached approximately 100 chondroitin sulfate chains, 15 to 50 keratan sulfate chains, and numerous N- and O-linked oligosaccharides. The complete amino acid sequence of the large aggregating proteoglycan from the rat chondrosarcoma (Doege, et al., 1987), and human cartilage (Doege et al., 1991) as well as partial sequences from human (Baldwin et al., 1989), bovine (Oldberg et al., 1987), and chicken (Sai et al., 1986), have been deduced from cDNA clones. Studies on the rat chondrosarcoma indicate that a continuous sequence of 6.55 kb encodes for a 2124-amino acid-long core protein (220,952 Da), including a 19-amino acid signal peptide (Doege et al., 1987). Analysis of the cDNA sequence, together with past biochemical studies, suggest that the core protein contains some 8 structural domains. These domains are numbered from the NH_2 terminus and are designated based on their structure or function. Domains 1 (G1), 3 (G2), and 8 (G3)

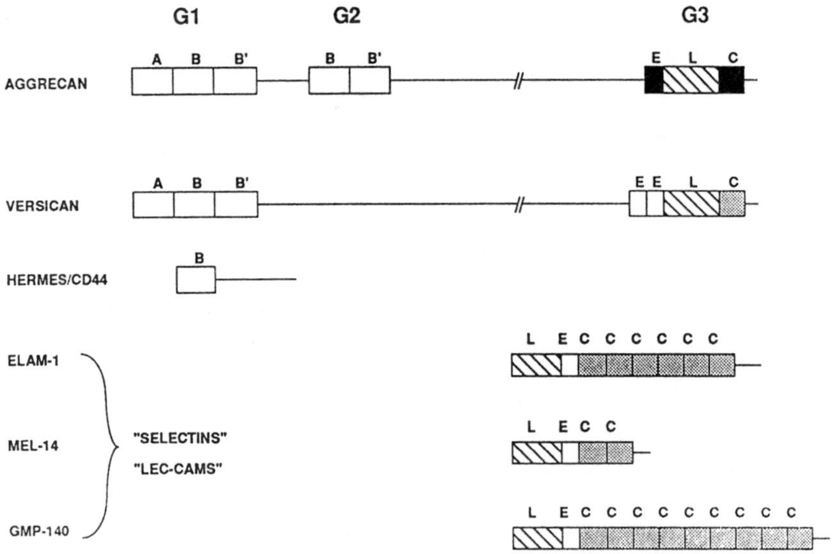

Figure 5. Schematic of related structural features among proteins and proteoglycans sharing homologous domains with aggrecan. Related domains are shown as *boxes* and are labeled A, B, and B′ for the link motifs; E,L, and C for the EGF-like, lectin-like, and complement regulatory protein-like domains, respectively. The E and C domains of aggrecan are blackened to indicate their apparent regulation by alternative splicing. (From Doege et al., 1991).

are globular. Domain 1 is composed of two structural motifs, an Ig variable region fold and a tandem repeat of two homologous loops (B and B′). This domain can interact noncovalently with hyaluronic acid and has thus been termed the hyaluronic acid binding region (HABR) (Heinegard and Hascall, 1974; Hardingham et al., 1986). The tandem repeat motif, but not the Ig fold, is also found in domain 3. Domain 2 is an interglobular domain (IGD); domain 4 (KS) is highly substituted with keratan sulfate; and domains 5 to 7 (CS1–CS3) make up the chondroitin sulfate-bearing portion of the core protein and are distinguishable on the basis of internal sequence repeats. The spacing of the glycosaminoglycan side chains, their distribution along the core protein, and the size and nature of the side chains differ between species. Typically, those from rat contain no, or very small amounts of keratan sulfate compared with those from larger mammals like cow and man (Heinegard and Oldberg, 1989). At the protein level this depends on the presence of a region-contained hexapeptide repeated 23 times in the proteoglycans, which is not represented in the proteoglycans from rat (Antonsson et al., 1989).

Due to the large number of structurally different domains in aggrecan, several homologies can be identified between regions of aggrecan and known proteins. Domains 1 and 3 have extensive sequence homology with link protein, a 39-kDa protein which interacts with domain 1 of aggrecan and hyaluronic acid to stabilize the interaction. Domain 1 is a copy of the complete link protein sequence, while domain 3 is a copy of the COOH-terminal two-thirds portion of link protein. However, electron microscopical data have demonstrated that domain 3 (G2) is not involved in the interactions of aggrecan with hyaluronic acid or link protein (Morgelin et al., 1988). Domains 1 and 3 contain two homologous loops of a tandem repeat structure which share significant sequence homology with a human lymphocyte homing receptor, CD44. CD44 is a polymorphic integral membrane glycoprotein with a postulated role in matrix adhesion and has been identified as the hyaluronate receptor on mouse and hamster fibroblast cell lines (see below). Domain 8 (G3) has sequence homology with a family of animal lectins which can bind specifically to galactose and fucose (Halberg et al., 1988). Domain 8 possesses two additional disulfide-containing domains: one with homology to the EGF repeat, and one related to complement regulatory proteins (CRP). Both the EGF-like domain and CRP-like domains have been determined to be alternatively expressed in the human proteoglycan (Doege et al., 1991) (Figure 5).

Under normal physiological conditions, aggrecan exists as large multimolecular aggregates bound to hyaluronic acid and link protein. Link protein is thought to stabilize the proteoglycan-hyaluronate interaction (Hardingham, 1979) and may direct the arrangement of proteoglycans along the hyaluronic acid chain. Due to the presence of numerous glycosaminoglycan side chains, aggrecan has a high fixed-charge density which attracts water and provides a swelling pressure (Muir, 1982). In cartilage, this pressure functions to resist compression in joints, as well as provides growth of skeletal rudiments during development. Defects in aggrecan structure or synthesis lead to abnormalities in the cartilage matrix and result in altered physical properties of the tissue. In the avian mutant, nanomelia (Argraves et al., 1981), as well as in the murine cartilage matrix deficiency (cmd/cmd) mutant (Kimata et al., 1981), the synthesis of aggrecan core protein is dramatically reduced, resulting in pronounced defects in cartilaginous tissues and cartilage-derived structures. An increase in aggrecan synthesis and accumulation has been shown to occur within the posterior sclera of the chick eye during the development of experimental myopia, resulting in exaggerated axial elongation of the eye globe (Rada et al., 1991b).

Aggrecan shows an age-related decrease in size and an enrichment in keratan sulfate relative to chondroitin sulfate (Thonar and Kuettner, 1987). These reflect changes in the synthesis of the glycosaminoglycan and also cleavage of the core protein by proteases (Sweet et al., 1979; Roughley and White, 1983). Furthermore, in osteoarthritis, a degenerative joint disease, there is a generalized loss of aggrecan from the tissues, resulting eventually in the failure of the tissue as a load-bearing material (Carney and Muir, 1988).

Although typically found in cartilage, a large (300 kDa) chondroitin/keratan sulfate-rich aggregating proteoglycan with EGF and lectin-like domains similar to those described for aggrecan has been identified in neuroglial cells of the central nervous system (Norling et al., 1978; Rauch et al., 1991). Furthermore, the core proteins of large proteoglycans from both sclera and tendon have the same general domain organization as described for aggrecan (Morgelin et al., 1989). Large proteoglycans resembling cartilage proteoglycans have also been isolated from aorta (McMurtry et al., 1979; Morgelin et al., 1989) which contain only two globular domains, but which can aggregate with hyaluronic acid.

Versican

A large aggregating chondroitin sulfate proteoglycan from human fibroblasts with a 2389-amino acid-long core protein has been named versican ("versatile proteoglycan") because it shares structural motifs with growth factors and adhesion molecules (Zimmermann and Ruoslahti, 1989). Based on amino acid sequence deduced from cDNA clones, the versican core protein is proposed to contain several distinct domains (Figure 5). The amino-terminal domain of the versican core protein is similar to the 3-loop structure of link protein, and to the link protein-like sequences in aggrecan (Doege, 1987). The amino acid sequences of the B and B′ loops, which are known to mediate the hyaluronic acid binding of link protein (Goetinck et al., 1987) are especially well conserved. Furthermore, hyaluronic acid-binding activity has been reported for what seems to be the same proteoglycan produced by human embryonic skin fibroblasts (Johansson et al., 1985). The versican sequence adjacent to the link protein-like domain includes an unusual polyglutamic acid stretch. Similar sequences occur in a number of DNA binding proteins as well as in the cell adhesion protein, bone sialoprotein, where they are thought to interact with histones. A glycosaminoglycan attachment domain consists of a 1750-amino acid stretch which contains attachment sites for 12 to 15 chondroitin sulfate side chains, 20 N-linked oligosaccharides, and a number of O-linked oligosaccharides. At the carboxy-terminal portion of the protein core, the versican proteoglycan sequence contains two EGF-like repeats, that do not appear to be present in the cartilage core protein. A lectin-like domain, similar to that of aggrecan, is located on the c-terminal side of the EGF domain which has homologies with mammalian lectins specific for glycans with terminal galactosyl, N-acetylglucosaminyl, and mannosyl residues (Krusius et al., 1987). The carboxy-terminal portion of the core protein contains a domain similar to the repeats identified in several compliment regulatory proteins (Chung et al., 1985; Kristensen and Tack, 1986).

We can only speculate about the function of versican. EGF-like repeats are known to participate in a number of protein-protein interactions (Appella et al., 1988). Furthermore, cell growth may be promoted by these EGF-like sequences similar to that reported for laminin (Panayotou et al., 1989). The lectin-like domain in versican may interact with carbohydrates, similar to that of aggrecan which has been demonstrated to bind fucose and galactose. The carboxy-terminal portion of both versican and aggrecan contains an EGF-like, a lectin-like, and a complement regulatory protein-like motif. This motif is also present on a group of cell adhesion molecules, termed LEC-CAMs, which, in turn, are present in lymphocyte-homing and adhesion molecules (Bevilacqua et al., 1989; Johnston et al., 1989; Lasky et al., 1989; Siegelman et al., 1989). The similarity to adhesive proteins in the carboxy-terminal portion and the occurrence of a hyaluronic acid-binding element at its amino-terminus suggest that versican could also connect hyaluronic acid with the cell membrane (Zimmermann and Ruoslahti, 1989).

Versican has been isolated and cloned from a human lung fibroblast cell line (Krusius et al., 1987) as well as from human placenta (Zimmerman and Ruoslahti, 1989). Another large, chondroitin sulfate proteoglycan (500–550 kDa), termed PG-M, has been isolated from the mesenchyme of chick embryo limb buds (Kimata et al., 1986). Recent sequence analyses indicate that PG-M is closely related to human versican. PG-M has been shown to appear preferentially in areas of precartilage mesenchymal condensation, and then disappears from the cartilage upon the completion of differentiation (Kimata et al., 1986; Shinomura et al., 1990).

CD44

CD44, also known as Pgp-1 (Hughes et al., 1981; Trowbridge et al., 1982), ECMR III (Carter and Wayner, 1988; Gallatin et al., 1989), or Hermes antigen (Jalkanen et al., 1986), has been described as a polymorphic family of related membrane proteoglycans and glycoproteins possessing extensive diversity in both glycosylation and core protein sequence. Although originally associated with hemopoietic cell types, the distribution of CD44 is now known to be much more widespread (for review see Haynes et al., 1989). Recently CD44 was identified as the principal cell surface receptor for hyaluronate on mouse and hamster fibroblast cell lines (Aruffo et al., 1990). The predominant mRNA transcripts of CD44 in baboon (Idzerda et al., 1989) and human (Goldstein et al., 1989; Stamenkovic et al., 1989) lymphoid cells codes for a 37-kDa core protein with 75, 7, and 20% of the polypeptide designated as extracellular, transmembrane, and cytoplasmic domains, respectively. The NH_2-terminal portion of the extracellular domain has homology to the HA-binding regions of link protein and cartilage PG (Goldstein et al., 1989; Stamenkovic et al., 1989). The protein core of CD44 is modified by extensive N- and O-glycosylation to an 80–90-kDa form by addition of chondroitin sulfate to a 180–200-kDa form (Carter and Wayner, 1988; Jalkanen, 1988; Omary et al., 1988; Gallatin et al., 1988; Stamenkovic et al., 1989). However, glycosylation of these forms can be unique for each cell type with heparan sulfate identified as the major carbohydrate derivative of epidermal cells (Brown et al., 1991).

The diverse roles proposed for CD44 can be generalized to an overall function of cell adhesion. CD44 has been shown to localize in filopodia which are involved in cell-cell and cell-substratum contacts (Albrecht-Buehler et al., 1976). Furthermore, based on its homology to link protein, CD44 may function in linking extracellular matrix proteins such as collagen or hyaluronic acid with the cytoskeleton.

C. Perlecan

Structure and Distribution

Perlecan was first isolated and characterized from the murine basement membrane producing Engelbreth-Holms-Swarm (EHS) tumor (Hassell et al., 1980). It consists of a core protein of 400,000 Da containing three heparan sulfate (HS) side chains of 65,000 Da each (Ledbetter et al., 1985). Antibodies to the core protein of perlecan localize to all basement membranes (Hassell et al., 1980; Ekblom, 1981; Laurie et al., 1982; Charonis et al., 1983; Klein et al., 1983; Laurie et al., 1984; Laurie et al., 1988). Kleinschmidt spreading, negative staining, and rotary shadowing electron microscopy of perlecan, elucidate a core protein with a series of six to seven globular domains separated by thin segments (Paulsson et al., 1987; Laurie et al., 1988). The heparan sulfate side chains appear as three long filaments attached to one end. Antiperlecan antibodies that recognize epitopes on the core protein (Ledbetter et al., 1987; Klein et al., 1988) localize mainly to the lamina densa part of basement membranes, with very little staining of the lamina lucida (Laurie et al., 1982, 1984; Grant and Leblond, 1988); whereas cationic markers that bind to the HS side chains are found primarily in the lamina lucida (Farquhar, 1981). Based on these staining patterns and the structure of the molecule, the core protein is believed to be inserted vectorally in the lamina densa where it interacts with other components while the end carrying the side chains protrudes in the lamina lucida (Laurie et al., 1988; Heremans et al., 1989).

The primary structure of perlecan was first derived in its entirety from murine cDNA clones (Noonan et al., 1991), and more recently from human cDNA clones (Kallunki and Tryggvasson, 1992). λgt11 expression vector libraries were screened with polyclonal anti-perlecan antibodies to obtain the first two clones of murine perlecan (Noonan et al., 1988). The perlecan message is 12,685-bp long with a 5′ untranslated region of 578 bp, and encodes for a protein of 3707 amino acids (396,000 Da). The core protein contains 5 distinct domains (Figure 6). The N-terminal, Domain 1, is a 172-amino acid-long segment that is unique to perlecan. It is rich in acidic amino acids, aspartic, and glutamic, as well as leucine and valine, but lacks cysteines. As in several other proteoglycans, domain 1 has three well conserved ser-gly-asp sequences that can serve as attachment sites for the three HS side chains. Adjacent Domain 2 (211 amino acids) has four LDL receptor-like cysteine repeats and may similarly serve a ligand-binding function in perlecan. Domain 3 with 1172 amino acids, bears remarkable resemblance to domains 3, 4, and 5 of the short arms of laminin. Domain 3 consists of three cysteine free globular structures separated by segments of cysteine-rich repeats. Domain 4 contains 14 93-amino acid-long repeats that are closely similar to those found in the cell adhesion molecule N-CAM. Each repeat contains two cysteines and a tryptophan at highly conserved positions. These repeats are typically found in members of the immunoglobulin super-family. As in N-CAM and other homophillic binding pro-

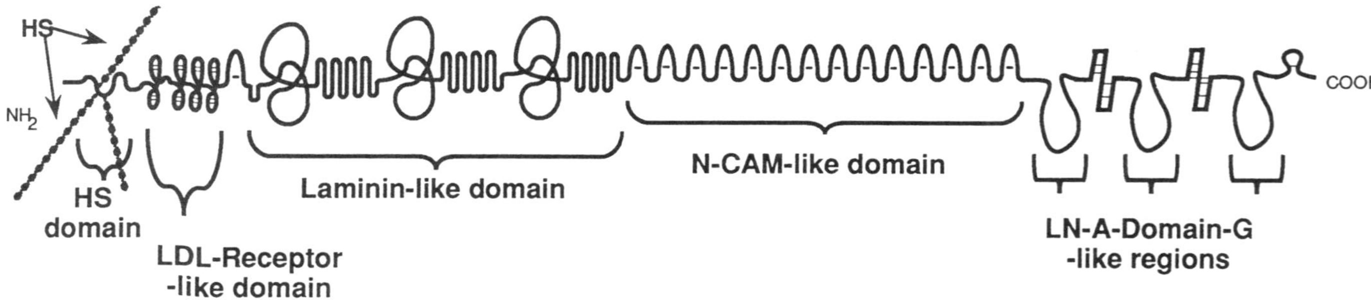

Figure 6. Diagrammatic representation of the domain structure of perlecan. The domains are numbered I through V as follows: Domain I (HS-domain); Domain II (LDL-receptor-like domain); Domain III (Laminin-like domain); Domain IV (N-CAM-like domain); Domain V (Laminin A domain G-like regions).

teins, in perlecan these repeats may be capable of self-interaction. Domain 5 consists of three cysteine-free globules separated by cysteine-rich stretches. The globular subdomains are very similar to those of the G-domain of laminin A chain. As evident for other extracellular matrix proteins, the different modular units of perlecan are derived from several eukaryotic proteins, making this another example of a mosaic protein (Doolittle, 1986).

A cDNA sequence for the human perlecan (Kallunki and Tryggvasson, 1992) reveals a core protein structure that is highly homologous to murine perlecan (Noonan and Hassell, 1993). Each of the 5 domains of the human perlecan is 80 to 90% identical to its mouse counterpart with a few notable differences: (1) in human perlecan domain 1, the position of the first ser-gly-asp does not conform to a canonical acidic-X-S-G-acidic glycosaminoglycan attachment site; (2) in domain 3 the spacer amino acid sequence between the first and second cysteine repeat is poorly conserved; and (3) domain 4 of human perlecan core has seven extra N-CAM-like repeats. It is believed that the position of the extra N-CAM repeats in human mark an alternate splice site in both the human and mouse sequence (Noonan and Hassell, in press; Kallunki and Trygvasson, 1992). The shorter isoform is more common in the mouse EHS-tumor. The gene for perlecan is localized in human chromosome 1p36.1 and to a syntenic region on mouse chromosome 4 (Wintle et al., 1990; Chakravarti et al., 1991; Iozzo et al., 1991; Kallunki et al., 1991).

Identical and Related Forms

In northern analysis, perlecan cDNA clones hybridize to one major 12- to 13-kb mRNA species, and antibodies to murine perlecan detect one major precursor protein of 400,000 Da. However, in previous studies, the EHS-tumor proteoglycan extract, on CsCl density centrifugations, yielded a low-buoyant and a high-buoyant density form (LD and HD form). These two forms contain different amounts of protein, but similar • mounts of glycosaminoglycans (Hassell et al., 1985). Antibodies raised against the two forms recognize the same precursor protein. Also, pulse-chase experiments with EHS tumor cells suggest a precursor product relationship between the HD and LD form. The HD form is believed to be generated from the LD form by variable clipping of the core protein with the GAG side chains still intact at one end. The high GAG-to-protein ratio of the HD form could explain its higher buoyant density in CsCl density gradient centrifugation.

Currently several studies document the presence of perlecan related HSPG in normal tissues. The extracellular matrix of cultured human lung fibroblasts contain one major HSPG with a core protein of 400,000 Da (Heremans et al., 1988). This core protein is identical to perlecan as shown by partial amino acid sequence and immunochemical data (Heremans et al., 1989). According to another study, low passage normal bovine aortic endothelial cells synthesize and secrete a HSPG with a core protein that is clearly related to perlecan. The similarities between their core proteins are evident from their size, peptide maps, rotary shadowing casting images, and immunochemical data (Saku and Furthmayr, 1989). In normal rat glomerular cells, anti-perlecan antibodies recognize a precursor protein that is similar in size to the perlecan precursor (Klein et al., 1988). In glomerular extracts, on the other hand, the same antibodies detect processed core proteins of smaller size (150,000 Da, 250,000 Da) as well as a core of 400,000 Da (Kato et al., 1988). Therefore, peptide mapping, immunochemical, and in the case of the human lung fibroblast form, limited protein sequence data indicate that these core proteins are products of the same gene, perlecan. Consequently, some of the observed variability in these products may be due to posttranslational modification and proteolytic processing.

In addition to the related HSPG discussed above, there may be products of other genes that give rise to HSPGs in the EHS tumor as well as normal tissues. Kato et al. (1987) isolated HD forms of HSPGs from the EHS tumor with core proteins ranging in size from 21,000 to 34,000 Da that may be derived from a gene(s) unrelated to perlecan. These authors show differences in two-dimensional tryptic peptide maps of the two forms and sensitivity of the LD and HD form to heparitinase but HD only to chondroitinase ABC. According to another study, normal calf lens epithelial cells contain, in addition to minor amounts of a large protein immunoprecipitable by anti-perlecan, other low molecular weight core proteins not recognized by this antibody (Mohan and Spiro, 1991). In competitive inhibition assays performed by the same study, there is very little inhibition of binding between perlecan and its antibodies by lens capsule proteoglycans. In similar assays, bovine glomerular basement membranes also show very limited affinity for murine anti-perlecan antibodies. These observations suggest that their HSPGs may be products of different genes. On the other hand, if the low molecular weight forms are proteolytic degradation products of a larger perlecan-like form, these would be carrying all the GAG chains and only domains 1 or domains 1 and 2 of the core protein. The low immune cross-reactivity of these forms could then be explained by their low antigenicity, as suggested by computer antigenicity analysis for the GAG-carrying domain 1 of the human and murine perlecan (Noonan et al., 1991; Kallunki and Tryggvasson, 1992).

Functions

Basement membrane HSPGs have three types of functions. First, these play an important role in the selective filtration of macromolecules. Second, they interact with various macromolecules which include cell surface proteins to modify their functions. Third, their interactions with other basement membrane components which provide structural support to the matrix and may be necessary intermediate steps in the matrix-assembly process. The last two functions are shared by other basement membrane components as well.

The polyanionic side chains play a major role in the selective filtration of macromolecules in glomerular basement membrane (GBM) and basement membranes of other vascular beds. Classic studies by Kanwar et al. (1978) and Farquhar

(1981) demonstrate increased permeability of the GBM to ferritin after enzymatic removal of the side chains.

At neuromuscular junctions, distribution of a basement membrane HSPG shows high correlation with the organization of acetylcholine receptors and may play a role in nerve-muscle adhesion and regulation of molecular organization of the synapse (Anderson and Fambrough, 1983). However, it is not clear at the moment as to whether the HSPG is perlecan and whether these interactions involve the core protein or the side chains.

Studies on the functions of the core protein of perlecan have just begun. The need for high salt or denaturing buffers to extract the proteoglycan (Hassell et al., 1985) and the failure to solubilize it from the matrix by heparatinase treatment alone (Heremans et al., 1990) suggest that core protein interactions stabilize the macromolecule in the matrix.

Heremans et al. (1990) document a high affinity binding of perlecan to fibronectin. ^{125}I-labeled perlecan or core protein alone from human lung fibroblast bound specifically to fibronectin blotted onto nitrocellulose or coupled to sepharose. Binding experiments with proteolytic fragments of the heparatinase digested core protein localize the activity to a 120-kDa doublet and traces of a 62-kDa fragment. The *in vitro* binding occurs at physiological salt concentrations. Furthermore, the estimated dissociation concentration of 10^{-9} M falls within the range of those reported for the interactions between other ECM components supporting the relevance of this interaction *in vivo*. It is likely that incorporation of perlecan into the extracellular matrix occurs after its binding with the matrix-bound fibronectin. This is supported by the observation that perlecan is capable of binding to fibronectin still complexed to other components in the protein-enriched fraction of the fibroblast matrix. Second, in the *in vitro* binding assays, the core protein shows a preference for insolubilized sepharose-coupled fibronectin.

Like other ECM proteins, perlecan shows self-assembly into dimers, trimers, and tetramers (Yurchenco et al., 1987). These higher order forms of perlecan were detected by sucrose density gradient velocity sedimentation of the EHS tumor proteoglycan. Since the core protein without the side chains also form dimers and trimers, the binding site(s) are believed to reside in the core. The noncovalent and weakly ionic nature of these interactions is apparent from the ease with which these higher order forms can be dissociated by low to moderate concentrations of urea and guanidine. Evidence from binding experiments using tryptic fragments and visualization of dimers and trimers by rotary-shadowing electron microscopy localize the binding activity to the C-terminal domain. Since the self-binding interactions *in vitro* occur under physiological ionic conditions, similar binding is possible *in vivo*. Binding at the C-terminal domain could orient the perlecan molecules so that the charged side chains are maximally separated. Such self-binding at the C-termini and interactions of HS side chains with other molecules like laminin may provide a well extended and relatively immobile mesh that could serve as a size and charge barrier.

Perlecan is also implicated in cell adhesion, spreading (Izzard et al., 1986; Gill et al., 1986), and in stabilizing focal contacts (Singer et al., 1987). A recent study (Clement et al., 1989) shows attachment of the core protein of perlecan to hepatocytes through interactions with two cell surface proteins (38,000 Da and 26,000 Da).

Pathology

The HS side chains of GBM proteoglycans are known to play a major role in the selective permeability of the GBM. In many renal diseases and ultimate renal failure, the permselectivity of the basement membrane is altered or lost. Because many of these studies concentrated on the side chains without clear characterization of the core proteins, the identity of the proteoglycan was unclear. However, some recent studies specifically look at the status of perlecan in these pathological conditions.

Puromycin amino nucleoside (PAN) induced nephrosis in rats is a model of human minimal change nephrotic syndrome, and often used to study experimental proteinurea. Immunoelectron microscopic studies showed reduced staining of PAN glomeruli with antibodies to perlecan (Mynderse et al., 1983; Rada and Carlson, 1991). Perlecan cDNA was used in northern blot analysis of mRNA from nephrotic and normal glomeruli (Nakamura et al., 1991). On day-8 after PAN injection, PAN-injected glomeruli show a remarkably low level of perlecan mRNA compared to the control glomeruli. On day-20 the PAN-injected glomeruli show somewhat higher levels of perlecan which are still lower than the corresponding control. Messages for laminin A and the $\alpha 1$ chain of collagen type IV, on the other hand in PAN nephrotic glomeruli show an increase up to day-8 and, then a decrease with levels consistently higher than control. The lowest level of perlecan message on day-8 also coincides with the peak of proteinurea suggesting an association between these observations.

Perlecan is believed to play a role in diabetes related alterations of the basement membrane. Thickening of the basement membranes in diverse tissues and increased permeability are major features in chronic diabetes. Studies using various animal models report reduction in HSPG (Cohen et al., 1981b; Brown et al., 1982; Hasslacher et al., 1982; Parthasarthy and Spiro, 1982; Rohrbach et al., 1982; 1983) and increase in type IV collagen and laminin in certain basement membranes (Cohen et al., 1977; 1981a; Brownlee and Spiro, 1979). Other studies investigate the effect of insulin and glucose levels on basement membrane producing cells (Hasslacher et al., 1979; Pihlajaniemi et al., 1982) and EHS-tumor tissues in organ culture (Ledbetter et al., 1987). In all of these studies either high levels of insulin or glucose or both alter the production of glycoproteins and proteoglycans. Therefore, in poorly controlled diabetes, severe fluctuations in glucose and insulin levels may alter responsiveness of the basement membrane-producing cells. A recent study (Ledbetter et al., 1990) using perlecan cDNA probes, examines levels of its message in the kidney cortex of KKAy mice, a genetically transmitted model for diabetes and compares it to levels of α1-IV collagens and laminin β1 messages. The results show constant levels of perlecan and laminin messages and increased levels of α1-IV collagen. Thus, it appears that increased type IV collagen and the relatively decreased perlecan and laminin lead to altered basement membrane structure and poor filtration properties.

There is some preliminary data that suggest perlecan is involved in the pathogenesis of autoimmune diseases affecting glomeruli. Injection of anti-HSPG IgG into rats result in the binding of the IgG to the glomerular basement membrane, C3 deposition, leukocyte infiltration, and proteinurea (Miettinen et al., 1986). Interestingly, domain 2 of perlecan containing the LDL receptor-like repeats bears a striking similarity with a domain of the Heymann nephratis antigen GP330 (Raychowdhury et al., 1989; Noonan et al., 1991). However, the exact structural relationship between these two proteins will only be apparent after elucidation of the entire primary structure of GP330.

VI. CONCLUDING REMARKS

Molecular biology technology has radically changed our perception of proteoglycans. cDNA clones to over 16 different proteoglycans has revealed a kaleidoscope of core protein domains. Most all of the proteoglycans have one or more domains that has binding activity for either matrix components or cell surfaces. This would suggest that the principal function of core proteins is to target a certain number of GAG chains to a specific location in the extracellular matrix or on the cell surface. cDNA clones to core proteins are now just starting to be used in expressing vector systems. This approach has already illustrated the dramatic effect of syndecan expression on cell shape (Leppà et al., 1992). The expression of recombinant core proteins will likely lead to a greater understanding of domain functions.

ACKNOWLEDGMENTS

This work was supported by National Institutes of Health grants 5 R01 EY08104, 5 R01 GM45380, and 5 R01 EY09391; CNR grants Biotechnologiae Biologia Molecolare 90.01299.CT14 and Ingeneria Genetica 91.0061.PF99; and Research to Prevent Blindness, Inc. Dr. Hassell is a Doris & Jules Stein RPB Professor.

REFERENCES

Ala-Kapee, M., Nevanlinna, H., Mali, M., Jalkanen M., & Schröder, J. (1990). Localization of gene for human syndecan, an integral membrane proteoglycan and a matrix receptor, to chromosome 2. Somatic Cell Mol. Gen. 16, 501–505.

Albrecht-Buehler, G. (1976). The function of filopodia in spreading 3T3 mouse fibroblasts. In: Cell Motility, pp. 247–264. Cold Spring Harbor Laboratories, Cold Spring Harbor, NY.

Alliel, P.M., Pèrin, J., Maillet, P., Bonnet, F., Rosa, J., & Jollès, P. (1988). Complete amino acid sequence of a human platelet proteoglycan. FEBS Lett. 236, 123–126.

Anderson, J.C. (1982). Bovine flexor tendon contains aggregating proteoglycan similar to that of cartilage. Biochem. Biophys. Res. Commun. 107, 1390–1394.

Anderson, M.J. & Farmbrough, D.M. (1983). Aggregates of acetylcholine receptors are associated with plaques of a basal lamina heparan sulfate proteoglycan on the surface of skeletal muscle fibers. J. Cell Biol. 97, 1396–1411.

Andres, J.L., DeFalcis, D., Noda, M., & Massaguè, J. (1992). Binding of two growth factor families to separate domains of the proteoglycan betaglycan. J. Biol. Chem. 267, 5927–5930.

Andres, J.L., Stanley, K., Cheifetz, S., & Massaguè, J. (1989). Membrane-anchored and soluble forms of betaglycan, a polymorphic proteoglycan that binds transforming growth factor-β. J. Cell Biol. 109, 3137–3145.

Angerth, T., Huang, R., Aveskogh, M., Pettersson, I., Kjellèn, L., & Hellman, L. (1990). Cloning and structural analysis of a gene encoding a mouse mastocytoma proteoglycan core protein; analysis of its evolutionary relation to three cross-hybridizing regions in the mouse genome. Gene 93, 235–240.

Antonsson, P., Heinegård, D., & Oldberg, Å. (1989). The keratan sulfate-enriched region of bovine cartilage proteoglycan consists of a consecutively repeated hexapeptide motif. J. Biol. Chem. 264, 16170–16173.

Antonsson, P., Heinegård, D., & Oldberg, Å. (1991). Posttranslational modifications of fibromodulin. J. Biol. Chem. 266, 16859–16861.

Appella, E., Weber, I.T., & Blasi, F. (1988). Structure and function of epidermal growth factor-like regions in proteins. FEBS Lett. 231, 1–4.

Argraves, W.S., McKeown-Longo, P.J., & Goetinck, P.F. (1981). Absence of proteoglycan core protein in the cartilage mutant nanomelia. FEBS Lett. 131, 265–268.

Aruffo, A., Stamenkovic, I., Melnick, M., Underhill, C.B., & Seed, B. (1990). CD44 is the principal cell surface receptor for hyaluronate. Cell 61, 1303–1313.

Avraham, S., Austen K.F., Nicodemus, C.F., Gartner, M.C., & Stevens, R.L. (1989a). Cloning and characterization of the mouse gene that encodes the peptide core of secretory granule proteoglycans and expression of this gene in transfected rat-1 fibroblasts. J. Biol. Chem. 264, 16719–16726.

Avraham, S., Stevens, R.L., Gartner, M.C., Austen, K.F., Lalley, P.A., & Weis, J.H. (1988). Isolation of a cDNA that encodes the peptide core of the secretory granule proteoglycan of rat basophilic leukemia-1 cells and assessment of its homology to the human analogue. J. Biol. Chem. 263, 7292–7296.

Avraham, S., Stevens, R.L., Nicodemus, C.F., Gartner, M.C., Austen, K.F., & Weis, J.H. (1989b). Molecular cloning of a cDNA that encodes the peptide core of a mouse mast cell secretory granule proteoglycan and comparison with the analogous rat and human cDNA. Proc. Natl. Acad. Sci. USA 86, 3763–3767.

Baldwin, C.T., Reginato, A.M., & Prockop, D.J. (1989). A new epidermal growth factor-like domain in the human core protein for the large cartilage-specific proteoglycan. Evidence for alternative splicing of the domain. J. Biol. Chem. 264, 15747–15750.

Benedek, G.B. (1971). Theory of transparency of the eye. Appl. Optics 10, 459.

Bentz, H., Chang, R.J., Thompson, A.Y., Glaser, C.B., & Rosen, D.M. (1990). Amino acid sequence of bovine osteoinductive factor. J. Biol. Chem. 265, 5024–5029.

Bentz, H., Nathan, R.M., Rosen, D.M., Armstrong, R.M., Thompson, A.Y., Segarini, P.R., Matthews, M.C., Dasch, J.R., Piez, K.A., & Seyedin, S.M. (1989). Purification and characterization of a unique osteoinductive factor from bovine bone. J. Biol. Chem. 264, 20805–20810.

Bettelheim, F.A. & Plessy B. (1975). The hydration of proteoglycans of bovine cornea. Biochim. Biophys. Acta 381, 203–214.

Bevilaequa, M.P., Stengelin, S., Gimbrone, M.A., Jr., & Seed, B. (1989). Endothelial leukocyte adhesion molecule 1: an inducible receptor for neutrophils related to complement regulatory proteins and lectins. Science 243, 1160–1165.

Bionco, P., Fisher, L.W., Young, M.F., Termine, J.D., & Robey, P.G. (1990). Expression and localization of the two small proteoglycans biglycan and decorin in developing human skeletal and nonskeletal tissues. J. Histochem. Cytochem. 38, 1549–1563.

Blochberger, T.C., Vergnes, J.P., Hempel, J., & Hassell, J.R. (1992). cDNA to chick lumican (corneal keratan sulfate proteoglycan) reveals homology to the small interstitial proteoglycan gene family and expression in muscle and intestine. J. Biol. Chem. 267, 347–352 .

Bourdon, M.A., Oldberg, Å., Pierschbacher, M., & Ruoslahti, E. (1985). Molecular cloning and sequence analysis of a chondroitin sulfate proteoglycan cDNA. Proc. Natl. Acad. Sci. USA 82, 1321–1325.

Bourdon, M.A., Shiga, M., & Ruoslahti, E. (1986). Identification from cDNA of the precursor form of a chondroitin sulfate proteoglycan core protein. J. Biol. Chem. 261, 12534–12537.

Boutin E.L., Sanderson, R.D., Bernfield, M., & Cunha, G.R. (1991). Epithelial-mesenchymal interactions in uterus and vagina alter the expression of the cell surface proteoglycan, syndecan. Devel. Biol. 148, 63–74.

Brauker, J.H., Trautman, M.S., & Bernfield, M. (1991). Syndecan, a cell surface proteoglycan, exhibits a molecular polymorphism during lung development. Devel. Biol. 147, 285–292.

Brown, D.M., Andres, G.A., Hostetter, T.H., Mauer, R., Price, R., & Venkatachalam, M.A. (1982). Kidney complications. Diabetes 31 (Suppl. 1), 71–81.

Brown, T.A., Bouchard, T., St. John, T., Wayner, E., & Carter, W.G. (1991). Human keratinocytes express a new CD44 core protein (CD44E) as a heparan-sulfate intrinsic membrane proteoglycan with additional exons. J. Cell Biol. 113, 207–221.

Brown, D.C., & Vogel K.G. (1989). Characteristics of the *in vitro* interaction of a small proteoglycan (PG II) s bovine tendon with type I collagen. Matrix 9, 468–478.

Brownlee, M. & Spiro, R.G. (1979). Glomerular basement membrane metabolism in the diabetic rat: *in vivo* studies. Diabetes 28, 121–125.

Bumol, T.F., Walker, L.E., & Reisfeld, R.A. (1984). Biosynthetic studies of proteoglycans in human melanoma cells with a monoclonal antibody to a core glycoprotein of chondroitin sulfate proteoglycans. J. Biol. Chem. 259, 12733–12741.

Carey, D.J. & Evans D.M. (1989). Membrane anchoring of heparan sulfate proteoglycans by phosphatidylinositol and kinetics of synthesis of peripheral and detergent-solubilized proteoglycans in schwann cells. J. Cell Biol. 108, 1891–1897.

Carney, S.C. & Muir, H. (1988). The structure and function of cartilage proteoglycans. Physiol. Rev. 68, 858–910.

Carter, W.G. & Wayner, E.A. (1988). Characterization of the class III collagen receptor, a phosphorylated transmembrane glycoprotein expressed in nucleated human cells. J. Biol. Chem. 263, 4193–4201.

Chajek-Shaul, T., Halimi, O., Ben-Naim, M., Stein, O., & Stein, Y. (1989). Phosphatidylinositol-specific phospholipase C releases lipoprotein lipase from the heparin releasable pool in rat heart cell cultures. Biochim. Biophys. Acta. 1014, 178–183.

Chakravarti, S., Phillips, S.L., & Hassell, J.R. (1991). Assignment of the perlecan (heparan sulfate proteoglycan) gene to mouse chromosome 4. Mammalian Genome 1, 270–272.

Chang, Y., Yanagishita, M., Hascall, V.C., & Wight, T.N. (1983). Proteoglycans synthesized by smooth muscle cells derived from monkey (macaca nemestrina) aorta. J. Biol. Chem. 258, 5679–5688.

Charonis, A.S., Tsilibary, P.C., Kramer, R.H., & Wissig, S.L. (1983). Localization of heparan sulfate proteoglycan in the basement membrane of continuous capillaries. Microvasc. Res. 26, 108–115.

Cheifetz, S., Andres, J.L., & Massaguè, J. (1988). The transforming growth factor-β receptor type III is a membrane proteoglycan. Domain structure of the receptor. J. Biol. Chem. 263, 16984–16991.

Cheifetz, S. & Massaguè, J. (1989). The TGF-β receptor proteoglycan. Cell surface expression and ligand binding in the absence of glycosaminoglycan chains. J. Biol. Chem. 264, 12025–12028.

Choi, H.V., Johnson, T.L., Pal, S., Tang, L.H., Rosenberg, L.C., & Neame, P.J. (1989). Characterization of the dermatan sulfate proteoglycans, DS-PGI and DS-PGII, from bovine articular cartilage and skin isolated by octyl-sepharose chromatography. J. Biol. Chem. 264, 2876–2884.

Chopra, R.K., Pearson, C.H., Pringle, G.A. Fackre, D.S., & Scott, P.G. (1985). Dermatan sulfate is located on Serine-4 of bovine skin proteodermatan sulfate. Biochem. J. 232, 277–279.

Chung, L.P., Bentley, D.R., & Reid, K.B.M. (1985). Molecular cloning and characterization of the cDNA coding for C4b-binding protein, a regulatory protein of the classical pathway of the human complement system. Biochem. J. 230, 133–141.

Clement, B., Bartolome, S.-R., Hassell, J.R., Martin, G.R., & Yamada, Y. (1989). Identification of a cell surface-binding protein for the core protein of the basement membrane proteoglycan. J. Biol. Chem. 264, 12467–12471a.

Cohen, M.P., Dasmahapatra, A., & Wu, V.Y. (1981a). Deposition of basement membrane *in vitro* by normal and diabetic rat renal glomeruli. Nephron. 27, 146–151.

Cohen, M.P. & Khalifa, A. (1977). Renal glomerular collagen synthesis in streptozotocin diabetes: reversal of increased basement membrane synthesis with insulin therapy. Biochim. Biophys. Acta 500, 395–404.

Cohen, M.P. & Surma, M.L. (1981b). ^{35}S-sulfate incorporation into glomerular basement membrane glycosaminoglycans is decreased in experimental diabetes. J. Lab. Clin. Med. 98, 715–722.

Cöster, L., Carlstedt, I. & Malmström, A. (1979). Isolation of ^{35}S- and ^{3}H- labelled proteoglycans from cultures of human embryonic skin fibroblasts. Biochem. J. 183, 669–681.

David, G., Lories, V., Decock, B., Marynen, P., Cassiman, J.J., & Van den Berghe, H. (1990). Molecular cloning of a phosphatidylinositol-anchored membrane heparan sulfate proteoglycan from human lung fibroblasts. J. Cell Biol. 111, 3165–3176.

Day, A.A., McQuillan, C.I., Termine, J.D., & Young, M.R. (1987). Molecular cloning and sequence analysis of the cDNA for small proteoglycan II of bovine bone. Biochem. J. 248, 801–805.

Doege, K., Sasaki, M., Horigan, E., Hassell, J.R., & Yamada, Y. (1987). Complete primary structure of the rat cartilage proteoglycan core protein deduced from cDNA clones. J. Biol. Chem. 262, 17757–17767.

Doege, K.J., Sasaki, M., Kimura, T., & Yamada, Y. (1991). Complete coding sequence and deduced primary structure of the human cartilage large aggregating proteoglycan, aggrecan-human-specific repeats, and additional alternatively spliced forms. J. Biol. Chem. 266, 894–902.

Doolittle, R.F., Feng, D.F., Johnson, M.S., & McClure, M.A. (1986). Relationship of human protein sequences to those of their organisms. In: Cold Spring Harbor Symposium on Quantitative Biology. 51, 447–455.

Dreher, K.L., Asundi, V., Matzura, D., & Cowan, K. (1990). Vascular smooth muscle biglycan represents a highly conserved proteoglycan within the arterial wall. Eur. J. Cell Biol. 53, 296–304.

Ekblom, P. (1981). Formation of basement membranes in the embryonic kidney: an immunohistological study. J. Cell Biol. 91, 1–10.

Elenius, K., Maata, A., Salmivirta, M., & Jalkanen, M. (1992). Growth factors induce 3T3 cells to express bFGF-binding syndecan. J. Biol. Chem. 267, 6435–6441.

Farquhar, M.G. (1981). The glomerular basement membrane. A selective macromolecular filter. In: Cell Biology of the Extracellular Matrix (Hay, E.D., Ed.), pp. 335–378. Plenum, New York.

Fisher, L.W., Hawkins, G.R., Tuross, N., & Termine, J.D. (1987). Purification and partial characterization of small proteoglycans I and II, bone sigloproteins I and II, and osteonectin from mineral compartment of developing human bone. J. Biol. Chem. 262, 9702–9708.

Fisher, L.W., Termine, J.D., & Young, M.F. (1989). Deduced protein sequence of bone small proteoglycan I (biglycan) shows homology with proteoglycan II (decorin) and several nonconnective tissue proteins in a variety of species. J. Biol. Chem. 264, 4571–4576.

Fisher, L.W., Heegaard, A.M., Vetter, V., Vogel, W., Just, W., Termine, J.D., & Young, M.F. (1991). Human biglycan gene. J. Biol. Chem. 266, 14371–14377.

Fransson, L.-Å. (1985). The Polysaccharides (Aspinall, G.D., Ed.), Vol. III, pp. 337–415. Academic Press, New York.

Fresco, L.D., Harper, D.S., & Keene, J.D. (1991). Leucine periodicity of U2 small nuclear ribonucleoprotein particle (snRNP) A′ protein is implicated in snRNP assembly via protein-protein interactions. Mol. Cell Biol. 11, 1578–1589.

Friederich, E., Fritz, H.J., & Huttner, W.B. (1988). Inhibition of tyrosine sulfation in the trans-Golgi retards the transport of a constitutively secreted protein to the cell surface. J. Cell Biol. 107, 1655–1667.

Funderburgh, J.L., Caterson, B., & Conrad, G.W. (1987). Distribution of proteoglycans antigenically related to corneal keratan sulfate proteoglycan. J. Biol. Chem. 262, 11634–11640.

Funderburgh, J.L. & Chandler, J.W. (1989). Proteoglycans of rabbit corneas with nonperforating wounds. Invest. Ophthalmol. Vis. Sci. 30, 435–442.

Funderburgh, J.L., Funderburgh, M.L., Rodriques, M.M., Krachmer, J.H., & Conrad, G.W. (1990). Altered antigenicity of keratan sulfate proteoglycan in selected corneal diseases. Invest. Ophthalmol. Vis. Sci. 31, 419–428.

Funderburgh, J.L., Funderburgh, M.L., Mann, M.M., & Conrad, G.W. (1991). Physical and biological properties of keratan sulfate proteoglycan. Biochem. Soc. Trans. 19, 871–876.

Funderburgh, J.L., Funderburgh, M.L., Mann, M.M., & Conrad, G.W. (1991). Arterial lumican. J. Biol. Chem. 266, 24773–24777.

Gallatin, W.M., Wayner, E.A., Hoffman, P.A., St. John, T., Butcher, E.C., & Carter, W.G. (1989). Structural homology between lymphocyte receptors for high endothelium and class III extracellular matrix receptor. Proc. Natl. Acad. Sci. USA 86, 4654–4658.

Gill, P.J., Silbert, K., & Silbert, J.E. (1986). Effects of heparan sulfate removal on attachment and reattachment of fibroblasts and endothelial cells. Biochem. 25, 405–410.

Giorda, R., Chambers, W.H., Dahl, C.A., & Trucco, M. (1990). Isolation and characterization of a cDNA that encodes the core protein of the cytolytic granule protein In rat natural killer cells. Nat. Immun. Cell Growth Regul. 9, 91–102.

Goetinck, P.F., Stirpe, N.S., Tsonis, P.A., & Carlone, D. (1987). The tandemly repeated sequences of cartilage link protein contain the sites for interaction with hyaluronic acid. J. Cell. Biol. 105, 2403–2408.

Goldstein, L.A., Zhou, D.F.H., Picker, L.J., Minty, C.N., Bargatze, R.F., Ding, J.F., & Butcher, E.C. (1989). A human lymphocyte homing receptor, the Hernes antigen, is related to cartilage proteoglycan core and link proteins. Cell 56, 1063–1072.

Gougos, A. & Letarte, M. (1990). Primary structure of endoglin, an RGD-containing glycoprotein of human endothelial cells. J. Biol. Chem. 265, 8361–8364.

Grant, D.S. & Leblond, C.P. (1988). Immunogold quantitation of laminin, type IV collagen and heparan sulfate proteoglycan in a variety of basement membranes. J. Histochem. 36, 271–283.

Halberg, D.F., Proulx, G., Doege, K., Yamada, Y., & Drickamer, K. (1988). A segment of the cartilage proteoglycan core protein has lectin-like activity. J. Biol. Chem. 263, 9486–9490.

Hardingham, T.E. (1979). The role of link-protein in the structure of cartilage proteoglycan aggregates. Biochem. J. 177, 237–247.

Hardingham, T.E., Beardmore-Gray, M., Dunham, D.G., & Ratcliffe, A. (1986). Cartilage proteoglycans. Ciba Found. Symp. 124, 30–46.

Harper, J.R. & Reisfeld, R.A. (1987). Cell-associated proteoglycans in human malignant melanoma. In: Biology of Proteoglycans (Wight, T.N. & Mecham, R.P., Eds.). Academic Press, New York.

Hascall, V.C., Heinegård, D.K., & Wight, T.N. (1991). Proteoglycans: metabolism and pathology. In: Cell Biology of Extracellular Matrix, second edition (Hay, E., Ed.), pp. 149–175. Plenum Press, New York.

Hashimoto, C., Hudson, K.L., & Anderson K.V. (1988). The toll gene of drosophila, required for dorsal-ventral embryonic polarity, appears to encode a transmembrane protein. Cell 52, 269–279.

Hassell, J.R., Kimura, J.H., & Hascall, V.C. (1986). Proteoglycan core protein families. Ann. Rev. Biochem. 55, 539–567.

Hassell, J.R., Ley Shon, W.C., Ledbetter, S.R., Tyree, B., Suzuki, S., Kato, M., Kimata, K., & Kleinman, H.K. (1985). Isolation of two forms of basement membrane proteoglycans. J. Biol. Chem. 260, 8098–8105.

Hassell, J.R., Newsome, D.A., Krachmer, J.H., & Rodriques, M.M. (1980). Macular corneal dystrophy: failure to synthesize a mature keratan sulfate proteoglycan. Proc. Natl. Acad. Sci. USA 77, 3705–3709 .

Hassell, J.R., Robey, P.G., Barrach, H.J., Wilczek, J., Rennard, S.L., & Martin, G.R. (1980). Isolation of a heparan sulfate containing proteoglycan from basement membrane. Proc. Natl. Acad. Sci. USA 77, 4494–4498

Hassell, J.R., SundarRaj, N., Cintron, C., Midura, R., & Hascall, V. (1989). Alterations in the synthesis of keratan sulfate proteoglycan in corneal wound healing and in macular corneal dystrophy. In: Keratan Sulfate (Greiling, H. & Scott, E., Eds.), pp. 215–225. The Biochemical Society, London.

Hasslacher, C., Kopischke, H.G., Burklin, E., Gechter, F., & Reichenbacher, R. (1982). *In vivo* studies on biosynthesis in diabetic and nondiabetic mice. Res. Exp. Med. 181, 245–251.

Hasslacher, C., Munderloh, R.H., & Wahl, P. (1979). Effect of glucose and insulin on protein and basement membrane synthesis in isolated glomeruli of diabetic rats. Acta Diabetol. Lat. 16, 219–226.

Haynes, B.F., Telen, M.J., Hale, L.P., & Denning, S.M. (1989). CD44: a molecule involved in leukocyte adherence and T-cell activation. Immunol. Today 10, 423–428.

Hedbom, E. & Heinegård, D. (1989). Interaction of a 59-kDa connective tissue matrix protein with collagen I and collagen II. J. Biol. Chem. 264, 6898–6905.

Heinegård, D., Bjorne-Bersson, A., Cöster, L., Franzén, A., Gardell, S., Malmström, A., Paulsson, M., Sandfalk, R., & Vogel, K. (1985). The core proteins of large and small interstitial proteoglycans from various connective tissues form distinct subgroups. Biochem. J. 230, 181–194.

Heinegård, D. & Hascall, V.C. (1974). Aggregation of cartilage proteoglycans. 3. Characteristics of the proteins isolated from trypsin digests of aggregates. J. Biol. Chem. 249, 4250–4256.

Heinegård, D., Hedbom, E., Antonsson, P. & Oldberg, A. (1990). Structural variability of large and small chondroitin sulphate/dermatan sulphate proteoglycans. Biochem. Soc. Trans. 18, 209–217.

Heinegård, D., Larsson, T., Sommarin, Y., Franzén, A., Paulsson, M., & Hedbom, E. (1986). Two novel matrix proteins isolated from articular cartilage show wide distributions among connective tissues. J. Biol. Chem. 261, 13866-13872.

Heinegård, D. & Oldberg, A. (1989). Structure and biology of cartilage and bone matrix noncollagenous macromolecules. FASEB J. 3, 2042–2051.

Heine, U.I., Munoz, E.F., Flanders, K.C., Ellinsworth, L.R., Lam, H.Y.P., Thompson, N.L., Roberts, A.B., & Sporn, M.B. (1987). Role of transforming growth factor-β in the development of the mouse embryo. J. Cell Biol. 105, 2861–2876.

Heremans, A., Cassiman, J.J., Vanten Berghe, H., & David, G. (1988). Heparan sulfate proteoglycan from the extracellular matrix of human lug fibroblasts. J. Biol. Chem. 263, 4731–4739.

Heremans, A., DeCock, B., Cassiman, J.-J., Den Berghe, H.V., & David, G. (1990). The core proteins of the matrix-associated heparan sulfate proteoglycan binds to fibronectin. J. Biol. Chem. 265, 8716–8724.

Heremans, A. B. Vander Schueran, B., DeCock, Paulsson, M., Cassiman J.-J., H.V.D. Berghe, & David, G. (1989). Matrix-associated heparan sulfate proteoglycan: core protein-specific monoclonal antibodies decorate the pericellular matrix of connective tissue cells and the stromal side of basement membranes. J. Cell Biol. 109, 3199–3211.

Huang, S.S., Huang, J.S., & Deuel, T.F. (1982). Proteoglycan carrier of human platelet factor 4. J. Biol. Chem. 257, 11546–11550.

Hughes, E.N., Mengod, G., & August, J.T. (1981). Murine cell surface glycoproteins. Characterization of a major component of 80,000 daltons as a polymorphic differentiation antigen of mesenchymal cells. J. Biol. Chem. 256, 7023–2027.

Idzerda, R.L., Carter, W.G., Nottenburg, C., Wayner, E.A., Gallatin, W.M., & St. John, T. (1989). Isolation and DNA sequence of a cDNA clone encoding a lymphocyte adhesion receptor for high endothelium. Proc. Natl. Acad. Sci. USA 86, 4659–4663.

Iozzo, R.V. (1991). Biosynthesis of heparan sulfate proteoglycan by human colon carcinoma cells and its localization at the cell surface. J. Cell Biol. 99, 403–417.

Iozzo, R.V. & Wight, T.N. (1982). Isolation and characterization of proteoglycans synthesized by human colon and colon carcinoma. J. Biol. Chem. 257, 11135–11144.

Ishihara, M., Fedarko, N.S., & Conrad, H.E. (1987). Involvement of phosphidylinositol and insulin in the coordinate regulation of proteoheparan sulfate metabolism and hepatocyte growth. J. Biol. Chem. 262, 4708–4716.

Izzard, C.S., Radinsky, R., & Culp, L.A. (1986). Substratum contacts and cytoskeletal reorganization of BALB/C 373 cells on a cell-binding fragment and heparin-binding fragments of plasma fibronectin. Exp. Cell Res. 165, 320–326.

Jalkanen, S., Bargatze, R.F., Herron, C.R., & Butcher, E.C. (1986). A lymphoid cell surface protein involved in endothelial cell recognition and lymphocyte homing in man. Eur. J. Immunol. 16, 1195–1202.

Jalkanen, S., Jalkanen, M., Bargatze, R., Tammi, M., & Butcher, E.C. (1988). Biochemical properties of glycoproteins involved in lymphocyte recognition of high endothelial venules in man. J. Immunol. 141, 1615–1623.

Johansson, S., Hedman, K., Kjellén, L., Christner, J., Vaheri, A., & Hook, M. (1985). Structure and interactions of proteoglycans in the extracellular matrix produced by cultured human fibroblasts. Biochem. J. 232, 161–168.

Johnston, G.I., Cook, R.G., & McEver, R.P. (1989). Cloning of GMP-140, a granule membrane protein of platelets and endothelium: sequence similarity to proteins involved in cell adhesion and inflammation. Cell 56, 1033–1044.

Kallunki, P., Eddy, R.L., Byers, M.G., Kestila, M., Shows, T.B., & Tryggvason, K. (1991). Cloning of human heparan sulfate proteoglycan core protein assignment of the gene HSPG1 to lp36.1>p35 and identification of a BamH1 restriction fragment length polymorphism. Genomics 11, 399–396.

Kallunki, P. & Tryggvason K. (1992). Human basement membrane heparan sulfate proteoglycan core protein: A 467-kDa protein containing multiple domains resembling elements of the low density lipoprotein receptor, laminin, neural cell adhesion molecules and epidermal growth factor. J. Cell Biol. 116, 559–571.

Kanwar, Y.S., Linker, A., & Farquhar, M.G. (1978). Increased permeability of the glomerular basement membrane to ferritin after removal of glycosaminoglycans (heparan sulfate) by enzyme digestion. J. Cell Biol. 86, 688–693.

Kato, M.Y., Koike, Y., Ito, Y., Suzuki, S., & Kimata, K. (1987). Multiple forms of heparan sulfate proteoglycans in the Engelbreth-Holm-Swarm mouse tumor. J. Biol. Chem. 262, 7180–7188.

Kato, M., Koike, Y., Suzuki, S., & Kimata, K. (1988). Basement membrane proteoglycan in various tissues: Characterization using monoclonal antibodies to the Engelbreth-Holm-Swarm mouse tumor low tensity heparan sulfate proteoglycan. J. Cell Biol. 106, 2203–2210.

Kiang, W., Margolis, R.U., & Margolis, R.K. (1981). Fractionation and properties of a chondroitin sulfate proteoglycan and the soluble glycoproteins of brain. J. Biol. Chem. 256, 10529–10537.

Kiefer, M.C., Stephans, J.C., Crawford, K., Okino, K., & Barr, P.J. (1990). Ligand-affinity cloning and structure of a cell surface heparan sulfate proteoglycan that binds basic fibroblast growth factor. Proc. Natl. Acad. Sci. USA 87, 6985–6989.

Kimata, K., Barrach, H.-J., Brown, K.S., & Pennypacker, J.P. (1981). Absence of proteoglycan core protein in cartilage from the cmd/cmd (cartilage matrix deficiency) mouse. J. Biol. Chem. 256, 6961–6968.

Kimata, K., Oike, Y., Tani, K., Shinomura, T., Yamagata, M., Uritani, M., & Suzuki, S. (1986). A large chondroitin sulfate proteoglycan (PG-M) synthesized before chondrogenesis in the limb bud of chick embryo. J. Biol. Chem. 261, 13517–13525.

Kjellèn, L. & Lindahl, U. (1991). Proteoglycans: Structures and Interactions. Annu. Rev. Biochem. 60, 443–475.

Kjellèn, L., Pettersson, I., Lillhager, P., Steen, M., Pettersson, U., Lehtonen, P., Karlsson, T., Ruoslahti, E., & Hellman, L. (1989). Primary structure of a mouse mastocytoma proteoglycan core protein. Biochem. J. 263, 105–113.

Klagsbrun, M. & Baird A. (1991). A dual receptor system is required for basic fibroblast growth factor activity. Cell 67, 229–231.

Klein, D.J., Brown, D.M., Oegama, T.R., Brenchley, P.E., Anderson, J.C., Dickinson, M.A.J., Horigan, E.A., & Hassell, J.R. (1988). Glomerular basement membrane proteoglycans are derived from a large precursor. J. Cell Biol. 106, 963–970.

Klein, D.J., Oegema, T.R., Eisentein, R., Furcht, L., Michael, A.F., & Brown, D.M. (1983). Renal localization of heparan sulfate proteoglycan by immunochemistry. Am. J. Pathol. 111, 323–330.

Kojima, T., Leone, C.W., Marchidon, G.A., Marcum, J.A., & Rosenberg, R.D. (1992a). Isolation and characterization of heparan sulfate proteoglycans produced by cloned rat microvascular endothelial cells. J. Biol. Chem. 267, 4859–4869.

Kojima, T., Shworak, N.W., & Rosenberg, R.D. (1992b). Molecular cloning and expression of two distinct cDNA-encoding heparan sulfate proteoglycan core proteins from a rat endothelial cell line. J. Biol. Chem. 267, 4870–4877.

Krantz, D.D., Zidovetzki, R., Kagan, B.L., & Zipursky, S.L. (1991). Amphipathic β structure of a leucine-rich repeat peptide. J. Biol. Chem. 266, 16801–16807.

Kristensen, T., & Tack, B.F. (1986). Murine protein H is comprised of 20 repeating units, 61 amino acids in length. Proc. Natl. Acad. Sci. USA 83, 3963–3967.

Krusius, T., Gehlsen, K.R., & Ruoslahti, E. (1987). A fibroblast chondroitin sulfate proteoglycan core protein contains lectin-like and growth factor-like sequences. J. Biol. Chem. 262, 13120–13125.

Krusius, T. & Ruoslahti, E. (1986). Primary structure of an extracellular matrix proteoglycan core protein deduced from cloned cDNA. Proc. Natl. Acad. Sci. USA 83, 7683–7687.

Kukita, A., Bonewald, L., Rosen, D., Seyedin, S., Mundy, G.R., & Roodman, G.D. (1990). Osteoinductive factor inhibits formation of human osteoclast-like cells. Proc. Natl. Acad. Sci. USA 87, 3023–3026.

Lasky, L.A., Singer, M.S., Yednock, T.A., Dowbenko, D., Fennil, C., Rodrigues, H., Nguyen, T., Stachel, S., & Rosen, S.D. (1989). Cloning of a lymphocyte homing receptor reveals a lectin domain. Cell 56, 1045–1055.

Laurie, G.W., Inove, S., Bing, J.T., & Hassell, J.R. (1988). Visualization of the large heparan sulfate proteoglycan from basement membrane. Am. J. Anat. 181, 320–326.

Laurie, G.W., Leblond, C.P., Inove, S., Martin, G.R., & Chung, A. (1984). Fine structure of the glomerular basement membrane and immunolocalization of five basement membrane components to the lamina densa (basal lamina) and its extensions in both glomeruli and tubules of the rat kidney. Am. J. Anat. 169, 463–481.

Laurie, G.W., Leblond, C.P., & Martin, G.R. (1982). Localization of type IV collagen, laminin, heparan sulfate proteoglycan and fibronectin to the basal lamina of basement membranes. J. Cell Biol. 95, 340–344.

Ledbetter, S.R., Tyree, B., Hassell, J.R., & Horigan, E.A. (1985). Identification of the precursor protein to basement membrane heparan sulfate proteoglycans. J. Biol. Chem. 260, 8106–8113.

Ledbetter, S., Copeland, E.J., Noonan, D., Vogeli G., & Hassell, J.R. (1990). Altered steady-state mRNA levels of basement membrane proteins in diabetic mouse kidneys and thromboxane synthase inhibition. Diabetes 39, 196–203.

Ledbetter, S.R., Fisher, L.W., & Hassell, J.R. (1987). Domain structure of the basement membrane heparan sulfate proteoglycan. Biochemistry 26, 988–995.

Ledbetter, S.R., Wagner, C.W., Martin, G.R., Rohrbach, D.H., & Hassell, J.R. (1987). Response of diabetic basement membrane-producing cells to glucose and insulin. Diabetes 36, 1029–1034.

Lee, F.S. & Vallee, B.L. (1990). Modular mutagenesis of human placental ribonuclease inhibitor, a protein with leucine-rich repeats. Proc. Natl. Acad. Sci. USA 87, 1879–1883.

Leppà, S., Mali, M., Miettinen, H.M., & Jalkanen, M. (1992). Syndecan expression regulates cell morphology and growth of mouse mammary epithelial tumor cells. Proc. Natl. Acad. Sci. USA 89, 932–936.

Li, W., Vergnes, J.P., Cornuet, P.K., & Hassell, J.R. (1992). cDNA clone to chick corneal chondroitin/dermatan sulfate proteoglycan reveals identity to decorin. Arch. Biochem. Biophys. (In press).

Lòpez-Casillas, F., Cheifetz, S., Doody, J., Andres, J.L., Lane, W.S., & Massaguè, J. (1991). Structure and expression of the membrane proteoglycan betaglycan, a component of the TGF-β receptor system. Cell 67, 785–795.

Madisen, L., Neubauer, M., Plowman, G., Rosen, D., Segarini, P., Dasch, J., Thompson, A., Ziman, J., Bentz, H., & Purchia, A.F. (1990). Molecular cloning of a novel bone-forming compound: osteoinductive factor. DNA Cell Biol. 9, 303–309.

Mali, M., Jaakkola, P., Arvilommi, A., & Jalkanen, M. (1990). Sequence of human syndecan indicates a novel gene family of integral membrane proteoglycans. J. Biol. Chem. 265, 6884–6889.

Mann, D.M., Yamaguchi, Y., Bourdon, M.A., & Ruoslahti, E. (1990). Analysis of glycosaminoglycan substitution in decorin by site-directed mutagenesis. J. Biol. Chem. 265, 5317–5323.

Marynen, P., Zhang, J., Cassiman, J.J., Van den Berghe, H., & David, G. (1989). Partial primary structure of the 48- and 90-kDa core proteins of cell surface-associated heparan sulfate proteoglycans of lung fibroblasts. Prediction of an integral membrane domain and evidence for multiple distinct core proteins at the cell surface of human lung fibroblasts. J. Biol. Chem. 264, 7017–7024.

McBride, O.W., Fisher, L.W., & Young, M.F. (1990). Localization of PGI (biglycan, BGN) and PGII (decorin, DCN, PG-40) genes on human chromosomes Xql3-qter and 12q, respectively. Genomics 6, 219–225.

McMurtrey, J., Radhakrishnamurthy, B., Dalferes, E.R., Jr., Berenson, G.S., & Gregory, J.D. (1979). Isolation of proteoglycanhyaluronate complexes from bovine aorta. J. Biol. Chem. 254, 1621–1626.

Midura, R.J. & Hascall, V.C. (1989). Analysis of the proteoglycans synthesized by corneal explants from embryonic chicken. J. Biol. Chem. 264, 1423–1430.

Miettinen, A., Stow, J.L., Mentone, S., & Farquhar, M.G. (1985). Antibodies to basement membrane heparan sulfate proteoglycans bind to the laminae rarae of the glomerular basement membrane (GBM) and induce subepithelial GBM thickening. J. Exp. Med. 163, 1064–1084.

Mohan, P.S. & Spiro, R.G. (1991). Characterization of heparan sulfate proteoglycan from calf lens capsule and proteoglycans synthesized by cultured lens epithelial cells. Geriatrics 266, 8567–8575.

Mörgelin, M., Paulsson, M., Hardingham, T.E., Heinegård, D., & Engel, J. (1988). Cartilage proteoglycans assembly with hyaluronate and link protein as studied by electron microscopy. Biochem. J. 253, 175–185.

Mörgelin, M., Paulsson, M., Malmström, A., & Heinegård, D. (1989). Shared and distinct structural features of interstitial proteoglycans from different bovine tissues revealed by electron microscopy. J. Biol. Chem. 264, 12080–12090.

Muir, H. (1982). Proteoglycans as organizers of the intercellular matrix. Biochem. Soc. Trans. 11, 613–622.

Mynderse, L.A., Hassell, J.R., Kleinman, H.K., Martin, G.R., & Martinez-Hernandez, A. (1983). Loss of heparan sulfate proteoglycan from glomerular basement membrane of nephrotic rats. Lab. Invest. 48, 292–302.

Nakamura, T., Isao, E., Isao, S., Tomino, Y., & Koide, H. (1991). Modulation of basement membrane component gene expression in glomeruli of aminonucleoside nephrosis. Lab. Invest. 64, 640–647.

Neame, P.J., Choi, H.U., & Rosenberg, L.C. (1989). The primary structure of the core protein of the small, leucine-rich proteoglycan (PGI) from bovine articular cartilage. J. Biol. Chem. 264, 8653–8661.

Nicodemus, C.F., Avraham, S., Austen, K.F., Purdy, S., Jablonski, J., & Stevens, R.L. (1990). Characterization of the human gene that encodes the peptide core of secretory granule proteoglycans in promyelocytic leukemia HL-60 cells and analysis of the translated product. J. Biol. Chem. 265, 5889–5896.

Niehrs, C. & Huttner, W.B. (1990). Purification and characterization of tyrosylprotein sulfotransferase. EMBO J. 9, 35–42.

Nishiyama, A., Dahlin, K.J., Prince, J.T., Johnstone, S.R., & Stallcup, W.B. (1991). The primary structure of NG2, a novel membrane-spanning proteoglycan. J. Cell Biol. 114, 359–371.

Noonan, D.M., Fulle, A., Valente, P., Cai, S., Horigan, E., Sasaki, S., Yamada, Y., & Hassell, J.R. (1991). The complete sequence of perlecan, a basement membrane heparan sulfate proteoglycan, reveals extensive similarity with laminin A chain, LDL-receptor, and the N-CAM. J. Biol. Chem. 266, 22939–22947.

Noonan, D.M. & Hassell, J.R. (1993). Perlecan, the large low density proteoglycan of basement membranes: Structure and variant forms. Kidney International 43, 53–60.

Noonan, D.M., Horlgan, E.A., Ledbetter, S.R., Vogeli, G., Sasaki, M., Yamada, Y., & Hassell, J.R. (1988). Identification of cDNA clones encoding different domains of the basement membrane heparan sulfate proteoglycan. J. Biol. Chem. 263, 16379–16387.

Norling, B., Glimelius, B., Westermark, B., & Wasteson, A. (1978). A chondroitin sulphate proteoglycan from human cultured glial cells aggregates with hyaluronic acid. Biochem. Biophys. Res. Commun. 84, 914–921.

Oettinger, H.F., Streeter, H., Edward, L., Copeland, N.G., Gilbert, D.J., Justice, M.J., Jenkins, N.A., Mohandas, T., & Bernfield, M. (1991). Chromosome mapping of the murine syndecan gene. Genomics 11, 334–338.

Oldberg, Å., Antonsson, P., & Heinegård, D. (1987). The partial amino acid sequence of bovine cartilage proteoglycan, deduced from a cDNA clone, contains numerous Ser-Gly sequences arranged in homologous repeats. Biochem. J. 243, 255–259.

Oldberg, Å., Antonsson, P., Lindblom, K., & Heinegård, D. (1989). A collagen-binding 59-kDa protein (fibromodulin) is structurally related to the small interstitial proteoglycans PG-S1 and PG-S2 (decorin). EMBO J. 8, 2601–2604.

Omary, M.B., Trowbridge, I.S., Letarte, M., Kagnoff, M.F., & Isacke, C.M. (1988). Structural heterogeneity of human Pgp-1 and its relationship with p85. Immunogenetics 27, 460–464.

Parthasarthy, N. & Spiro, R.G. (1982). Effect of diabetes on the glycosaminoglycan component of the human glomerular basement membrane. Diabetes 31, 738–741.

Panayotou, G., End, P., Aumailley, M., Timpl, R., & Engel, J. (1989). Domains of laminin with growth-factor activity. Cell 56, 93–101.

Paulsson, M., Yurchenco, P.D., Ruben, G.C., Engel, J., & Timpl, R. (1987). Structure of low density heparan sulfate proteoglycan isolated from a mouse tumor basement membrane. J. Mol. Biol. 197, 297–313.

Pauwels, S., Dockray, G.J., & Walker, R. (1987). Comparison of the metabolism of sulfated and unsulfated heptadecapeptide gastrin in humans. Gastroenterology 92, 1220–1225.

Pierce, A., Lyon, M., Hampson, I.N., Cowling, G.J., & Gallagher, J.T. (1992). Molecular cloning of the major cell surface heparan sulfate proteoglycan from rat liver. J. Biol. Chem. 267,3894–3900.

Pihlajaniemi, T.R., Myllyla, K.I., Kivirikko, & Tryggvason, K. (1982). Effects of streptozotocin diabetes, glucose and insulin on the metabolism of type IV collagen and proteoglycan in murine basement membrane-forming EHS tumor tissues. J. Biol. Chem. 257, 14914–14920.

Plaas, A.H.K., Neame, P.J., Nivens, C.M., & Reiss L. (1990). Identification of the keratan sulfate attachment sites on bovine fibromodulin. J. Biol. Chem. 265, 20634–20640.

Rada, J.A. & Carlson, E.C. (1991). Anionic site and immunogold quantitation of heparan sulfate proteoglycans in glomerular basement membranes of puromycin aminonucleoside nephrotic rats. Anat. Records 231, 35–47.

Rada, J.A., Schrecengost, P.K., & Hassell, J.R. (1993). Regulation of corneal collagen fibrillogenesis *in vitro* by corneal (lumican and decorin) proteoglycan core protein. Exp. Eye Res. (In press).

Rada, J.a., Thoft, R.A., & Hassell, J.R. (1991b). Increased aggrecan (cartilage proteoglycan) production in the sclera of myopic chicks. Devel. Biol. 147, 303–312.

Radhakrishnamurthy, B., Smart, F., Dalferes, E.R., & Berenson, G.S. (1980). Isolation and characterization of proteoglycans from bovine lung. J. Biol. Chem. 255, 7575–7582.

Rapraeger, A. (1989). Transforming growth factor (type β) promotes the addition of chondroitin sulfate side chains to the cell surface proteoglycan (syndecan) of mouse mammary epithelia. J. Cell Biol. 109, 2509–2518.

Rapraeger, A., Jalkanen, M., & Bernfield, M. (1986). Cell surface proteoglycan associates with the cytoskeleton at the basolateral cell surface of mouse mammary epithelial cells. J. Cell Biol. 103, 2683–2696.

Rapraeger, A.C., Krufka, A., & Olwin, B.B. (1991). Requirement of heparan sulfate for bFGF-mediated fibroblast growth and myoblast differentiation. Science 252, 1705–1708.

Rasmussen, S. & Rapraeger, A. (1988). Altered structure of the hybrid cell surface proteoglycan of mammary epithelial cells in response to transforming growth factor-β. J. Cell Biol. 107, 1959–1967.

Rauch, U., Gao, P., Janetzko, A., Flaccus, A., Hilgenberg, L., Tekotte, H., Margolis, R.K., & Margolis, R.U. (1991). Isolation and characterization of developmentally regulated chondroitin sulfate and chondroitin/keratan sulfate proteoglycans of brain identified with monoclonal antibodies. J. Biol. Chem. 266, 14785–14801.

Raychowdbury, R., Niles, J.L., McCluskey, R.T., & Smith, J.A. (1989). Autoimmune target in Heymann Nephritis is a glycoprotein with homology to the LDL-receptor. Science 244, 1163–1165.

Reinke, R., Krantz, D.E., Yen, D., & Zipursky, S.L. (1988). Chaoptin, a cell surface glycoprotein required for drosophila photoreceptor cell morphogenesis, contains a repeat motif found in yeast and human. Cell 52, 291–301.

Rothenberg, M.E., Pomerantz, J.L., Owen, J.W.F., Avraham, S., Soberman, R.J., Austen, K.F., & Stevens, R.I. (1988). Characterization of a human eosinophil proteoglycan, and augmentation of its biosynthesis and size of interleukin 3, interleukin 5, and granulocyte/macrophage colony stimulating factor. J. Biol. Chem. 263, 13901–13906.

Rohrbach, D.H., Hassell, J.R., Kleinman, H.K., & Martin, G.R. (1982). Alterations in the basement membrane (heparan sulfate) proteoglycan in diabetic mice. Diabetes 31, 185–188.

Rohrbach, D.H., Wagner, C.W., Star, V.L., Martin, G.R., Brown, K.S., & Yoon, J.W. (1983). Reduced synthesis of basement membrane heparan sulfate proteoglycan in streptozotocin-induced diabetic mice. J. Biol. Chem. 258, 11672–11677.

Roughley, P.J. & White, R.J. (1983). The use of caesium sulfate density gradient centrifugation to analyze proteoglycans from human articular cartilages of different ages. Biochim. Biophys. Acta 759, 58–66.

Sai, S., Tanaka, T., Kosher, R.A., & Tanzer, M.L. (1986). Cloning and sequence analysis of a partial cDNA for chicken cartilage proteoglycan core protein. Proc. Natl. Acad. Sci. USA 83, 5081–5085.

Saku, T. & Furthmayr, H. (1989). Characterization of the major heparan sulfate proteoglycan secreted by bovine aortic endothelial cells in culture. J. Biol. Chem. 264, 3514–3523.

Sanderson, R.D. & Bernfield, M. (1988). Molecular polymorphism of a cell surface proteoglycan: Distinct structures on simple and stratified epithelia. Proc. Natl. Acad. Sci. USA 85, 9562–9566.

Sawaguchi, S., Yue, B.Y., Chang, I., Sugar, J., & Robin, J. (1991). Proteoglycan molecules in keratoconus corneas. Invest. Ophthalmol. Vis. Sci. 32, 1846–1853.

Saunders, S., Jalkanen, M., O'Farrell, S., & Bernfield, M. (1989). Molecular cloning of syndecan, an integral membrane proteoglycan. J. Cell Biol. 108, 1547–1556.

Sawhney, R.S., Hering, T.M., & Sandell, L.S. (1991). Biosynthesis of small proteoglycan II (decorin) by chondrocytes and evidence for a procore protein. J. Biol. Chem. 266, 9231–9240.

Schrecengost, P.K., Blochberger, T.C., & Hassell, J.H. (1992). Identification of chick corneal keratan sulfate proteoglycan precursor protein in whole corneas and in cultured corneal fibroblasts. Arch. Biochem. Biophys. 292, 54–61.

Scott, J.E. (1991). Proteoglycan: collagen interactions and corneal ultrastructure. Biochem. Eye 19, 877–881.

Scott, J.E., Cummings, C., Greiling, H., Stuhlsatz, H., Gregory, J.D., & Damle, S.P. (1990). Examination of corneal proteoglycans and glycosaminoglycans by rotary shadowing and electron microscopy. Int. J. of Biol. Macromol. 12, 180–184.

Scott, P.G. & Dodd, C.M. (1990). Self aggregation of bovine skin proteodermatan sulfate promoted by removal of three N-linked oligosaccharides. Connective Tissue Res. 24, 1.

Segarini, P.R. & Seyedin, S.M. (1988). The high molecular weight receptor to transforming growth factor-β contains glycosaminoglycan chains. J. Biol. Chem. 263, 8366–8370.

Shinomura, T., Jensen, K.L., Yamagata, M., Kimata, K., & Solursh, M. (1990). The distribution of mesenchyme proteoglycan (PG-M) during wing bud outgrowth. Anat. & Embryol. 181, 227–233.

Shinomura, T. & Kimata, K. (1992). Proteoglycan-Lb, a small dermatan sulfate proteoglycan expressed in embryonic chick epiphyseal cartilage, is structurally related to osteoinductive factor. J. Biol. Chem. 267, 1265–1270.

Shinomura, T., Kimata, K., Oike, Y., Maeda, N., Yano, S., & Suzuki, S. (1984). Appearance of distinct types of proteoglycan in a well-defined temporal and spatial pattern during early cartilage formation in the chick limb. Devel. Biol. 103, 211–220.

Shimomura, H. & Spiro, R.G. (1986). Studies on macromolecular components of human glomerular basement membrane and alterations in diabetes. Diabetes 36, 374-381.

Siegelman, M.H., van de Rijn, M., & Weissman, I.L. (1989). Mouse lymph node homing receptor cDNA clone encodes a glycoprotein revealing tandem interaction domains. Science 243, 1165–1172.

Singer, I.I., Scott, S., Kawka, D.W., & Hassell, J.R. (1987). Extracellular matrix fibers containing fibronectin and basement membrane heparan sulfate proteoglycan coalign with focal contacts and microfilament bundles in stationary fibroblasts. Exp. Cell Res. 173, 558–571.

Soberman, R.J., Austen, K.F., & Stevens, R.I. (1988). Characterization of a human eosinophil proteoglycan, and augmentation of its biosynthesis and size by interleukin 3, interleukin 5, and granulocyte/macrophage colony stimulating factor. J. Biol. Chem. 263, 13901–13906.

Stamenkovic, I., Amiot, M., Resando, J.M., & Seed B. (1989). A lymphocyte molecule implicated in lymph node homing is a member of the cartilage link protein family. Cell 56, 1057–1062.

Stellrecht, C.M., Mars, W.M., Miwa, H., Beran, M., & Saunders, G.F. (1991). Expression pattern of a hemopoietic proteoglycan core protein gene during human hematopoiesis. Differentiation 48, 127–135.

Stellrecht, C.M. & Saunders, G.F. (1989). Nucleotide sequence of a cDNA encoding a hemopoietic proteoglycan core protein. Nucl. Acids Res. 17, 7623.

Stevens, R.L. (1987). Intracellular proteoglycans in cells of the immune system. In: Biology of Proteoglycans (Wight, T.N. & Mecham, R.P., Eds.). Academic Press, New York.

Stevens, R.I., Avraham, S., Gartner, M.C., Bruns, G.A.P., Austen, K.F., & Wels, J.H. (1988). Isolation and characterization of a cDNA that encodes the peptide core of the secretory granule proteoglycan of human promyelocytic leukemia HL-60 cells. J. Biol. Chem. 263, 7287–7291.

Suzuki, N., Choe, H.R., Yoshimitsu, N., Yamawaki-Kataoka, Y., Ohnishi, S., Tamaoki, T., & Kataoka, T. (1990). Leucine-rich repeats and carboxyl terminus are required for interaction of yeast adenylate cyclase with RAS proteins. Proc. Natl. Acad. Sci. USA 87, 8711–8715.

Sweet, M.B.E., Thonar, E.J.-M.A., & Marsh, J. (1979) Age-related changes in proteoglycan structure. Arch. Biochem. Biophys. 198, 439–448.

Thesleff, I., Jalkanen, M., Vainio, S., & Bernfield, M. (1988). Cell surface proteoglycan expression correlates with epithelialmesenchymal interaction during tooth morphogenesis. Dev. Biol. 129, 565–572.

Thonar, E.J.-M.A., & Kuettner, K.E. (1987). Biochemical basis of age-related changes in proteoglycans. In: Biology of Proteoglycans (Wight, T.N. & Mecham, R.P., Eds.), pp. 211–246. Academic Press, Orlando, FL.

Toole, B.P. (1991). Proteoglycans and hyaluronan in morphogenesis and differentiation. In: Cell Biology of the Extracellular Matrix, second edition (Hay, E.D., Ed.), pp. 305–341. Plenum Press, New York.

Trautman, M.S., Kimelman, J., & Bernfield, M. (1991). Developmental expression of syndecan, a proteoglycan matrix receptor in epithelial cell development. Development 111, 213–220.

Trowbridge, I.S., Lesley, J., Schulte, R., Hyman, R., & Trotter, J. (1982). Biochemical characterization and cellular distribution of a polymorphic murine cell-surface glycoprotein expressed on lymphoid tissues. Immunogenetics 15, 299–312.

Vogel, K.G., Paulsson, M., & Heinegård, D. (1984). Specific inhibition of type I and type II collagen fibrillogenesis by the small proteoglycan of tendon. Biochem. J. 223, 587–597.

Wight, T.N., Heinegård, D.K., & Hascall, V.C. (1991). Proteoglycans: structure and function. In: Cell Biology of Extracellular Matrix, second edition (Hay, E., Ed.). Plenum Press, New York, pp. 45–78.

Wilkinson, D.G., Bhatt, S., & McMahon, A.P. (1989). Expression pattern of the FGF-related proto-oncogene int-2 suggests multiple roles in fetal development. Development 105, 131–136.

Wintle, R.F., Kisilevsky, R., Noonan, D., & Duncan, A.M.V. (1990). In situ hybridization to human chromosome 1 of a cDNA probe for the gene encoding the basement membrane heparan sulfate proteoglycan (HSPG). Cytogenet. Cell Genet. 54, 60–61.

Yamaguchi, Y. & Ruoslahti, E. (1988). Expression of human proteoglycan in Chinese hamster ovary cells inhibits cell proliferation. Nature 336, 244–246.

Yamaguchi, Y., Mann, D.M., & Ruoslahti, E. (1990). Negative regulation of transforming growth factor-β by the proteoglycan decorin. Nature 346, 281–284.

Yanagashita, M. & McQuillan, D.J. (1989). Two forms of plasma membrane-intercalated heparan sulfate proteoglycans in rat ovarian granulosa cells. J. Biol. Chem. 264, 17551–17558.

Yanon, A., Klagsbrun, M., Esko, J.D., Leder, P., & Ornitz, D.M. (1991). Cell surface heparin-like molecules are required for binding of basic fibroblast growth factor to its high affinity receptor. Cell 64, 841–848.

Yurchenco, P.D., Cheng, Y.-S., & Ruben, G.C. (1987). Self-assembly of a high molecular weight basement membrane heparan sulfate proteoglycan into dimers and oligomers. J. Biol Chem. 262, 17668–17676.

Zimmermann, D.R. & Ruoslahti, E. (1989). Multiple domains of the large fibroblast proteoglycan, versican. EMBO J. 8, 2975–2981.

STRUCTURE-FUNCTION OF THROMBOSPONDINS:
REGULATION OF FIBRINOLYSIS AND CELL ADHESION

Deane F. Mosher, Xi Sun, Jane Sottile, and Philip J. Hogg

I. INTRODUCTION

Thrombospondin (TSP) is a 450-kDa trimer of identical 150-kDa subunits (Mosher, 1990; Frazier, 1991). The subunit contains lengthy unique sequences at its amino-

Advances in Molecular and Cell Biology
Volume 6, pages 115–132

ISBN: 1-55938-515-4

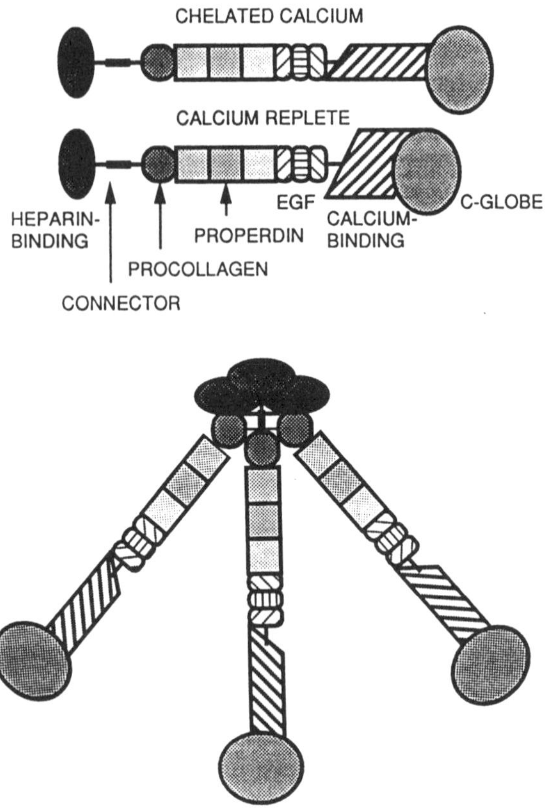

Figure 1. Model of TSP. The subunit (*top*) contains an unique heparin-binding domain at its amino-terminus (darkly shaded oval), a connecting region that contains the cysteines that participate in interchain disulfides (thin and thick lines), a procollagen module (circle), 3 properdin or type 1 modules (squares), 3 epidermal growth factor or type 2 modules (small ovals), unique calcium-binding or type 3 repeats (heavily cross-hatched, different in Ca^{2+}-depleted versus Ca^{2+}-replete TSP), and an unique carboxyl-terminal globular (C-globe) domain (large oval). The subunits are together in the connector by disulfide linkages to form the trimeric molecule shown at the *bottom*.

and carboxyl-terminus; three different types of well-recognized modules of the procollagen, properdin (type 1), and epidermal growth factor (type 2) varieties; and an unique array of calcium-binding (type 3) repeats (Figure 1). The subunits are connected by disulfide linkages (Figure 1). TSP binds to cells, platelets, and numerous extracellular matrix and blood proteins including collagen, fibronectin, heparan sulfate proteoglycan, laminin, fibrinogen, plasminogen, histidine-rich glycoprotein, and transforming growth factor-β (Mosher, 1990;

Frazier, 1991; Murphy-Ullrich et al., 1992). Platelet TSP is secreted from α-granules upon activation and is normally present at very low concentration (approximately 0.1 μg/ml or 0.5% of the serum concentration) in plasma (Mosher, 1990). Normal and transformed cells in culture also produce TSP in a regulated fashion.

Recently, a second gene for TSP was described in the mouse (Bornstein et al., 1991a) and chicken (Lawler et al., 1991). Thus, there are at least two genes for TSPs: *Thbs 1* encoding TSP1, and *Thbs 2* encoding TSP2. The two genes also are present in humans (unpublished studies mentioned in Bornstein et al., 1991a). The expression of the *Thbs 2* gene in various tissues of the mouse differs from that of the *Thbs 1* gene, and the *Thbs 2* gene is not subject to as much upregulation in cultured cells exposed to serum growth factors as the *Thbs 1* gene (Bornstein et al., 1991a, 1991b). Comparison of the two TSPs reveals a gradient of sequence identity in which the aminoterminal regions are less homologous than the carboxyterminal

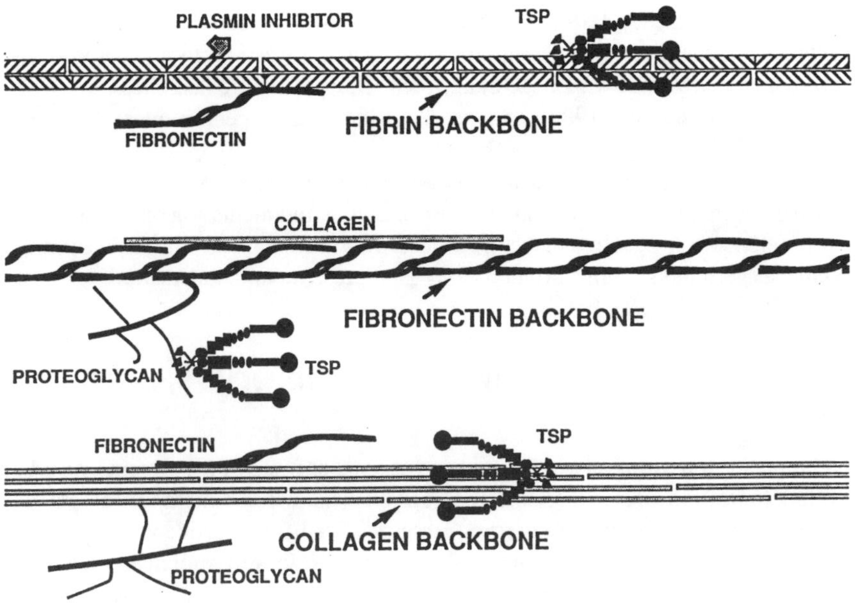

Figure 2. Interactions of TSP with extracellular matrices. The three matrices that form sequentially during wound healing, the provisional fibrin clot (*top*), the early fibronectin-rich matrix (*middle*), and the definitive collagen matrix (*bottom*), all have been shown to contain TSP. Fibronectin, α_2-antiplasmin (plasmin inhibitor), and TSP are cross-linked to fibrin by blood coagulation factor XIIIa (Bale and Mosher, 1986). The major interaction with fibronectin ma+trix is shown as being with proteoglycan because incorporation of TSP into matrix is blocked by heparin (McKeown-Longo et al., 1984). As discussed in the text, TSP may also interact with fibronectin and collagen. The same uncertainty exists for the collagen matrix.

regions. For instance, mouse TSP1 is 38% identical to mouse TSP2 in its amino-terminal heparin-binding domain, and 65 to 75% identical in the type 1 properdin-like modules (Bornstein et al., 1991a; also compare Figures 3 and 5). The functions of both forms of TSP remain to be elucidated.

TSP complexes to fibrinogen (Leung and Nachman, 1982) and copolymerizes with fibrin in clots (Bale, 1987). In addition, TSP binds to extracellular matrix of cultured fibroblasts (McKeown-Longo et al., 1984), probably by interacting with proteoglycans, especially heparan sulfate proteoglycan (McKeown-Longo et al., 1984), and collagen, especially type V collagen (Mumby et al., 1984; Galvin et al., 1987). Thus, the three matrices that form sequentially during wound healing—the provisional fibrin clot, the early fibronectin-rich matrix, and the definitive collagen matrix—all potentially contain TSP (Figure 2). TSP, however, is a temporary component of the matrices as ascertained by immunostaining of TSP in blood clots formed *in situ* (Murphy-Ullrich and Mosher, 1985), immunohistochemistry of wound biopsies (Raugi et al., 1987), and metabolic studies in cell culture (McKeown-Longo et al., 1984). Reasons for the transient nature of TSP include regulated expression, endocytosis, and degradation mediated by cell surface heparan sulfate proteoglycan (McKeown-Longo et al., 1984; Murphy-Ullrich and Mosher, 1987a; Murphy-Ullrich et al., 1988), and, as discussed below, specific interactions with proteases.

The purposes of this review are to describe and discuss two activities of TSP1 purified from platelets, modulation of fibrinolysis, and mediation of cell adhesion. Some attention is paid to whether these activities may be present in TSP2, although actual experiments on this protein await its expression and purification.

II. EFFECTS OF TSP ON FIBRINOLYSIS

A number of observations, mostly made with one or more of the proteins adsorbed to microtitration plates, indicate that TSP enhances the generation of plasmin through its interactions with various proteins of the fibrinolytic system. TSP has been demonstrated to form bimolecular complexes with plasminogen or plasmin (Silverstein et al., 1984; DePoli et al., 1989) and a trimolecular complex with plasminogen and histidine-rich glycoprotein (Silverstein et al., 1985a). Activation of plasminogen by tissue plasminogen activator (tPA) is facilitated by these interactions, and most of the plasmin generated remains bound in the complex and protected from inactivation by α_2-antiplasmin (Silverstein et al., 1985b). The binding of plasminogen to TSP is enhanced by tPA and urokinase (uPA) (Silverstein et al., 1986). This enhancement is dependent on the generation of plasmin, which is postulated to enzymatically modify TSP and plasminogen and thereby enhance the formation of the TSP-plasminogen complexes. TSP can also form a complex with uPA, but not tPA; the TSP-uPA complex has plasminogen activating activity equivalent to fluid phase uPA, but is protected from inhibition by plasminogen activator inhibitor type 1 (Harpel et al., 1990).

We have found that TSP in solution has an opposite effect; that is, it inhibits fibrinolysis (Hogg et al., 1992). Platelet TSP is a slow tight-binding inhibitor of plasmin as determined by the loss of plasmin's amidolytic, fibrinolytic, and fibrinogenolytic activities (Hogg et al., 1992). The second order rate constant for inhibition is 6.3×10^3 $M^{-1}sec^{-1}$. This rate constant is roughly comparable to the rate of inhibition of thrombin by antithrombin in the absence of heparin. The stoichiometry of the inhibition is approximately a mol of Glu- or Lys-plasmin to a mol of trimeric TSP, and the dissociation constant for plasmin-TSP is <1 nM (Hogg et al., 1992). There is limited degradation of TSP into monomeric amino-terminal heparin-binding "heads" and trimeric carboxyl-terminal "tails", but not formation of a complex stable in sodium dodecyl sulfate. Plasmin in complex with streptokinase or ε-aminocaproic acid is protected from inhibition by TSP, thereby implicating the lysine-binding kringles of plasmin in the inhibition process. However, the protease part of plasmin must be of primary importance, because TSP is a good inhibitor of trypsin also (P.J. Hogg, manuscript in preparation). TSP also inhibits uPA, albeit, much more slowly than it inhibits plasmin, has no effect on thrombin or factor Xa, and has a stimulatory effect on the amidolytic activity of tPA (Hogg et al., 1992).

At least 10 families of protein inhibitors of serine proteases have been identified, most of which follow the "standard mechanism" of inhibition (Laskowski and Kato, 1980). The mechanism involves the binding of the inhibitor to the protease followed by the formation of a stable complex. During complex formation, a peptide bond at the reactive site of the inhibitor is hydrolyzed very slowly by the protease. Although the overall structures of the inhibitors are not conserved, the inhibitors share a common loop structure at their reactive centers (Carrell et al., 1987). Small inhibitors have numerous disulfide bonds that, along with hydrophobic interactions, stabilize the tertiary structure of the proteins and provide a "scaffold" for the loop containing the reactive center.

Protease inhibitor modules have been detected in complex extracellular proteins (Engel, 1991) in the form of the Kunitz inhibitor modules of β-amyloid precursor protein (Kido et al., 1990; Sinha et al., 1990), type VI collagen (Chu et al., 1990), and agrin (Rupp et al., 1991). None of the protein modules within TSP, however, have previously been implicated in inhibition of proteases (Laskowski and Kato, 1980; Carrell et al., 1987). Due to the one-to-one stoichiometry of TSP to plasmin (Hogg et al., 1992), likely candidates for the inhibitory domain of TSP are the properdin modules and the procollagen module, because these are located adjacent to the intermolecular disulfide bridge region, and the bridge region itself.

A leech protein, antistasin, which shares a short sequence similarity with part of a properdin module of TSP (Holt et al., 1989), has recently been identified as a potent inhibitor of Factor Xa and exhibits many functional similarities to other serine protease inhibitors (Dunwiddie et al., 1989). The properdin module in TSP,

```
                                                      *
hTSP1     CYHNGVQYRNNEEWTVDSCTECHCQNSVTICKKVSCPIMP  CSNATVPDGECCPRC
mTSP1     -F------K-------------------------------  ---------------
                                                      *
                  *
mTSP2     -VQE-RIFAE--T-V-----T-T-KKFK-V-HQIT-SPAT  -A-PFSVE-----S-
cTSP2     -WQD-RVFADS-S-I-----K-T-QDSKIV-HQIT--PVS  -ADPSFIE-----V-


hα1(I)    -S-L-LL-HDRDV-LP-P-QI-V-D-GKVL-DD-T-DETK N-PG-D--E-----I-
hα1(II)   -VQD-QR-NDKDV-KPEP-RI-V-DTGTVL-DDII-EDVK D-LSPEI-L-----I-
hα1(III)  -S-L-QS-ADRDV-KPEP-QI-V-DSGSVL-DDII-DDQELD-P-PEI-F-----A-
```

Figure 3. Amino acid sequences of procollagen modules. The procollagen domains of human (h), mouse (m), and chicken (c) TSPs and human interstitial procollagen chains are aligned. The thrombospondin sequences are from Lawler et al. (1991) and Bornstein et al. (1991 a). The procollagen sequences are from Sandell and Boyd (1990). Identity with the sequence of human TSP1 is indicated by a dash (–). Asterisks (*) indicate potential sites of N-linked glycosylation. Spaces are left to accommodate the extra residues in human α1(III) procollagen. Note the conserved cysteines (C) and glycines (G).

however, does not contain the lysine residue identified as the reactive site for Factor Xa in antistasin (Dunwiddie et al., 1989), nor does it contain any other lysine or arginine residue in an obvious reactive site. Moreover, the sequence similarity between antistasin and the properdin modules does not extend to all of the cysteines in the properdin modules, making it unlikely that the properdin modules can adopt an antistasin-like "scaffold". TSP1 contains a short basic sequence (DKRFKQ) inserted between the first and second properdin modules (Lawler and Hynes, 1986; Bornstein et al., 1991a). This sequence is presumably the site where trypsin cleaves Ca^{2+}-replete TSP to yield the 85 kDa monomeric carboxyl-terminal fragment described by Lawler et al. (1985), and therefore is a candidate for the reactive loop even though it is not within a module.

The procollagen modules of TSP and of the procollagens themselves form two subfamilies (Figure 3). The function of the N-propeptide of interstitial collagens is obscure (Sandell and Boyd, 1990). If the procollagen module does account for the inhibitory activity of TSP1 towards plasmin and trypsin, the inhibitory activity may not extend to TSP2 or the procollagens, because the distributions of Lys and Arg are different among the several examples of the module (Figure 3).

The bridge region is thought to be stabilized by the three subunits coming together in an α-helical coiled-coil (Sottile et al., 1991). Changes in the presumptive coiled-coil region are conservative between TSP1 and TSP2, so both TSPs likely form the same type of structure and may form heterotrimers (Figure 4). A major difference between the two forms of TSP is five extra residues inserted between the end of the presumptive coiled-coil and the procollagen module of TSP2 (Bornstein et al., 1991a; Lawler et al., 1991). This is very near the site of the cleavage, probably

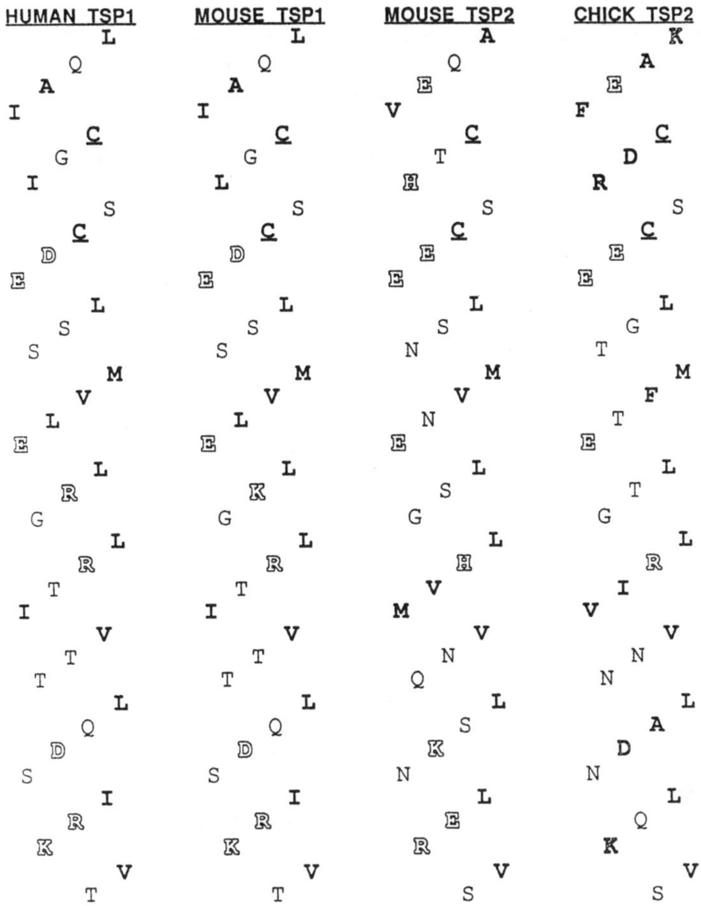

Figure 4. Depiction of the sequences of TSP1 and TSP2 about the interchain disulfides as α-helical surfaces. The sequence from Leu-248 to Thr-284 of human TSP1 and homologous sequences of other TSPs are shown. Hydrophobic residues are in boldface type. Charged residues are outlined. Cys-252 and Cys-256 that participate in interchain disulfides are underlined. Note the hydrophobic surface on the right-hand side of the surface below the cysteines. This surface is preserved in TSP1 and TSP2.

by a Glu-specific protease, that gives rise to the monomeric 140 kDa fragment of TSP isolated by Good et al. (1990).

TSP released from activated platelets and incorporated into the fibrin clot may play an important role in fibrinolysis. The kinetics of activation of plasminogen by tPA are unfavorable in circulation, but upon binding of tPA and plasminogen to fibrin, plasmin is readily formed and clot lysis can occur (Silverstein and Nachman,

1987). Platelet-rich clots, however, have been observed to be much more resistant to lysis with tPA than erythrocyte-rich clots for reasons that are unclear (Jang et al., 1989). It is likely that this resistance is due, at least partially, to TSP.

Because plasmin activity is important for phenomena such as inflammation and tumor cell implantation and migration that are unrelated to lysis of fibrin (Saksela and Rifkin, 1988), TSP has the potential to play important roles in many processes. For example, WI-38 embryonic lung fibroblasts secrete substantial quantities of plasminogen activator and lyse a fibrin clot formed over the cell layer without digesting the extracellular matrix (Mosher et al., 1977). TSP binds to the extracellular matrix of WI-38 fibroblasts and is selectively removed in a process that requires intact cells (McKeown-Longo et al., 1984). One may hypothesize a pathway in which matrix-bound TSP forms a complex with plasmin, being cleaved and consumed in the process but protecting and preserving remaining matrix elements, such as plasmin-sensitive fibronectin (Petersen et al., 1983). As another example, uPA and TSP colocalize to invading mammary carcinoma cells (Harpel et al., 1990). Studies have linked increased production of uPA by tumor cells with a more malignant phenotype (Ossowski et al., 1991). The biology of plasminogen activation is complex, however, and involves, in addition to uPA and its activators, the cell surface receptor for uPA and specific activator inhibitors (Ossowski et al., 1991; Pyke et al., 1991). One of the striking differences between MCF-7 tumorigenic human breast cancer cells and MCF-7 cells, suppressed in tumorigenicity by hybridization with normal human mammary epithelial cells, is increased expression of TSP by the nontumorigenic hybrids (Zajchowski et al., 1990). An effect of increased TSP may be to regulate plasmin activity. Studies on the inhibition of plasmin by TSP in more complex systems are required to test these speculations and learn the relative importance of TSP in plasmin-related processes such as tumorigenesis.

III. EFFECTS OF TSP ON CELL ADHESION

A number of extracellular proteins containing the sequence Arg-Gly-Asp (RGD), such as vitronectin, fibronectin, and fibrinogen, unambiguously promote cell adhesion when coated onto substrata in a process mediated by cell surface integrin receptors (Ruoslahti and Pierschbacher, 1987). Both TSP1 and TSP2 contain the RGD sequence (Lawler and Hynes, 1986; Lawler et al., 1991). The adhesive activity of TSP1, however, is most ambiguous. Different groups have tested TSP purified from platelet releasate for adhesive activity on various cell lines with contradictory and controversial results.

For the same cell type, opposite results have been observed. Lahav (1988a) reported that TSP inhibits endothelial cell adhesion, while others found that TSP promotes endothelial cell adhesion (Lawler et al., 1988; Murphy-Ullrich and Höök, 1989; Taraboletti et al., 1990). Lahav (1988b) also reported that TSP inhibits platelet adhesion, whereas Tuszynski et al. (1987) and Tuszynski and Lowalska (1991) found that TSP supports platelet adhesion.

Among the pro-adhesion papers, the proposed cell surface receptors that mediate cell adhesion to TSP-coated substrate are controversial. Candidates include the GPIIb/IIIa ($\alpha_{IIb}\beta_3$) (Karczewski et al., 1989; Tuszynski et al., 1989) and vitronectin receptor ($\alpha_v\beta_3$) (Lawler et al., 1988; Lawler and Hynes, 1989; Tuszynski et al., 1989) integrins. Candidates also include heparan sulfate proteoglycan (Roberts et al., 1987; Kaesberg et al., 1989), GPIV (also known as CD36) (Asch et al., 1991), and unidentified receptors that recognize the carboxyl-terminal region (Roberts et al., 1987; Kosfeld et al., 1991), or the type 1 repeat (Prater et al., 1991) of TSP. For adhesion of C32 human melanoma cells to TSP-coated substrate, Tuszynski et al. (1989) proposed that the adhesion was mediated by GPIIb/IIIa-like vitronectin receptor, inasmuch as the adhesion is sensitive to RGD-containing peptide and polyclonal antibodies against GPIIb/IIIa or vitronectin receptor. Asch et al. (1991), in contrast, proposed that the adhesion of C32 cells to TSP was mediated by the combination of heparan sulfate proteoglycan and GPIV, because heparin and monoclonal antibody against GPIV together, but not alone, inhibited the adhesion.

Even for the same cell type adhering to TSP by the same presumptive receptor, different results have been obtained. Lawler et al. (1988) found that TSP supported endothelial cell attachment but not spreading, and that the cell attachment was blocked by RGD-containing peptide. Taraboletti et al. (1990), in contrast, observed that TSP strongly supported endothelial cell spreading as well as attachment and that RGD peptide mainly inhibited cell spreading rather than cell attachment.

We undertook studies to try to reconcile some of the reported findings (Sun et al., 1992). The inconsistencies in the literature on the adhesive activity of TSP were reproduced in our experiments. Most preparations of our TSP did not have RGD-inhibitable adhesive activity for trypsinized cells, although these preparations did have heparin-inhibitable adhesive activity for cells that are allowed to recover from trypsinization, and thus, to display cell surface proteoglycans (Kaesberg et al., 1989). However, a single preparation of our TSP and TSP purified following Lawler's procedure did have the RGD-inhibitable adhesive activity. These findings indicate that at least part of the controversies in the literature are due to differences in TSPs, with some TSPs being adhesive and some not, suggesting that, as proposed by Lawler et al. (1988), exposure of the RGD sequence in TSP is regulated.

Because the RGD sequence of TSP is located in the type 3 repeat region, which contains multiple Ca^{2+}-binding sites, and because TSP adsorbed to plastic in the presence of EGTA did not support cell adhesion upon Ca^{2+} restoration, Lawler et al. (1988) concluded that exposure of the RGD sequence is dependent on Ca^{2+}. This region of TSP undergoes a major conformational change upon chelation of Ca^{2+} as ascertained by rotary shadowing (Galvin et al., 1985; Lawler et al., 1985; Dixit et al., 1986; see Figure 1), immunological reactivity (Dixit et al., 1986), susceptibility to proteolysis (Lawler et al., 1982; Lawler and Simons, 1983; Lawler et al., 1985), circular dichroism spectroscopy (Lawler and Simons, 1983), electron spin resonance spectroscopy (Slane et al., 1988), and other techniques (Lawler et al., 1982). A Hill plot of ellipticity versus [Ca^{2+}] revealed a Hill coefficient of 12.3, indicative of a highly cooperative interaction (Lawler and Simons, 1983). A possible difference between Lawler's TSP purification procedures and ours is Lawler's addition of 2-mM Ca^{2+} to the platelet suspension before activation by thrombin and 2-mM rather than 0.1-mM Ca^{2+} to all the buffers used in later purification steps. However, our platelet releasate contained >1-mM Ca^{2+} when measured by Ca^{2+}-ion electrode (Sun et al., 1992) due to endogenous Ca^{2+} released from platelets upon activation (Detwiler and Feinman, 1973). Incubation of our nonadhesive TSPs with 2-mM Ca^{2+} before it was coated on plastic well did not make it adhesive (Sun et al., 1992). These results suggest that Ca^{2+} alone is not sufficient to regulate the exposure of the RGD sequence.

Recently, Speziale and Detwiler (1990) studied the number and position of free thiols in TSP freshly released from platelet by derivatization of thiols with radiolabeled N-ethylmaleimide or iodoacetamide, followed by digestion and sequencing of peptides containing derivatized thiols. Although there was a stoichiometry of s single free thiol per TSP subunit, the derivatized cysteine residues were mapped at 12 different positions (Figure 5). Of the 12 radiolabeled thiols, 10 were in the type 3 repeat region, which contains 17 of the total of 69 cysteines in each subunit of TSP. The remaining two-labeled thiols were in the adjacent C-globe region. The number of reactive thiols was a function of [Ca^{2+}]. With increased [Ca^{2+}], the number of derivatized thiols decreased in the same cooperative manner as described above for the change in circular dichroism. Thus, the type 3 repeats and adjacent C-globe of TSP participate in a complex intramolecular thiol-disulfide isomerization that is regulated by Ca^{2+}. In our purified nonadhesive TSP, only Cys^{974} in the C-globe region was labeled (Sun et al., 1992). These observations led us to hypothesize that the exposure of the RGD sequence in TSP is controlled by disulfides.

Figure 5. Sequences and schematic diagram of the type 3 repeat and adjacent C-globe regions. The sequence of human TSP1 from Asp-674 to Pro-1152 (Lawler and Hynes, 1986) is given in the upper lines. The 12 variants of the Ca^{2+}-binding loop sequence are numbered. Asterisks (*) indicate potential sites of N-linked glycosylation. The sequence of chicken TSP2 (Lawler et al., 1991) is shown below the sequence of human TSP1; identity is indicated by a dash (–). Note the well-conserved sequences compared to the sequences for the TSP procollagen modules shown in Figure 3. A schematic diagram of this region is also shown. The 12 loops are numbered. Small circles represent the 19 cysteines. Open circles, lightly shaded circles, and heavily shaded circles represent cysteines not labeled, lightly labeled, and heavily labeled in the experiments of Speziale and Detwiler (1990). Note that the three heavily labeled cysteines are in proximity to the RGD adhesive sequence in loop 11. The uncertain arrangements of disulfides in this region of TSP are indicated by the question mark for the pairing of Cys-1149 at the extreme carboxyl-terminus and are discussed in the text.

```
              *
DTDLDGWPNENLVCVANATYNCKKDNCPNLPNSGQE
-S-------N----A-------V-----L-------

DYDKDGIGDACDD                               1
-F----K-----E

DDDNDKIPDDRDNCPFHYNPAQY                     2
-----GVE--K----LLF--R-F

DYDRDDVGDRCDNCPYNHNPDQA                     3
---K-E----------V---A-I

DTDNNGEGDACAA                               4
---------S--V

DIDGDGILNERDNCQYVYNVDQR                     5
-----D-F------P----T--S

DTDMDGVGDQCDNCPLEHNPDQL                     6
---G------------M-----T

DSDSDRIGDTCDNNQ                             7
-A-N-LV--Q----E

DIDEDGHQNNLDNCPYVPNANQA                     8
----------Q-----I------

DHDKDGKGDACDH                               9
------------P

DDDNDGIPDDKDNCRLVPNPDQK                     10
----------R-----RY--E-E

DSDGDGRGDACKD                               11
---------I---

DFDHDSVPDIDDICP                             12
---D-N----F-V--

ENVDISETDFRRFQMIPLDPKGTSQNDPNWVVRHQGKELVQTVNCDPGLAVGYDEFNAVD
--NA-------K---V------A-I------I----------A-S---I-------SS--
                                                           *
FSGTFFINTERDDDYAGFVFGYQSSSRFYVVMWKQVTQSYWDTNPTRAQGYSGLSVKVVN
-----YV--D--------------------L-------T--EDK----Y----V-L----

STTGPGEHLRNALWHTGNTPGQVRTLWHDPRHIGWKDFTAYRWRLSHRPKTGFIRVVMYE
----T-------------------------KN-----Y-----H-I------L-K-LV--

GKKIMADSGPIYDKTYAGGRLGLFVFSQEMVFFSDLKYECRDP
--QV-V-------T-F---------------Y----------A
```

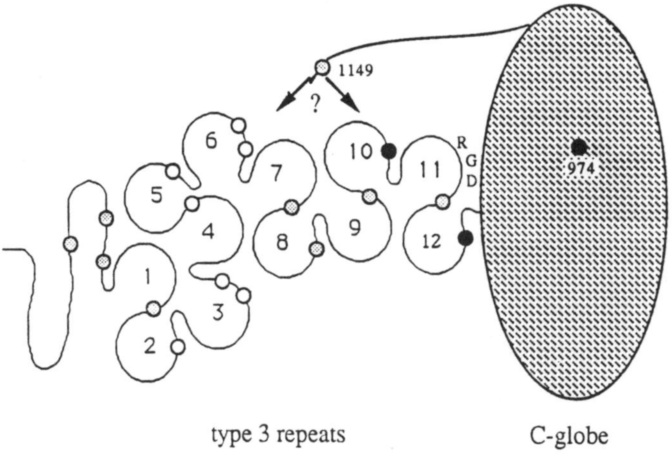

To test this hypothesis, TSP was reduced with dithiothreitol (DTT) to break disulfides and tested for its adhesive activity. Several interesting phenomena were observed (Sun et al., 1992).

First, DTT-treated TSP acted as a potent adhesive molecule. In the previous work when TSP was reported to be adhesive, it was coated to plastic wells at ≥ 20 μg/ml (Tuszynski et al., 1987; Lawler et al., 1988) and supported cell attachment but not robust cell spreading (Lawler et al., 1988). In contrast, cells were well spread on TSP coated at concentrations as low as 1 μg/ml and then treated with DTT.

Second, the adhesion of cells to DTT-reduced TSP was inhibited by RGD peptide and sensitive to a monoclonal antibody which recognizes a functional site of the $\alpha_v\beta_3$ vitronectin receptor as previously reported (Lawler et al., 1988).

Third, the DTT concentrations needed to generate adhesive activity were 10-fold higher in the presence of Ca^{2+} than in the presence of EDTA. Ca^{2+} also protected disulfides from reduction as ascertained by gel electrophoresis and quantification of free thiol. Such a protective effect has been observed previously by Turk and Detwiler (1986) and Speziale and Detwiler (1990).

Finally, cell blots of reduced TSP fragments localized the RGD-inhibitable adhesive activity at its Ca^{2+}-sensitive and RGD-containing type 3 repeat region.

The possibility that the exposure of RGD sequence is controlled by disulfide(s) in the type 3 repeat prompted us to analyze closely the amino acid sequence in this region. When amino acid sequence and the homologous alignment in this region was first described by Lawler and Hynes (1986), residues were aligned into eight repeats to maximize the homology among the repeats. A Ca^{2+}-binding loop sequence (DXDXDGXXDXXD or slight modifications thereof) appeared twice in each full type 3 repeat of TSP. This sequence has been noted in several Ca^{2+}-binding proteins (Dang et al., 1985). These proteins adopt a characteristic helix-loop-helix conformation (Kretsinger, 1976; Sudhakar Babu et al., 1985), with the Ca^{2+} bound in the loop between the helices. The coordination of Ca^{2+} can be represented by an octahedron with six vertices. Five vertices are coordinated with oxygen atoms from side chains, and a peptide oxygen coordinates at the sixth (Kretsinger, 1976, 1980).

If the 12 loop sequences in this region are aligned, a striking pattern emerges (Figure 5). DXDXDGXXDXXDX loop sequences are followed 6 times by a 10-amino acid extension, tentatively termed the break sequence. The boundaries of both loop and break sequences line up, with the exception at line 7, in which the extension of the loop sequence is only two amino acids. The cysteines are aligned at either in the loop sequence (short lines) or in the break sequence (long lines), with the exceptions of lines 3 and 6, in which the cysteines are vicinal. The alignment suggests that of the 12 loops, 10 are in pairs forming 5 loop-loop structures that are followed by a break region whereas loops 3 and 6 are separated by break region after a single loop.

Inasmuch as there are homologies among the break sequences, it is likely that these regions share a common secondary structure. As pointed out by Lawler et al. (1988), the algorithm of Chou and Fasman (1978) predicts that the conserved

NPDQ sequence has a strong tendency to form β-turn, with a calculated probability 15-fold greater than average. Thus, we propose that this multiple Ca^{2+}-binding region is composed of loop(loop)–β-turn repeats (Figure 5). Loop 11, containing the RGD, and loop 12, as Lawler et al. (1988) proposed, form the last of the double loop structures.

Figure 5 also shows the relative labeling of cysteines in the type 3 repeat and C-globe regions (Speziale and Detwiler, 1990). The cysteines that are fractionally labeled may be free or participate in more than one type of disulfide. For instance, Cys^{912} in loop 11 may form a disulfide with Cys^{928} next to loop 12, or with Cys^{892} next to loop 10, Cys^{892} may form a disulfide with Cys^{912} or with Cys^{876} in loop 9, etc. Cys^{974}, which is the only labeled cysteine in our TSP and is replaced by a serine in TSP2 (Lawler et al., 1991), may play a critical role in the isomerization either by forming a disulfide with Cys^{1149} or with a cysteine in the type 3 repeats. Cys^{1149}, which is the fourth residue from the carboxy terminus of TSP and is thought to be involved in a disulfide (Dixit et al., 1985), may be important in maintaining the C-globe structure and the exposure of the RGD sequence. One or more of the many potential isomers of TSP may have an exposed RGD adhesive sequence and mimic our reduced TSP. Such isomers may be only a small percentage of the TSP molecules or subunits released from platelets. It will be of great interest to identify factors in addition to Ca^{2+} and reducing molecules that modulate exposure of the RGD sequence. One possible modifier of TSP is type V collagen, which interacts more strongly with Ca^{2+}-depleted TSP than with Ca^{2+}-replete TSP (Galvin et al., 1987). TSP2 has the loop-(loop)-break structure, the RGD sequence in loop 11, and the C-globe but a serine at 974 rather than cysteine. It will be of interest to learn if TSP2 has a regulated RGD or whether its adhesive sequence is constitutively exposed or constitutively cryptic.

IV. CONCLUDING REMARKS

Sage and Bornstein (1991) have grouped TSP, tenascin, and SPARC (secreted protein acidic and rich in cysteine, also known as osteonectin) into a group of nonhomologous proteins with related functions. Common attributes include "anti-adhesive" effects (at least in some experimental systems!), regulated secretion, affinity for extracellular matrix molecules, and failure to accumulate or function as structural components in the normal adult organism. The group of proteins, therefore, is proposed to play dynamic roles in embryogenesis and morphogenesis (Sage and Bornstein, 1991), as has been proposed previously for TSP (Murphy-Ullrich and Mosher, 1987b; Mosher, 1990). Nevertheless, TSP1 has a number of unique features that set it apart: its ability to inhibit plasmin and perhaps other trypsin-like proteases and thereby regulate fibrinolysis and tissue proteolysis; its variably exposed RGD adhesive sequence inbedded in an imposingly complex array of metal ion-binding repeats; and its related but quite distinct homolog, TSP2. By focusing on these unique features, it is hoped that this review will help the reader understand forthcoming information about TSPs.

NOTE ADDED IN PROOF

Three additional TSP-like proteins have been cloned (Bornstein, 1992; Lawler et al., 1992; Oldberg et al., 1992). These proteins lack the procollagen and properdin modules found in TSP1 and TSP2, contain four EGF-like modules rather than three, and, where the subunit structure is known (Mörgelin et al., 1992), are pentameric rather than trimeric. Thus, the three new proteins may be considered a subclass of TSPs. It is noteworthy that all contain the twelve calcium-binding sequences and the long carboxyl-terminal sequence.

ACKNOWLEDGMENTS

This research was supported by HL29586 and HD07118 from the National Institutes of Health.

REFERENCES

Asch, A.S., Tepler, J., Silbiger, S., & Nachman, R.L. (1991). Cellular attachment to thrombospondin. J. Biol. Chem. 266, 1740–1745.

Bale, M.D. (1987). Noncovalent and covalent interactions of thrombospondin with polymerizing fibrin. Semin. Thromb. Hemostasis 13, 326–334.

Bale, M.D. & Mosher, D.F. (1986). Thrombospondin is a substrate for blood coagulation factor XIIIa. Biochemistry 25, 5667–5673.

Bornstein, P. (1992). Thrombospondins: Structure and regulation of expression. FASEB J. 6, 3290–3299.

Bornstein, P., Devarayalu, S., Li, P., Disteche, C.M., & Framson, P. (1991b). A second thrombospondin gene in the mouse is similar in organization to thrombospondin 1 but does not respond to serum. Proc. Natl. Acad. Sci. USA 88, 8636–8640.

Bornstein, P., O'Rourke, K., Wikstrom, K., Wolf, F. W., Katz, R., Li, P., & Dixit, V.M. (1991a). A second, expressed thrombospondin gene (Thbs2) exists in the mouse genome. J. Biol. Chem. 266, 12821–12824.

Carrell, R.W., Pemberton, P.A., & Boswell, D.R. (1987). The serpins: Evolution and adaptation in a family of protease inhibitors, Cold Spring Harbor Symp. on Quant. Biol. 527–535.

Chou, P.Y. & Fasman, G.D. (1978). Prediction of the secondary structure of proteins from their amino acid sequence. Adv. Enzymol. 47, 45–148.

Chu, M.L., Zhang, R.Z., Pan, T., Stokes, D., Conway, D., Kuo, H.J., Glanville, R., Mayer, U., Mann, K., Deutzmann, R., & Timpl, R. (1990). Mosaic structure of globular domains in the human type VI collagen α3 chain: similarity to von Willebrand factor, fibronectin, actin, salivary proteins and aprotinin type protease inhibitors. EMBO J. 9, 385–393.

Dang, C.V., Ebert, R.F., & Bell, W.R. (1985). Localization of a fibrinogen calcium binding site between γ-subunit positions 311 and 336 by terbium fluorescence. J. Biol. Chem. 260, 9713–9719.

DePoli, P., Bacon-Baguley, T., Kendra-Franczak, S., Cederholm, M. T., & Walz, D.A. (1989). Thrombospondin interaction with plasminogen. Evidence for binding to a specific region of the kringle structure of plasminogen. Blood 73, 976–982.

Detwiler, T.C. & Feinman, R.D. (1973). Kinetics of the thrombin-induced release of calcium (II) by platelet. Biochemistry 12, 282–289.

Dixit, V.M., Galvin, N.J., O'Rourke, K.M., & Frazier, W.A. (1986). Monoclonal antibodies that recognize calcium-dependent structures of human thrombospondin. Characterization and mapping of their epitopes. J. Biol. Chem. 261, 1962–1968.

Dunwiddie, C., Thornberry, N.A., Bull, H.G., Sardana, M., Friedman, P.A., Jacobs, J.W., & Simpson, E. (1989). Antistasin, a leech-derived inhibitor of factor Xa: Kinetic analysis of enzyme inhibition and identification of the reactive site. J. Biol. Chem. 264, 16694–16699.

Engel, J. (1991). Common structural motifs in proteins of the extracellular matrix. Curr. Opinions Cell Biol. 3, 779–785.

Frazier, W.A. (1991). Thrombospondins. Curr. Opinion Cell. Biol. 3, 792–799.

Galvin, N.J., Vance, P.M., Dixit, V.M., Fink, B., & Frazier, W.A. (1987). Interaction of human thrombospondin with types I–V collagen: Direct binding and electron microscopy. J. Cell Biol. 104, 1413–1422.

Galvin, N.J., Dixit, V.M., O'Rourke, K. M., Santoro, S.A., Grant, G.A., & Frazier, W.A. (1985). Mapping of epitopes for monoclonal antibodies against human platelet thrombospondin with electron microscopy and high sensitivity amino acid sequencing. J. Cell Biol. 101, 1434–1441.

Good, D.J., Polverini, P.J., Rastinejad, F., Le Beau, M.M., Lemons, R.S., Frazier, W.A., & Bouck, N.P. (1990). A tumor suppressor-dependent inhibitor of angiogenesis is immunologically and functionally indistinguishable from a fragment of thrombospondin. Proc. Natl. Acad. Sci. USA 87, 6624–6628.

Harpel, P.C., Silverstein, R.L., Pannell, R., Gurewich, V., & Nachman, R.L. (1990). Thrombospondin forms complexes with single-chain and two-chain forms of urokinase. J. Biol. Chem. 265, 11289–11294.

Hogg, P.J., Stenflo, J., & Mosher, D.F. (1992). Thrombospondin is a slow tight-binding inhibitor of plasmin. Biochemistry 31, 265–269.

Holt, G.D., Krivan, H.C., Gasic, G.J., & Ginsburg, V. (1989). Antistasin, an inhibitor of coagulation and metastasis, binds to sulfatide (Gal(3-SO_4)β1-1Cer) and has a sequence homology with other proteins that bind sulfated glycoconjugates. J. Biol. Chem. 264, 12138–12140.

Kido, H., Fukutomi, A., Schilling, J., Wang, Y., Cordell, B., & Katunuma, N. (1990). Protease-specificity of Kunitz inhibitor domain of Alzheimer's disease amyloid protein precursor. Biochem. Biophys. Res. Commun. 167, 716–721.

Jang, I.K., Gold, H.K., Ziskind, A.A., Fallon, J.T., Holt, R.E., Leinbach, R.C., May, J.W., & Collen, D. (1989). Differential sensitivity of erythrocyte-rich and platelet-rich arterial thrombi to lysis with recombinant tissue-type plasminogen activator: A possible explanation for resistance to coronary thrombolysis. Circulation 79, 920–928.

Kaesberg, P.R., Ershler, W.B., Esko, J.D., & Mosher, D.F. (1989). Chinese hamster ovary cell adhesion to human platelet thrombospondin is dependent on cell surface heparan sulfate proteoglycan. J. Clin. Invest. 83, 994–1001.

Karczewski, J., Knudsen, K.A., Smith, L., Murphy, A., Rothman, V.L., & Tuszynski, G.P. (1989). The interaction of thrombospondin with platelet glycoprotein GPIIb-IIIa. J. Biol. Chem. 264, 21322–21326.

Kosfeld, M.D., Pavlopoulos, T.V., & Frazier, W.A. (1991). Cell attachment activity of the carboxyl-terminal domain of human thrombospondin express in *Escherichia coli*. J. Biol. Chem. 266, 24257–24259.

Kretsinger, R.H. (1980). Structure and evolution of calcium-modulated proteins. Crit. Rev. Biochem. 31, 119–174.

Kretsinger, R.H. (1976). Calcium-binding proteins. Ann. Rev. Biochem. 45, 239–266.

Lahav, J. (1988a). Thrombospondin inhibits adhesion of endothelial cell. Exp. Cell Res. 177, 199–204.

Lahav, J. (1988b). Thrombospondin inhibits adhesion of platelets to glass and protein-coated substrata. Blood 71, 1096–1099.

Laskowski, M., Jr. & Kato, I. (1980). Protein inhibitors of proteinases. Ann. Rev. Biochem. 49, 593–626.

Lawler, J., Duquette, M., Whittaker, C.A., Adams, J.C., McHenry, K., & DeSimone, D.W. (1992). Identification and characterization of thrombospondin-4, a new member of the thrombospondin gene family. J. Cell Biol. 120, 1059–1067.

Lawler, J. & Hynes, R.O. (1986). The structure of human thrombospondin, an adhesive glycoprotein with multiple calcium-binding sites and homologies with several different proteins. J. Cell Biol. 103, 1635–1648.

Lawler, J. & Hynes, R.O. (1989). An integrin receptor on normal and thrombasthenic platelets that binds thrombospondin. Blood 74, 2022–2027.

Lawler, J. & Simons, E.R. (1983). Cooperative binding of calcium to thrombospondin. J. Biol. Chem. 258, 12098–12101.

Lawler, J., Chao, F.C., & Cohen, C.M. (1982). Evidence for calcium-sensitive structure in platelet thrombospondin. Isolation and partial characterization of thrombospondin in the presence of calcium. J. Biol. Chem. 257, 12257–12265.

Lawler, J., Derick, L.H., Connolly, J.E., Chen, J-H., & Chao, F.C. (1985). The structure of human platelet thrombospondin. J. Biol. Chem. 260, 3762–3772.

Lawler, J., Duquette, M., & Ferro, P. (1991). Cloning and sequencing of chicken thrombospondin. J. Biol. Chem. 266, 8039–8043.

Lawler, J., Weinstein, R., & Hynes, R.O. (1988). Cell attachment to thrombospondin: the role of Arg-Gly-Asp, calcium, and integrin receptors. J. Cell Biol. 107, 2351–2361.

Leung, L.L.K. & Nachman, R.L. (1982). Complex formation of platelet thrombospondin with fibrinogen. J. Clin. Invest. 70, 542–549.

McKeown-Longo, P.J., Hanning, R., & Mosher, D.F. (1984). Binding and degradation of platelet thrombospondin by cultured fibroblasts. J. Cell Biol. 98, 22–28.

Mörgelin, M., Heinegård, D., Engel, J., & Paulsson, M. (1992). Electron microscopy of native cartilage oligomeric matrix protein purified from the Swarm rat chondrosarcoma reveals a five-armed structure. J. Biol. Chem. 267, 6137–6141.

Mosher, D.F. (1990). Physiology of thrombospondin. Annu. Rev. Med. 41, 85–97.

Mosher, D.F., Saksela, O., & Vaheri, A. (1977). Synthesis and secretion of alpha-2-macroglobulin by cultured adherent lung cells: Comparison with cell strains derived from other tissues. J. Clin. Invest. 60, 1036–1045.

Mumby, S.M., Raugi, G.J., & Bornstein, P. (1984). Interactions of thrombospondin with extracellular matrix proteins: Selective binding to type V collagen. J. Cell. Biol. 98, 646–652.

Murphy-Ullrich, J.E. & Höök, M. (1989). Thrombospondin modulates focal adhesions in endothelial cells. J. Cell Biol. 109, 1309–1319.

Murphy-Ullrich, J.E. & Mosher, D.F. (1985). Localization of thrombospondin clots formed *in situ*. Blood 66, 1098–1104.

Murphy-Ullrich, J.E. & Mosher, D.F. (1987a). Interaction of thrombospondin with endothelial cells: receptor-mediated binding and degradation. J. Cell Biol. 105, 1603–1611.

Murphy-Ullrich, J.E. & Mosher, D.F. (1987b). Interactions of thrombospondin with cells in culture: Rapid degradation of both soluble and matrix thrombospondin. Semin. Thromb. Hemost. 13, 343–351.

Murphy-Ullrich, J.E., Schultz-Cherry, S., & Höök, M. (1992). Transforming growth factor-β complexes with thrombospondin. Mol. Biol. Cell 3, 181–188.

Murphy-Ullrich, J.E., Westrick, L.G., Esko, J.D., & Mosher, D.F. (1988). Altered metabolism of thrombospondin by Chinese hamster ovary cells defective in glycosaminoglycan synthesis. J. Biol. Chem. 263, 6400–6406.

Oldberg, Å, Antonsson, P., Lindblom, K., & Heinegård, D. (1992). COMP (cartilage oligomeric matrix protein) is structurally related to the thrombospondins. J. Biol. Chem. 267, 22346–22350.

Ossowski, L., Clunie, G., Masucci, M.-T., & Blasi, F. (1991). *In vivo* paracrine interaction between urokinase and its receptor: Effect on tumor cell invasion. J. Cell Biol. 115, 1107–1112.

Petersen, T.E., Thøgersen, H.C., Skorstengaard, K., Vibe Pedersen, K., Sahl, P., Sottrup-Jensen, L., & Magnusson, S. (1983). Partial primary structure of bovine plasma fibronectin: Three types of internal homology. Proc. Natl. Acad. Sci. USA 80, 137–141.

Prater, C.A., Plotkin, J., Jaye, D., & Frazier, W.A. (1991). The properdin-like type I repeats of human thrombospondin contain a cell attachment site. J. Cell Biol. 112, 1031–1040.

Pyke, C., Kristensen, P., Ralfkiaer, E., Grøndahl-Hansen, J., Eriksen, J., Blasi, F., & Danø, K. (1991). Urokinase-type plasminogen activator is expressed in stromal cells and its receptor in cancer cells at invasive foci in human colon adenocarcinomas. Am. J. Pathol. 138, 1059–1067.

Raugi, G.J., Olerud, J.E., & Gown, A.M. (1987). Thrombospondin in early human wound tissue. J. Invest. Dermatol. 89, 551–554.

Ruoslahti, E. & Pierschbacher, M.D. (1987). New perspectives in cell adhesion: RGD and integrins. Science 238, 491–497.

Rupp, F., Payan, D.G., Magill-Solc, C., Cowan, D.M., & Scheller, R.H. (1991). Structure and expression of a rat agrin. Neuron 6, 811–823.

Roberts, D.D., Sherwood, J.A., & Ginsburg, V. (1987). Platelet thrombospondin mediates attachment and spreading of human melanoma cells. J. Cell Biol. 104, 131–139.

Sage, E.H. & Bornstein, P. (1991). Extracellular proteins that modulate cell-matrix interactions. J. Biol. Chem. 266, 14831–14834.

Saksela, O. & Rifkin, D.B. (1988). Cell-associated plasminogen activator: Regulation and physiological function. Ann. Rev. Cell Biol. 4, 93–126.

Sandell L.J. & Boyd, C.D. (1990). Conserved and divergent sequence and functional elements within collagen genes. In: Extracellular Matrix Genes (Sandell, L.J. & Boyd, C.D., Eds.), pp. 1–56. Academic Press, New York.

Silverstein, R.L., Harpel, P.C., & Nachman, R.L. (1986). Tissue plasminogen activator and urokinase enhance the binding of plasminogen to thrombospondin. J. Biol. Chem. 261, 9959–9965.

Silverstein, R.L., Leung, L.L.K., Harpel, P.C., & Nachman, R.L. (1984). Complex formation of platelet thrombospondin with plasminogen: Modulation of activation by tissue activator. J. Clin. Invest. 74, 1625–1633.

Silverstein, R.L., Leung, L.L.K., Harpel, P.C., & Nachman, R.L. (1985a). Platelet thrombospondin forms a trimolecular complex with plasminogen and histidine-rich glycoprotein. J. Clin. Invest. 55, 2065–2073.

Silverstein, R.L. & Nachman, R.L. (1987). Thrombospondin-plasminogen interactions: Modulation of plasmin generation. Sem. Thromb. Hem. 13, 335–342.

Silverstein, R.L., Nachman, R.L., Leung, L.L.K., & Harpel, P.C. (1985b). Activation of immobilized plasminogen by tissue activator: Multimolecular complex formation. J. Biol. Chem. 260, 10346–10352.

Sinha, S., Dovey, H.F., Seubert, P., Ward, P.J., Blacher, R.W., Blaber, M., Bradshaw, R.A., Arici, M., Mobley, W.C., & Lieberburg, I. (1990). The protease inhibitory properties of the Alzheimer's β-amyloid precursor protein. J. Biol. Chem. 265, 8983–8985.

Slane J.M.K., Mosher, D.F., & Lai, C-S. (1988). Conformation change in thrombospondin induced by removal of bound Ca^{2+}. A spin label approach. FEB 229, 363–366.

Sottile, J., Selegue, J., & Mosher, D.F. (1991). Synthesis of truncated amino-terminal trimers of thrombospondin. Biochemistry 30, 6556–6562.

Speziale, M.V. & Detwiler, T.C. (1990). Free thiols of platelet thrombospondin. J. Biol. Chem. 265, 17859–17867.

Sudhakar Babu, Y., Sack, J.S., Greenhough, T.J., Bugg, C.E., Means, A.R., & Cook, W.J. (1985). Three-dimensional structure of calmodulin. Nature 315, 37–40.

Sun, X., Skorstengaard, K., & Mosher, D.F. (1992). Disulfides modulate RGD-inhibitable adhesive activity of thrombospondin. J. Cell Biol. 118, 693–701.

Taraboletti, G., Roberts, D., Liotta, L. A., & Giavazzi, R. (1990). Platelet thrombospondin modulates endothelial cell adhesion, motility, and growth: a potential angiogenesis regulatory factor. J. Cell Biol. 111, 765–772.

Turk, J.L. & Detwiler, T.C. (1986). Thiol-disulfide exchange by thrombospondin: evidence for a thiol and a disulfide bond protected by calcium. Arch. Biochem. Biophys. 245, 446–454.

Tuszynski, G.P., Karczewski, J., Smith, L., Murphy, A., Rothman, V.L., & Knudsen, K.A. (1989). The GPIIb-IIIa-like complex may function as a human melanoma cell adhesion receptor for thrombospondin. Exp. Cell Res. 182, 473–481.

Tuszynski, G.P. & Kowalska, M.A. (1991). Thrombospondin-induced adhesion of human platelets. J. Clin. Invest. 87, 1387–1394.

Tuszynski, G.P., Rothman, V., & Murphy, A. (1987). Thrombospondin promotes cell-substratum adhesion. Science 236, 1570–1573.

Zajchowski D.A., Band, V., Trask, D.K., Kling, D., Connolly, J.L., & Sager, R. (1990). Suppression of tumor-forming ability and related traits in MCF-7 human breast cancer cells by fusion with immortal mammary epithelial cells. Proc. Natl. Acad. Sci. USA 87, 2314–2318.

ELASTIN

William C. Parks, Richard A. Pierce,
Katherine A. Lee, and Robert P. Mecham

Advances in Molecular and Cell Biology
Volume 6, pages 133–181.

ISBN: 1-55938-515-4

I. INTRODUCTION

The structural integrity of a tissue or organ is maintained by its extracellular matrix. Although matrix proteins participate actively in numerous aspects of cellular regulation, historically they have been defined by their physical and adhesive properties. For example, the connective tissue protein elastin, which is a resilient protein polymer, is so named because of its rubber-like physical properties. Consequently, elastin is found in significant quantities in tissues, such as skin, lung, and large arteries, in which reversible extensibility is required. The ability of elastin to restore tissue shape after removal of stretch or deformation defines the principal characteristic of the fiber. Thus, alveolar capacity during respiration, transduction of arterial blood pressure, and resiliency of skin are each maintained by elastic fibers in the extracellular matrix, but these fibers are also involved in cell:matrix and matrix:matrix interactions. The importance of this protein is indicated by its limited phylogenetic distribution to vertebrates which have an enclosed, pressurized circulatory system. Many features of elastin, such as its limited developmental expression, extensive alternative splicing, and its extreme durability and hydrophobicity, distinguish this protein from other connective tissue components. In this chapter, we review many of the characteristics of elastin and discuss several recent findings related to its regulation, interaction with cells, gene structure, and assembly. Other recent reviews on elastin structure (Mecham and Heuser, 1991), gene sequence (Indik et al., 1990), cross-link formation (Franzblau et al., 1969; Kagan, 1986; Kagan and Trackman, 1991), physicochemical properties (Gosline, 1976; Urry, 1983; Gosline and Rosenbloom, 1984), heterogeneity (Parks and Deak, 1990), degradation (Robert and Hornebeck, 1989), and role in disease (Uitto et al., 1982; Uitto and Ryhanen, 1987) can be consulted for additional information.

II. COMPOSITION OF ELASTIC FIBERS

In discussions of elastic tissues, a distinction is made among elastic fibers, elastin, and tropoelastin. The elastic fiber is a complex, extracellular structure comprised of elastin, microfibrillar proteins, lysyl oxidase, and possibly other matrix molecules. Elastin is the predominant protein of mature elastic fibers, and it is assembled in the extracellular space by the enzymatic crosslinking of individual tropoelastin molecules. In its cross-link form, elastin is a chemically inert and extremely hydrophobic protein that is very resistant to dissolution by chaotropes, bases, and weak acids. Because of its durability, elastin, prior to the discovery of the soluble precursor molecule, tropoelastin, was defined only in operational and qualitative terms as the insoluble protein residue remaining after all other tissue components had been solubilized. With the discovery of tropoelastin and subsequent sequencing of the mRNA and gene, the primary structure of elastin was revealed.

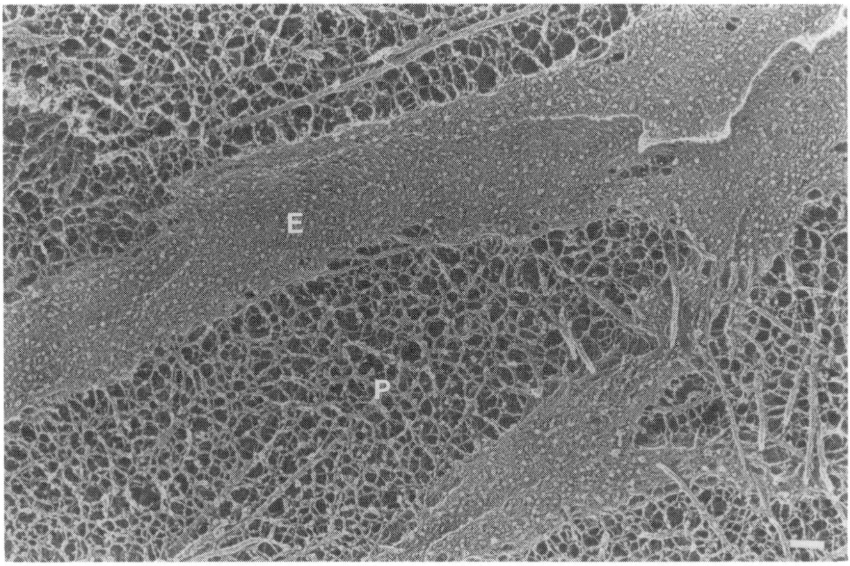

Figure 1. Ultrastructure of elastic fibers. The fine structure of an elastic fiber (E) in bovine nuchal ligament is seen by quick-freeze, deep-etch transmission electron microscopy. Typical of most elastic tissues, the fiber is surrounding by proteoglycans (P) which form a three-dimensional trabecular network. In this high resolution electron micrograph, the fiber is comprised of tightly packed globular units which resemble isolated tropoelastin monomers. (Bar = 0.2 μm, mag. 70,000×)

A. Elastin and Elasticity

Elastic fibers are comprised of two components that are morphologically and chemically distinct. Most of the mature fiber (>90%) is amorphous elastin (Figure 1), the insoluble product of cross-linked tropoelastin monomers. Tropoelastin is synthesized as a group of slightly different isoforms, ranging in size from 64 to 70 kDa, which result from alternative splicing of the pre-mRNA. In its secreted form, tropoelastin is nonglycosylated and is extremely rich in nonpolar amino acids, such as alanine, valine, and proline and the uncharged amino acid glycine (Table 1). Together, nonpolar and uncharged residues account for greater than 90% of the total amino acids of tropoelastin, making this one of the most hydrophobic proteins known. Consequently, charged residues are scarce, and the secreted protein is devoid of asparagine, histidine, glutamic acid, tryptophan, and methionine residues (Indik et al., 1990). The physical properties of elastic fibers are reflected in the primary sequence of tropoelastin. The protein is comprised of alternating hydrophobic and cross-linking domains which confer distinct physical properties to elastic fibers. The elasticity of the fiber is driven by hydrophobic interactions, while polymerization of elastin by covalent bonding of cross-link lysines provides durability and strength to the fiber. A characteristic of the hydrophobic domains is that they often contain repetitive sequences of three to nine uncharged amino acids, whereas cross-linking domains contain stretches of alanines interrupted by two or three lysines (Table 2).

Elastin is characterized by a high degree of reversible distensibility, including the ability to deform extensively with small forces. Biophysical studies of purified

Table 1. Amino Acid Composition of Secreted Bovine Tropoelastin*

Nonpolar		*Uncharged*		*Charged*	
Amino Acid	*Residues per Molecule*	*Amino Acid*	*Residues per Molecule*	*Amino Acid*	*Residues per Molecule*
Ala	154	Gly	232	Lys	38
Val	92	Gln	9	Arg	6
Pro/Hypro	89	Thr	8	Asp	3
Leu	44	Ser	8	Glu	0
Phe	23	Tyr	7	His	0
Ile	19	Cys	2		
Met	0	Asn	0		
Trp	0				
Subtotal	421		266		47
%Total	57.4%		36.2%		6.4%

Note: *Derived from cDNA sequence excluding amino acids from the signal peptide (Yeh et al., 1987).

Table 2. Domain Sequence of Mammalian Tropoelastin

Exon[1]	Type[2]	Species	Sequence[3]
		Bovine	MRSLTAAARRPEVLLLLLC-ILQPSQPG
1	SP	Human	AG P G S- H R
		Rat	VPQ G I LNL H A
		Bovine	GVPGAVPGGVP--------GGVFFP
2	H	Human	I -------- Y
		Rat	GGLPGGVP YY
		Bovine	GAGL-GGLGV
3	H	Human	- A G
		Rat	IG G
		Bovine	GGLGPGVKPAKP
4	CL	Human	A G L
		Rat	A G P
		Bovine	GVGGL----VGPG-LG-A
5	H	Human	VP ----A A - -
		Rat	A L GAFGA G G
		Bovine	GLGALPGA-F--PGA-LVPGGPAGAAAAYKAAAKA
6	CL	Human	F A T -- V D -
		Rat	P GLSY-ASR GV G
		Bovine	GAAGLGVGGIGGV----------------------------GGLGVST
7	H	Human	- - V-P ---------------------------- A
		Rat	- -- PGGVGVGGVPGAVGVGGVPGAVGGIGGI
		Bovine	GAVVPQLGAGVGAGVKPGKVP
8	CL	Human	P ----
		Rat	G
		Bovine	GVGLPGVYPGGVLPGAGA
9	H	Human	--
		Rat	T
		Bovine	RFPGIGVLPGVPTGAGVKPKAP
10	CL	Human	V
		Rat	V T A V
		Bovine	GGGG-AFAGIP
11	H	Human	V -
		Rat	G S
		Bovine	GVGPFGGQQPGVPLGYPIKAPKLP
12	CL	Human	P
		Rat	
		Bovine	AGYGLPYKTGKLPY
13	CL	Human	G T
		Rat	G TN
		Bovine	GFGPGGVAGSAGKAGYPTGT
14	H	Human	Y A
		Rat	----- AG
		Bovine	GVGPQAAAAA-AKAAAKL-
15	CL	Human	A F-
		Rat	S V - - YA
		Bovine	GAGGAGVLPGVGVGGAGIPGAPGAIPGIGGIA
16	H	Human	A -- V V
		Rat	G -- G GA T
		Bovine	GVGAPDAAAAAAAAAKAAKF
17	CL	Human	T A Y
		Rat	A T A K Y
		Bovine	GAAGGL-PGV-GVPGV-GVPGVGVPGVGVPGVGV-PGVGV-PGVGVPGVGVPGVGVPGVGVPG-VGVP
18	H	Human	A FV GP F V A A I- ----- V -- A I AA -----------
		Rat	-VPG- - -R A I I GI GI G----I G I G I
		Bovine	GAVSPAAAAKAAAKAAKF
19	CL	Human	V D Y
		Rat	Y
		Bovine	GARGGVGIGGIPTFGVGPGGFPGIG------------------------------
20	H	Human	P V Y A F VGVGGIPGVAGVPSVGGVPGVGGVPGVGIS
		Rat	--- Y A Y VGAGAGLG----------------------
		Bovine	-DAAAAQAAAAAKAAKI
21	CL	Human	P --- Y
		Rat	GASQ A Y
		Bovine	GAGGVGALGGLVPGA---------------------PGAIPGVPGVPGVGGVG
22	H	Human	AAA V P---------------------QA V T
		Rat	A T VPGALPGAVPGALPGAVPGAL V T
		Bovine	GVGIPAAAAA----KAAAKAAQF
23	CL	Human	T ----
		Rat	A T AAAA G Y

(*continued*)

Table 2. (*continued*)

Exon[1]	Type[2]	Species	Sequence[3]
		Bovine	GLGPGVGVAPG---VGVVPGVGVVP----GVGVAP------GIGLGPG-GV---I
24[4]	H	Human	V --- A A ---- L GVGVAP V VA V APG
		Rat	GV GVG GL - GPGGVT I TG ------ T V -DL---G
		Bovine	GAGVPAAAAKSAAKAAAKAQF
25	CL	Human	P GV V L
		Rat	
		Bovine	RAAAGLPAGVPGLGVGVGVPGLGVGVGVPGLGVGVGVPGFGA
26	H	Human	G I
		Rat	G A F --A AG F A
		Bovine	-------------------------------
26A[5]	M	Human	GADEGVRRSLSPELREGDPSSSQHLPSTPSS
		Rat	-------------------------------
		Bovine	--VPGTLAAAKAAKF
27	CL	Human	-- A Y
		Rat	GA S S Y
		Bovine	G--------------PGGV-GALGGVGDLGGA-GIPGGVA
28	H	Human	---------------AA P V L A V- V
		Rat	AAGGGLGGPGGLGG L- GP F GP LG V
		Bovine	GVGPAAAAAA-KAAAKAAQF
29	CL	Human	A A
		Rat	GA Y
		Bovine	GL-G--GVGGLGVGGLGA-----------------------VPGAVGLG
30	H	Human	V AA L ------------------------ VG
		Rat	- -- A A GGLGAGGLGAGGLGAGGLGAGGVI
		Bovine	GVSPAAAAKAAKF
31	CL	Human	IP Y
		Rat	Y
		Bovine	GAAGLGGVLGAGQPFPIG
32	H	Human	GAGQ- L
		Rat	-R G
		Bovine	GVAARPGFGLSPIFP
33	H	Human	
		Rat	Y
		Bovine	GG-AGGLGVG
34	H	Human	----------
		Rat	G
		Bovine	GKPPKPFGGALGALGFP
35	CL	Human	-----------------
		Rat	Y YQ
		Bovine	GGACLGKSCGRKRK
36	HL	Human	A
		Rat	G F

Notes: [1]The domain sequences are separated into the functional units which are encoded by separate exons (see text).

[2]The domains are classified as signal peptide (SP), hydrophobic (H), potential crosslinking (CL), mixed hydrophobic/hydrophilic (M) and hydrophilic (HL).

[3]The amino acid sequences are derived from the cDNA sequences for human, bovine (Indik et al., 1990), and rat (Pierce et al., 1991) tropoelastins. In this table, the human and rat sequences are compared to the bovine sequence. Homologies are represented by spaces; deletions are represented by dashes; and substitutions are listed with the replaced residue. For example, with the exception of a single proline substitution in the human, domain 12 is homologous among all species. Domains 34 and 35 are missing in the human sequence, and domain 7 in the rat has a large hydrophobic insert.

[4]Domain 24 in the bovine and human contains the repeated cell recognition domain VGVAPG.

[5]Domain 26A, which is only found rarely in human tropoelastin, is encoded by exon 26. The 26A sequences of the exon are usually alteratively spliced by use of an altemate donor splice site.

elastic fibers indicate that elastin behaves as a classical rubber polymer in that when it is relaxed, it is in a state of maximum disorder and, hence, maximum entropy (Hoeve, 1974). Stretching a rubber induces order in the direction of the extension, thereby decreasing entropy. When the external force is removed, elastic recoil occurs because the partially ordered chains return spontaneously to their initial, random state level of maximum entropy. Elastin exhibits other physical features characteristic of a polymeric rubber, including interchain cross-links and domains that behave as a kinetically free, random chain network (Fleming et al., 1980).

While the dominant pattern of elastin is that of a random structure, elastin is more complicated than a simple three-dimensional array of random chains. Several physicochemical studies suggest that regions of the molecule exhibit local order. For example, synthetic peptides that represent repeating hydrophobic tropoelastin sequences can assume a specific β-spiral (Venkatachalam and Urry, 1981). If the repeating peptides of tropoelastin conform to a regular structure, then the interactions among tropoelastin monomers in the extensively cross-linked fiber may result in a polymer with a regular structure. Indeed, using high-resolution transmission electron microscopy, an ordered, three-dimensional structure of elastic fibers and tropoelastin monomers is seen (Morocutti et al., 1988; Mecham and Heuser, 1990; Mecham and Heuser, 1991).

The extreme hydrophobicity of elastin has led to the generation of models of elasticity based on the possibility that nonpolar regions of the molecule are exposed to water when elastin is stretched (Partridge, 1969; Weis-Fogh and Andersen, 1970; Gray et al., 1973; and Urry, 1974). Recoil occurs after the distending force is removed, thereby allowing the nonpolar groups to reaggregate and expel the absorbed water (Gosline, 1978). The relative proportion of energy stored in the reversible exposure of hydrophobic groups to water versus entropy effects associated with ordering of random chains upon stretching, however, is still an open question. Regardless, the ability of elastic fibers to stretch and recoil by an energy-free mechanism provides an efficient means to confer reversible extensibility to tissues that undergo continued expansion and relaxation.

Studies of the structure of elastin are made difficult by the extensive cross-linking of the tropoelastin monomers. Moreover, the hydrophobic tropoelastin monomer is not amenable to crystallization. Therefore, structural and physicochemical studies on elastin have focused on elastin peptides, tropoelastin, and elastin cross-links. Synthetic elastin peptides—peptides generated from enzymatically degraded elastin and tropoelastin—undergo an unusual phenomenon when in solution. As the temperature of the solution is raised from 4 °C to about 25 °C, a phase separation occurs, and these components self-aggregate into a structure that resembles amorphous elastin *in situ*. This phase separation is called coacervation and demonstrates the tendency of elastin peptides and tropoelastin monomers to associate tightly with each other and to exclude water (Cox et al., 1974; Urry and Long, 1977). Findings from these models indicate that hydrophobic interactions are important in elastic recoil and possibly in the ordering of tropoelastin during fiber assembly. Since elastin polymerization *in vivo* is an enzymatic event, coacervated peptides probably do not completely reflect elastin structure or formation. Lacking in these physicochemical systems is the contribution of elastin cross-links, which are believed to be rigid, extended helical domains (Foster, 1982), and the influence of nonelastin proteins, such as microfibrils, on the final structure of elastic fibers.

B. Elastin Cross-Links

The potential cross-linking domains of tropoelastin contain two or three lysines interspersed among several alanine residues (Table 2). The lysines in these areas form inter- and intrachain covalent cross-links providing durability to the fiber. The principal cross-links found in elastin, desmosine, and isodesmosine, which are produced from four lysine residues, are unique to the elastic fibers and their presence serves as a diagnostic marker of elastin (Figure 2). Although intermediates in this process have been characterized, the precise chemical events have not been fully elucidated. In fact, new yet infrequent elastin cross-links continue to be

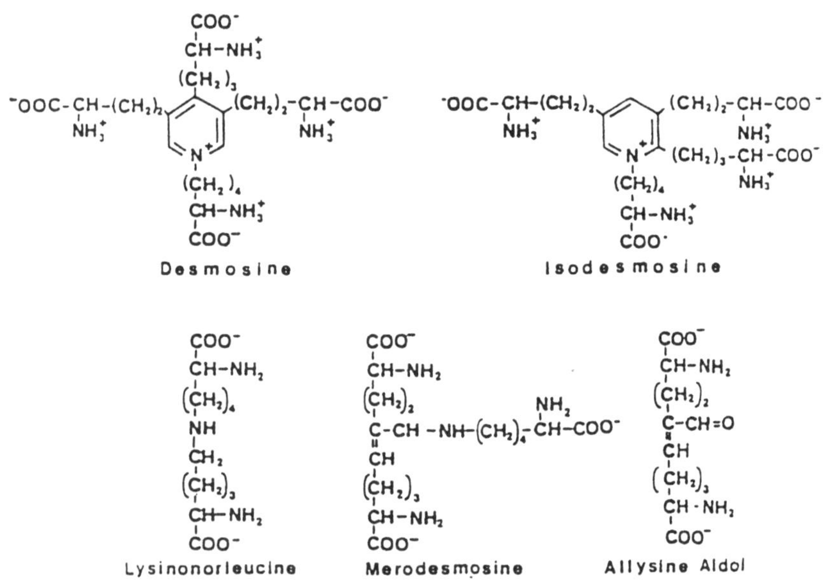

Figure 2. Elastin cross-links. Desmosine and isodesmosine cross-link isomers are produced from the enzymatic deamination of three lysines and spontaneous condensation with a fourth lysine residue. The other cross-links are condensation products of two or three lysines. Lysinonorleucine is an aldol condensation product of a deaminated and an unmodified lysine. Allysine aldol is a condensation product of two deaminated lysines, and merodesmosine is formed by the condensation of two deaminated lysines with one unmodified residue.

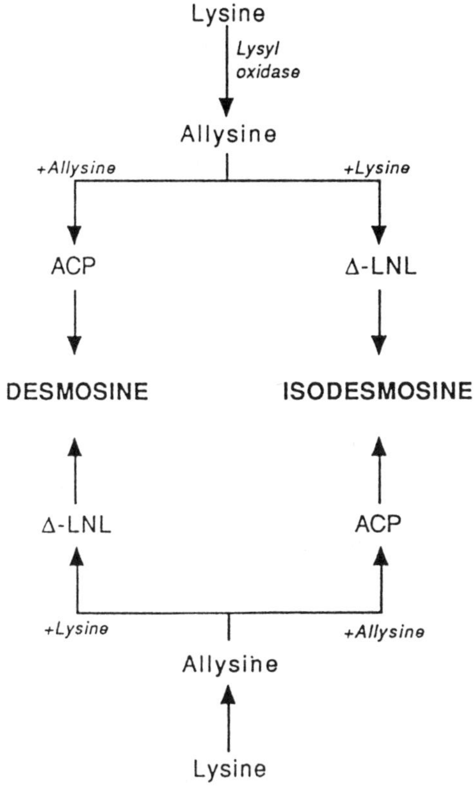

Figure 3. Elastin cross-link formation. Lysine is deaminated by the copper-requiring enzyme, lysyl oxidase, and the intermediate product, allysine, spontaneously condenses with another allysine to form an allysine aldol condensation product (ACP), or with an unmodified lysine to form dehydrolysinonorleucine (Δ-LNL). ACP and Δ-LNL then condense to form desmosine or isodesmosine cross-links. Reduction of Δ-LNL and ACP forms the bifunctional cross-link structures lysinonorleucine and allysine aldol.

discovered (Suyama and Nakamura, 1990). In spite of an incomplete understanding of the mechanism of cross-link formation, three steps are accepted as being key in the formation of desmosine and isodesmosine (Figure 3). The initial step in their formation is the deamination of the ε-amino group of lysine side chains by the copper-requiring enzyme lysyl oxidase. The reactive aldehyde that is formed (α-amino adipic δ-semi-aldehyde, or allysine) condenses spontaneously with either a second aldehyde residue to form allysine aldol or with an ε-amino group on lysine to form dehydrolysinonorleucine. Allysine aldol and dehydrolysinonorleucine then condense to form the pyridinium cross-links called desmosine and isodesmosine.

Also found in elastin are lysinonorleucine and merodesmosine cross-links which are the reduced forms of the original condensation products dehydrolysinonorleucine and dehydromerodesmosine (Figure 2), and these are probably formed by linkage of lysines from separate tropoelastin molecules rather than within a single cross-linking domain. The biochemistry and importance of cross-link reduction is obscure and nothing is known about the *in vivo* reducing agent.

Mammalian tropoelastin has about 34 lysine residues per molecule that may potentially participate in cross-link formation (Table 2). However, only about 30 lysines are modified and not all of these are accounted for as cross-links. In bovine nuchal ligament elastin, for example, there are about 2 desmosines, 1 isodesmosine, 2 lysinonorleucines, and 4 aldol condensation products per tropoelastin equivalent. This translates into 22 lysine equivalents per monomer, or about 75% of the available modified lysines. The difference may be that insoluble elastin contains as yet unidentified cross-links or that some regions are cleaved from tropoelastin before its incorporation into the fiber or that the primary sequence of some cross-linking domains is not fully conducive for condensation (see Tropoelastin Structure). Although desmosine and isodesmosine are tetrafunctional cross-links, current evidence suggests that they are produced from two cross-linking domains of separate tropoelastin monomers. An individual tropoelastin monomer can form cross-links with multiple adjacent tropoelastins providing for potential extensive networking among tropoelastins which may account for the significant growth and durability of elastic polymers.

C. Microfibrils

Microfibrils, the other component of elastic fibers, are a complex of glycoproteins organized as small, 10–12-nm diameter tubular fibrils (Cleary and Gibson, 1983). Microfibrils were initially identified as an integral component of elastin fibers that form a peripheral sheath around a core of amorphous elastin, but morphologically and immunologically similar fibrils are also seen without amorphous elastin. Microfibrils contain many charged and basic amino acid residues as well as numerous cysteines that may be important in promoting interactions with tropoelastin during fiber assembly. Microfibrils appear in the extracellular space in close association with the plasma membrane before fetal elastin production begins. As it is produced, tropoelastin is secreted and cross-linked onto the microfibrils, gradually obscuring them until, in mature elastin, they are visible only at the edges of the amorphous elastin core (see Figure 6).

Recently, substantial progress has been made toward identifying microfibrillar components. Gibson et al. (1986) isolated a 31-kDa glycoprotein from a reductive guanidine extract of microfibrils, called microfibril associated glycoprotein (MAGP) which they recently cloned and sequenced (Gibson et al., 1991). The amino acid sequence indicates that MAGP contains two structurally distinct regions. The amino-terminal half of the protein is rich in glutamine, proline, and acidic amino acids whereas the carboxy-terminal half contains all 13 of the cysteine residues and most of the basic amino acids. The high number of cysteine residues is consistent with the finding that MAGP forms intermolecular disulfide bonds and is suggestive that the protein interacts with other components. MAGP, however, does not bind tropoelastin (M. Gibson, personal communication), nor does it contain any arg-gly-asp (RGD) sequences.

Gibson et al. (1989) isolated four other proteins of 340-, 78-, 70-, and 25-kDa from fetal bovine nuchal ligament, and they demonstrated by immunostaining that two of these (340- and 78-kDa forms) localize to microfibrils. Sakai and co-workers identified a human 350-kDa glycoprotein called fibrillin that localizes to microfibrils (Sakai et al., 1986; Maddox et al., 1989). Like MAGP, fibrillin is rich in acidic amino acids and cysteine, but it is distinct in that it contains numerous EGF-like repeats that alternate with cysteine-rich domains (Malsen et al., 1991). It is now apparent that the 340-kDa protein reported by Gibson et al. (1989) is the bovine homologue of fibrillin (M. Gibson, personal communication).

Microfibrils distribute widely among both elastic and nonelastic tissues (Streeten and Licari, 1983; Sakai et al., 1986; Gibson and Cleary, 1987; Dahlbäck et al., 1990). Although morphologically and immunologically similar, it is unknown whether the components of microfibrils from elastic and nonelastic tissues are compositionally identical. Recently, genetic heterogeneity of fibrillin was shown by the partial characterization of two highly related yet distinct fibrillin genes (Lee et al., 1991). Like fibrillin, or fib-15, a name derived from the chromosomal localization of its gene, the gene for the new fibrillin, fib-5, encodes protein sequences that include repeated cysteine-rich, EGF-like motifs and a relatively high content of acidic amino acids. Although the tissue-specific expression of fib-5 is not known, this genomic heterogeneity of fibrillin suggests that differences may exist among microfibrils.

Several other proteins have been shown by antibody studies to be associated with elastic-microfibrils, including a 34-kDa glycoprotein with amine oxidase activity (Serafini-Fracassini et al., 1981) and lysyl oxidase (Kagan et al., 1986), but the precise relationship between these molecules and microfibrillar structure and function is yet to be resolved. Lysyl oxidase immunolocalizes to the interface between the amorphous core of mature elastin and its microfibrillar mantle, but the enzyme is not associated with microfibrils away from amorphous elastin (Kagan et al., 1986).

D. Association of Other Matrix Molecules with Elastin

Proteoglycans are seen in the extracellular matrix in close association with elastic fibers (Mecham and Heuser, 1990). In human dermis, core proteins of the small matrix proteoglycans, PG I and PG II (decorin) as well as to enzymatically digested forms of hyaluronic acid and dermatan sulfate side chains colocalize by immunostaining with elastin (Baccarani-Contri et al., 1985; Baccarani-Contri et al.,

1990). In addition, the intensity of elastin staining often increases markedly following brief digestion of the matrix with hyaluronidase suggesting that hyaluronic acid-containing sugars associate with the fiber (Wrenn and Mecham, 1987). The role of proteoglycans in the elastic fiber is not known, but they may provide hydration necessary for elastic recoil or prevent spontaneous aggregation of tropoelastin in the extracellular space thereby allowing fibrillogenesis to occur in a more orderly fashion (Fornieri et al., 1987). Other components, such as glycoprotein gp115 (Colombatti et al., 1988), vitronectin (Dahlbäck et al., 1989), amyloid P (Breathnach et al., 1981), and various complement factors (Niculescu et al., 1987; Werth et al., 1988; Dahlbäck et al., 1989), have been shown by immunostaining techniques to be associated with elastic fibers, but the significance of these interactions is not known.

III. TISSUE DISTRIBUTION OF ELASTIC FIBERS

Although elastic fibers are found in the connective tissue of most organs [see Farquharson and Robins (1988) and references within], they are particularly abundant in tissues that are subjected to repetitive deformation. Thus, blood vessel wall, alveolar septal interstices, and deep dermal layers each contain significant quantities of elastin. A high proportion of elastin are also found in elastic cartilage, such as in the external ear, the eustachian tube wall, and cartilages of the larynx, and in elastic ligaments, such as the ligamentum flava of the vertebral column and the nuchal ligament of certain grazing animals. The relative amount of elastin varies among tissues and reflects the physical demands placed on the tissue. In tissues in which the applied force is greater and more constant, such as the arteries and elastic ligaments, elastin comprises about 30 and 75%, respectively, of the dry weight of the tissue.

In skin, elastic fibers are confined to the reticular dermis and are relatively scarce comprising no more than 2% of the dry weight of the tissue (Uitto, 1987). In discussions of elastin in the skin, the terms "oxytalan fibers" and "elaunin fibers" are used to denote different forms of elastic fibers, but this nomenclature is imprecise and at times confusing. Oxytalan fibers are microfibrils with no associated amorphous elastin and are found extending vertically from the basal lamina to the upper portions of the papillary dermis (Figure 4). As these fibrils extend into the intermediate dermis, they become associated with small amounts of deposited elastin. These small elastic fibers are the so-called elaunin fibers and have often been thought of as "immature" elastic fibers. This classification implies that elastogenesis is incomplete in this area of the dermis, but these smaller elastic fibers probably are an endpoint product with a precise structural role. Deeper into the reticular dermis, the small, vertical elastic fibers anastomose with horizontally arranged cascades of large elastic fibers. Thus, the dermal elastic matrix can be considered as a continuous network of microfibrils that, as they extend from the

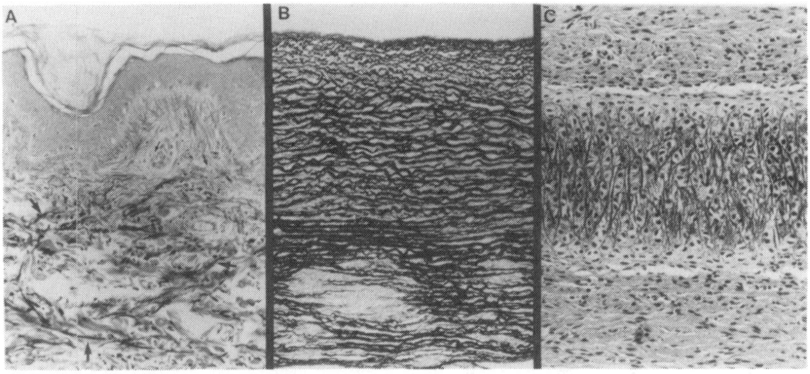

Figure 4. Tissue-specific organization of elastic fibers. The morphology of elastic fibers varies among tissues. **A.** In skin, large elastic fibers (large arrows) in the reticular dermis transverse parallel to the epidermis. In the papillary dermis, the fibers are thin (small arrows) and perpendicular to those in the deeper layers. As these fibers approach the surface, the amount of elastin diminishes. Shown is a section of human adolescent skin stained with Miller-Halmi (mag ×300). (Courtesy of Dr. Edward Clearly) **B.** Shown is a section of calf pulmonary artery stained with Verhoeff–Van Gieson stain which stains elastic fibers black. The photomicrograph is oriented with the vessel lumen at the top and the abluminal medial at the bottom; the adventitia has been stripped from this sample. In large arteries, elastic fibers are found in the medial layer and are predominantly organized in concentric sheets or laminae. **C.** In fetal bovine auricular cartilage, Verhoeff–Van Gieson stained elastic fibers are seen as an anastomotic trabecular network.

reticular to the papillary dermal layers, are associated with progressively less elastin. In this particular organization, microfibrils may have a structural role in anchoring the epidermis to the dermal elastin network (Cleary, 1987).

A. Developmental Expression

In most mammalian tissues, the bulk of elastin production occurs during fetal and neonatal periods and is essentially complete by the first decade of life (Figure 5). Thereafter, the rate of synthesis decreases sharply and is almost fully repressed by maturity, although production may be reinitiated in response to wounding or disease. The temporal sequence of elastogenesis is unique in that elastic fibers are deposited during a defined developmental window whereas many other extracellular matrix proteins are continually made and turned over. The lack of continued production reflects the extreme durability and extended half-life of elastic fibers. In most mammalian tissues, tropoelastin synthesis initiates near the start of the third trimester of gestation. During the last trimester of fetal development, the rate of

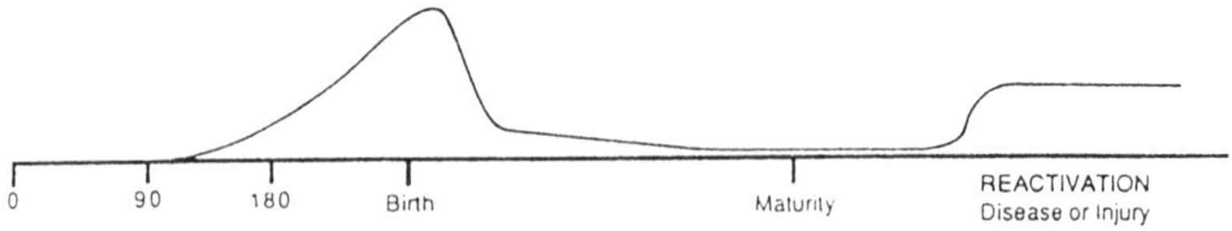

Figure 5. Developmental expression of tropoelastin. In mammals, elastic fibers first appear in the extracellular matrix near midgestation, and production is maximal in late fetal and neonatal periods. By maturity, elastogenesis is complete, and, concomitantly, the expression of tropoelastin drops to very low levels. Production may be reactivated, however, in response to disease or injury.

tropoelastin synthesis and elastin deposition increases sharply and reaches maximal levels during early neonatal periods. During peak periods of deposition, tropoelastin production may account for as much of 15% of total cell protein synthesis (Davidson et al., 1985), and tropoelastin mRNA may constitute about 1% of total RNA (Davidson et al., 1984). After this relatively short, yet intense period of production, tropoelastin synthesis begins to wane, and production markedly diminishes soon after birth. In the chick, the pattern of elastin production varies among tissues, but still the bulk of deposition occurs during the embryonic period (Foster et al., 1989). A notable exception to the general pattern of elastogenesis is the cyclic production and turnover of elastin in the gravid and postgravid uterus (Sharrow et al., 1989).

In some tissues, the precise postnatal time when tropoelastin synthesis is repressed has not been fully described, and in some tissues, low levels of production may persist. For example, the levels of tropoelastin synthesis and mRNA in human skin fibroblasts remain relatively constant between cell populations derived from fetal and adult donors (Sephel and Davidson, 1986), whereas in other tissues sharp differences are seen in tropoelastin expression between cells derived from fetal and adult samples (Parks et al., 1988b). In intact skin, however, the mass of deposited elastin does not change appreciably after maturity, and thus, if tropoelastin production is continuing, then some degradation of either the soluble or insoluble protein must be occurring to balance synthesis and assembly. Alternatively, the expression of tropoelastin in isolated skin cells may reflect the heterogeneity population of dermal fibroblasts or indicate that dermal tropoelastin production is regulated, in part, by its extracellular environment. In this context, removal of the cells from adult tissue may disrupt inhibitory controls resulting in a reactivation of tropoelastin expression.

B. Cellular Production

The cells responsible for elastin production and deposition are diverse and vary depending on anatomical location. As is common for interstitial connective tissue proteins, elastin producing cells in normal tissues are mostly of a mesodermal lineage. Fibroblasts produce elastin in the dermis, in elastic ligaments, and in the loose connective tissue of many soft tissues; smooth muscle cells produce elastin in the uterus; chondroblasts produce elastin found in cartilages; and mesothelial cells deposit elastin in the lung pleura.

In arteries, elastin is made by various cell types, including endothelial cells (Damiano et al., 1984), smooth muscle cells, and adventitial fibroblasts (Prosser et al., 1989). Although medial smooth muscle cells are responsible for the bulk of arterial elastin, the lumenal surface of the internal elastic lamina and the ablumenal side of the external elastic lamina are probably produced by endothelial cells and adventitial fibroblasts, respectively. In addition, some of the elastin in the wall of the great vessels may be synthesized by neural crest which migrate into the media (Rosenquist et al., 1990). Thus, not all elastogenic cells are mesodermally derived.

Cells responsible for elastin production in the lung parenchyma have been difficult to identify because this area is comprised of multiple cell types, and because the phenotype of the interstitial cell changes during development. As the alveolar septa begin to form, myofibroblasts are found in close association with capillary structures and are considered to the source of elastin at these sites (Collett and Des Biens, 1974; Fukuda et al., 1983; Campagnone et al., 1987; Noguchi et al., 1989). In rat lung, immature interstitial cells differentiate into lipid-filled and nonlipid-filled fibroblasts, the latter being alveolar septal myofibroblasts which are associated with elastic fibers (Vaccaro and Brody, 1978; Brody and Vaccaro, 1979). In culture, however, both cell subpopulations synthesize and deposit elastin (Campagnone et al., 1987; Berk et al., 1991) suggesting that multiple cell types are responsible for alveolar septal elastogenesis.

Elastic fibers are often found in desmoplastic areas associated with breast cancers, and this production has usually been attributed to stromal fibroblasts which received an inductive signal from the carcinoma cells (Parfey and Doyle, 1985). Recently, Krishnan and Cleary (1990) demonstrated that breast carcinoma cells themselves express tropoelastin, and these authors suggested that much of the elastin made during the desmoplastic response is a product of the cancer cells. The normal breast epithelial cell most likely does not make elastin since a intact basement membrane separates the ductal cells from the stroma. Thus, the acquisition of the elastogenic phenotype in the carcinoma cells may be dependent on the disruption of the basement membrane and contact with breast connective tissue. Indeed, tropoelastin production is increased in fibroblasts grown on matrix from breast tumor (Kao et al., 1986).

C. Morphology

Elastic fibers are organized in at least three distinct morphological forms (Figure 4). In elastic ligaments, skin and lung, the fibers are small and rope-like (Fahrenbach et al., 1966; Smith et al., 1982; Noguchi et al., 1989). In blood vessels, elastin

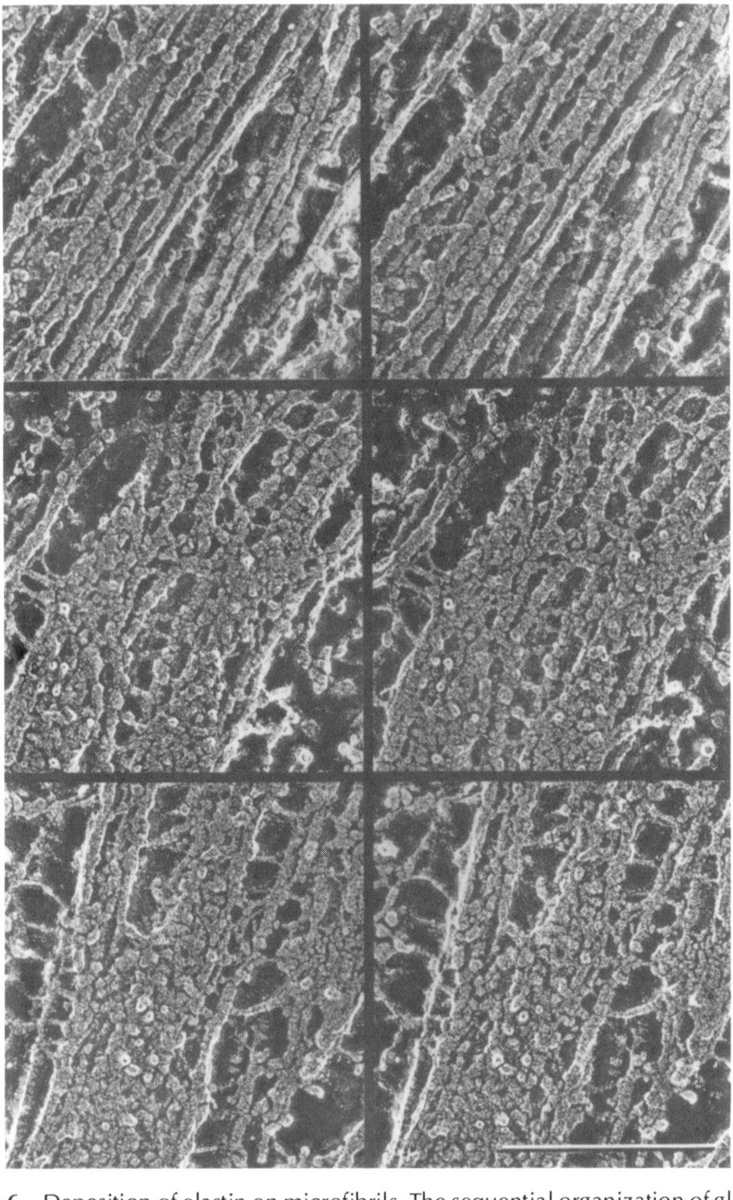

Figure 6. Deposition of elastin on microfibrils. The sequential organization of globular tropoelastin monomers on microfibrils during development is demonstrated in this series of freeze-etched stereo-electronmicrographs of nuchal ligament of 120-, 200- and 270-day gestation bovine fetuses. (Bar = 0.2 μm)

forms as concentric sheets or lamellae which are interconnected by fine elastin fibers (Crissman, 1986), and in elastic cartilage, the fibers organize an anastomotic trabecular network (Mecham and Heuser, 1990). In each case, the structure of the elastic fiber arises, in part, as a consequence of the direction and magnitude of the force placed on the tissue.

Although the histologic structure of elastic fibers differs among tissues, the fine detail of insoluble elastin is similar regardless of location. Ultrastructural analysis of mature elastic fibers reveals a core of lightly stained amorphous insoluble elastin surrounded by a mantle of electron-dense tubular-appearing microfibrils (Pasquali-Ronchetti and Fornieri, 1984). In its relaxed state, elastin appears disordered and granular by freeze-fracture electron microscopy. With stretching, however, anastomotic filaments, about 5 nm in diameter, organize relative to the applied force (Pasquali-Ronchetti et al., 1979). Using freeze-etch electron microscopy, fractured elastic fibers present a prominent three-dimensional face with much more detail than that seen by more conventional ultrastructural imaging techniques (Morocutti et al., 1988; Mecham and Heuser, 1990; Mecham and Heuser, 1991). These high resolution images suggest that elastin is composed of an array of densely packed filamentous structures 5 to 10 nm in diameter comprised of individual globular subunits (Figures 1 and 6). Rotary shadowing of purified tropoelastin shows it to be a globular protein with a diameter similar to that of filaments in the intact fiber suggesting that the filaments are indeed comprised of densely packed tropoelastin monomers (Mecham and Heuser, 1991). Of potential interest is that the globular subunits in elastic fibers display structural heterogeneity (Morocutti et al., 1988) which may reflect the presence of different tropoelastin isoforms.

IV. STRUCTURE OF TROPOELASTIN

Tropoelastin was discovered as the soluble precursor of elastin in the aorta of copper-deficient pigs (Sandberg and Davidson, 1984). Eventually, tryptic peptides of purified tropoelastin were sequenced but not ordered. This initial analysis demonstrated that the protein is mostly comprised of two distinct amino acid domains (Sandberg et al., 1969). These areas are characterized by sequences abundant in hydrophobic amino acids and by alanine-rich segments containing lysine residues which are potential cross-linking sites (Table 2). The hydrophobic domains often contain amino acid repeats which form β-helices that confer elasticity (Urry and Long, 1977), and the cross-linking domains are characterized by stretches of alanines interrupted by two or three lysines. The eventual characterization of tropoelastin cDNA confirmed the prediction that the primary structure of tropoelastin consists of alternating hydrophobic and cross-linking domains (see Figure 7). The sequence and organization of tropoelastin is highly conserved at both amino acid and nucleotide levels among mammals but differs significantly in some segments from the chick (Indik et al., 1990).

In general, the hydrophobic domains are less well conserved than the cross-linking domains in size and composition suggesting that tropoelastin can better tolerate variation of the hydrophobic amino acids without compromising their molecular role, and, in fact, these areas have diverged during evolution at twice the rate of the cross-link domains (Boyd et al., 1991). Overall, there is extensive homology among mammalian tropoelastins with most substitutions being conservative in nature. There are, however, some interesting differences. For example, the number of amino acids that may be found in a particular hydrophobic region may differ among species, suggesting that hydrophobic domains tolerate changes in either the number or sequence of amino acids. Chicken tropoelastin shows extensive homology with mammalian sequences for the first 300 residues and for the last 57 residues. Major differences exist in the central portion, however, with sequences that suggest duplication and deletion events (Bressan et al., 1987; Tokimitsu et al., 1987; Indik et al., 1990).

In contrast to the sequence variability in the hydrophobic domains, there appears to be absolute conservation in the number of alanine residues between lysines in the cross-linking domains. In these regions, two lysine residues are usually seen within a cluster of alanine residues, although in two domains near the center of the molecule, encoded by exons 19 and 25 (Table 2), three lysine residues are found. In addition, most potential cross-linking sequences in the first 200 residues frequently contain prolyl or other residues between the lysines instead of the usual alanines (exons 4, 8, 10, 12, and 13; Table 2). Typically, the two lysines are separated by two or three alanines, but never just one or more than three. This strict adherence to the spacing of lysine residues provides a conformation that is most conducive to the efficient formation of desmosine and isodesmosine cross-links. The alanine-rich domains form an α-helix with the side chains extended outward. Accounting for the spatial dimensions, the side chains of amino acids separated by two or three residues are relatively close to one another on the same side of the helix. Side chains of amino acids separated by one or four residues, on the other hand, would be situated on opposite sides of the helix, and this spacing is probably too great for the condensation reactions in the formation of elastin cross-links (Gray, 1977). In addition, substitution of bulkier amino acids between lysine residues may interfere with the steric requirements for cross-link formation. In light of this, the lysines encoded by exons 4, 8, 10, 12, 13, and 35 (Table 2) may not be fully modified to isodesmosine and desmosine cross-links, but rather they may be limited to the formation of lysinonorleucine and merodesmosine bonds between tropoelastin monomers.

In addition to the cross-linking and hydrophobic domains, a tyrosine-rich region downstream from the amino terminal end and the C-terminal amino acids are highly conserved among tropoelastins (Table 2). Although the role of the tyrosines is not known, they may affect oxidation of the ε-amino group of nearby lysines (Sandberg, 1976). The carboxy terminal sequence of tropoelastin is different from other domains in that it contains basic amino acids and, other than the signal peptide, is the only domain that contains cysteine residues. Significantly, the amino acid sequence of this segment is evolutionarily conserved, especially at the positions for cysteine, lysine, and arginine (Table 2). Recent studies have demonstrated that the cysteine residues in this domain form an interchain disulfide bond (Brown et al., 1992), and molecular modeling suggests that this linkage creates a loop of charged amino acids which may interact with other extracellular components such as microfibrils. As will be discussed later in Section IX, this potential interaction may serve in the initial steps of tropoelastin polymerization.

V. THE TROPOELASTIN GENE

As demonstrated by *in situ* hybridization and analysis of somatic cell hybrids, the tropoelastin gene localizes to chromosome 7q 11.2 in the human genome (Fazio et al., 1991). As for other genes coding for matrix proteins, the tropoelastin gene is separated into exons that code for distinct domains of the protein (Figure 7) (Indik et al., 1990). The human and bovine genes have been most thoroughly characterized, and these contain 33 and 36 exons, respectively, which have been demonstrated by cDNA sequence analysis to be present in the mRNA. Although 34 exons are shown in Figure 7 for the human mRNA, exon 22, which would encode a typical tropoelastin hydrophobic domain (Table 2), has never been found in any cDNA. Among species, the tropoelastin gene is highly conserved with respect to size, exon organization, 5′ flanking sequences, sequences within intron 1, and sequences coding for the 3′ untranslated region (UTR).

The human and bovine tropoelastin genes are unusually large, containing nearly 40 kb of sequence coding for a mRNA of about 3.5 kb. This large size results from an intron to exon ratio of approximately 15:1. In contrast, many collagen genes have an intron:exon ratio of about 8:1 (Sandell and Boyd, 1990). The exons of the tropoelastin gene are relatively small, ranging from about 25 to 200 bp, and most exons are arranged in a cassette-like fashion with sequences coding for either hydrophobic or cross-linking domains (Figure 7). There are, however, two exceptions to this general organization. The 5′ end of exon 26 in the human gene encodes a typical hydrophobic domain, but the 3′ end, called 26A, encodes an unusual mix of charged and nonpolar amino acids (Table 2). These amino acids probably comprise a minor component of human elastin since the 26A sequences are usually alternatively spliced from tropoelastin pre-mRNA. The second exception is exon 36 which is about 1.5 kb in mammals and which contains sequences that encode both the hydrophilic C-terminus and the large 3′ UTR of the mRNA (Figure 7).

A notable difference between the bovine and human tropoelastin genes is the absence in the human gene of cross-linking exon 34 and hydrophobic exon 35 (Figure 7). Deletion of these exons, however, is not characteristic of all primate DNA because a clone isolated from a baboon smooth muscle cell cDNA library contained sequences homologous to exon 34 (C.D. Boyd, personal communica-

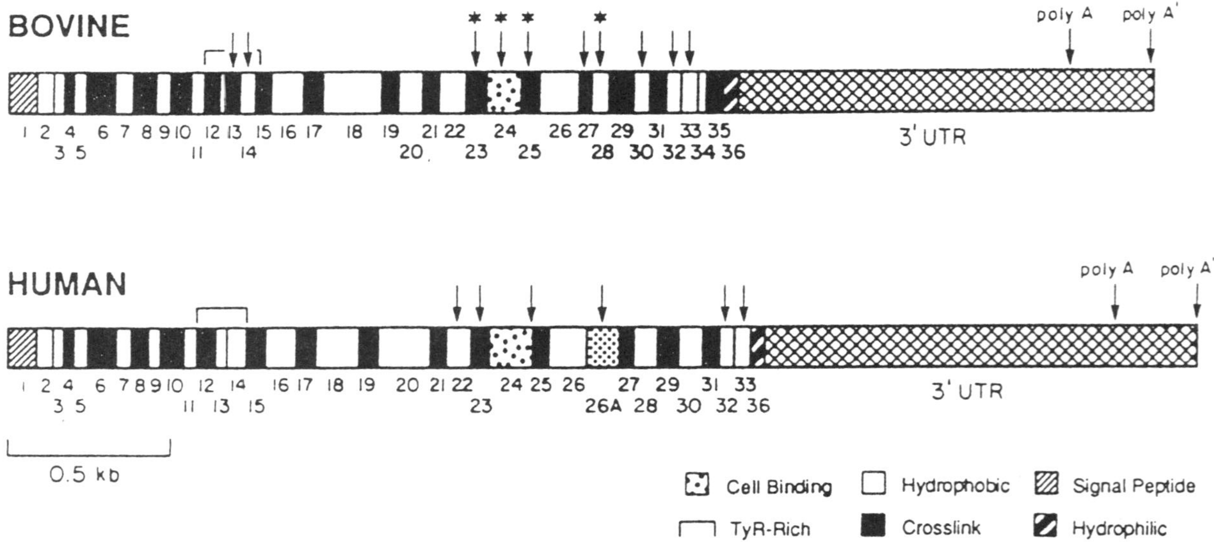

Figure 7. Alternating domain structure of tropoelastin. The structure of human and bovine tropoelastin mRNAs is divided into specific areas which are transcribed from discrete exons and which encode for distinct domains of the protein. Exons are numbered from the 5' end and are drawn approximately to scale. Human exons are numbered relative to the bovine exons; note in the human gene the deletion of exons 33 and 34 and the addition of 26A which codes for a domain containing both hydrophobic and hydrophilic amino acids. Exon 36 is divided to distinguish the 3' UTR from the sequences that code for the conserved, hydrophilic C-terminus. Arrows indicate exons that are alternatively spliced, and arrows with asterisks identify bovine exons that are alternatively spliced exclusively from adult bovine mRNA.

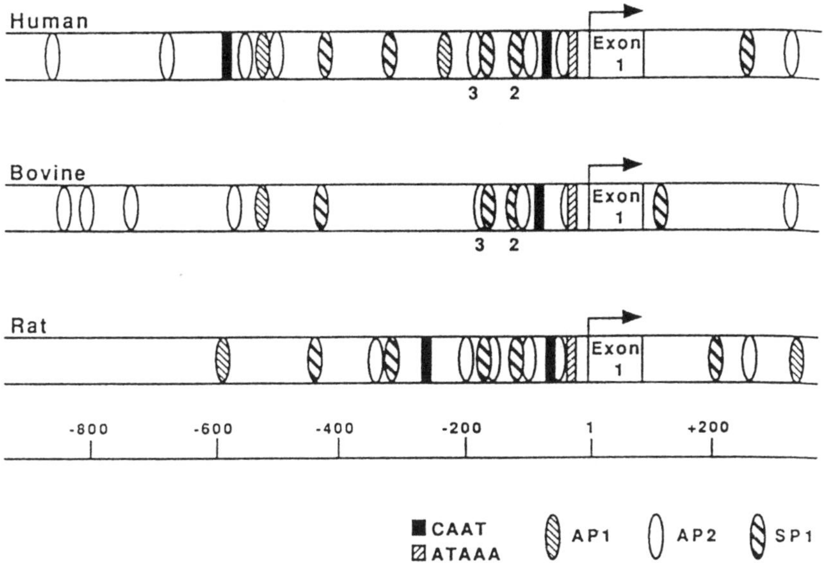

Figure 8. Transcription elements of the tropoelastin gene. The location and types of transcription elements is similar among tropoelastin genes, especially those sequences between –200 and 1. In the human and bovine genes, three AP2 and 2 SP1 sites are clustered together at about –170 and –100 bp, respectively. (Figure courtesy of Dr. Charles Boyd.)

tion). This variability of tropoelastin-coding sequences within such a closely related phylogenetic group such as primates suggests that considerable heterogeneity is tolerated in elastin without affecting the normal function of tissue.

Analogous to other genes, the 5′ flanking region of the tropoelastin gene contains several elements associated with the initiation of gene transcription, such as binding sites for SP1 and AP1 and a CAAT box (Bashir et al., 1989; Yeh et al., 1989), and the relative location of these is conserved among species (Figure 8). In contrast to many other genes, the promoter region of the tropoelastin gene does not contain a canonical TATA box but does have the element ATAAA. These sequences are important for basal transcription because, when mutated to AAAAA, transcriptional activity from the tropoelastin promoter is markedly decreased (J. Uitto, personal communication). Although involved in transcription, the ATAAA motif of the human tropoelastin gene may not act as a true TATA box since at least seven different transcription initiation sites are used (Bashir et al., 1989). The lack of a TATA box, along with several G/C-rich domains within the 5′ end of the gene, suggests that the promoter of the tropoelastin gene resembles regulatory regions of genes which are constitutively expressed. Regions of the tropoelastin gene that may be involved in control of gene transcription are discussed in Section VI.

VI. REGULATION OF TROPOELASTIN EXPRESSION

In essentially all models studied, the production of elastin or secretion of tropoelastin correlates with the steady-state levels of tropoelastin mRNA (for example, see Foster et al., 1980; Shibahara et al., 1981; Burnett et al., 1982; Davidson et al., 1984; Mecham et al., 1987; Parks et al., 1988a,b; Sharrow et al., 1989; Foster et al., 1990b; Hinek et al., 1991). These collective findings indicate that expression is regulated at a pretranslational step, but the contribution of transcriptional and posttranscriptional mechanisms to the control of elastogenesis is only now beginning to be delineated. As discussed, the promoter region of tropoelastin is similar to that of a constitutively expressed gene. This is somewhat surprising considering the precise temporal and tissue-specific production of elastin, and may indicate that unique regulatory mechanisms may control tropoelastin expression. Using multiple deletion constructs of the first 2.26 kb of the tropoelastin promoter and a chloramphenicol acetyltransferase (CAT) expression vector, Kähäri et al. (1992) have identified sequences within bases –128 to –1 that confer basal transcriptional activity, but no promoter sequences have yet been characterized as being important for age or tissue-specific production. In fact, these CAT constructs conferred similar transcriptional activity in both elastogenic and nonelastogenic cells, indicating that tissue and age specific elements lie elsewhere in the gene (Fazio et al., 1990; Kähäri et al., 1990). Similar to the first intron of the type II procollagen gene (Horton et al., 1987), it is possible that regions involved in temporal and tissue-specific expression of tropoelastin reside within the transcribed sequences of the gene.

Studies of cultured cells exposed to different modulators, or maintained under specific conditions, allow the determination of regulatory mechanisms that are active in controlling tropoelastin expression, and also indicate which extracellular factors may influence elastin production *in vivo*. From such studies, various cytokines, steroids, and culture conditions have been shown to affect tropoelastin production and elastin deposition (Table 3). Elastogenic cells are typically derived by explant culture from fetal or neonatal tissues, such as aorta, nuchal ligament, lung or skin, or by direct enzymatic digestion of fetal ear cartilage (Mecham, 1987b). The production of tropoelastin varies among these cells, paralleling the relative mass of elastin in the tissue from which the cells were derived as well as reflecting age-specific production. Thus, tropoelastin synthesis is highest in late fetal or early neonatal aortic smooth muscle cells and fetal auricular chondroblasts, and is low in cells derived from skin and from adult tissues. The low production of tropoelastin by dermal fibroblasts may be more reflective of a heterogeneous population rather than low expression by individual cells. A characteristic of elastin producing cells is that they lose their capacity to synthesize tropoelastin as they age in culture or are serially passed (Mecham et al., 1981; Sephel and Davidson, 1986). With explanted cells this problem is enhanced because numerous population doublings are needed to expand the primary population. Reflective of the *in vivo* developmental expression of elastin which occurs primarily during periods of rapid

Table 3. Modulators of Elastin Production[1]

Stimulators	*Repressors*
Insulin-like growth factor 1	1,25-$(OH)_2$ Vitamin D_3
Transforming growth factor β1	Tumor necrosis factor-α
SMEF	Interleukin 1
Elastic matrix	Phorbol ester
Glucocorticoids[2]	Epidermal growth factor
cGMP[3]	Interferon γ
	Monesin
	Ascorbate

Notes: [1]See text for references.
[2]As discussed in the text, glucocorticoid-mediated effects on elastin production may be age dependent.
[3]Mecham et al., 1985.

tissue growth, the synthesis of tropoelastin is higher in proliferating cultures than in nondividing cells, and is greatest during late log-phase growth (Mecham et al., 1981; Sjölund et al., 1986).

A. Effect of Matrix

The phenotypic instability of tropoelastin expression may be primarily a consequence of removal of the cells from their microenvironment *in vivo*, and may involve unique regulatory mechanisms. An interesting aspect of elastin regulation is that expression of tropoelastin is induced in preelastogenic fibroblasts grown on nuchal ligament elastic matrix from late gestation fetuses (Mecham et al., 1984a). This differentiation response apparently requires *de novo* gene induction, although it is not known if the tropoelastin gene is induced or if elastogenesis is dependent on transcriptional initiation of another gene product. Elastic matrix also stimulates tropoelastin synthesis by late fetal ligament fibroblasts (Mecham et al., 1984a) and elevates the very low levels of production by adult ligament fibroblasts (Parks et al., 1988b), and desmoplastic elastic matrix from malignant breast carcinoma stimulates production of elastin by dermal fibroblasts (Kao et al., 1986). Unlike the inductive effect in predifferentiated cells, the stimulation of tropoelastin production by ligament matrix does not require initiation of gene transcription, indicating that diverse mechanisms regulate tropoelastin expression at different stages of production. Furthermore, the stimulation of tropoelastin synthesis by adult cells grown on elastic matrix indicates that the tropoelastin gene remains in an operative state after elastogenesis has ceased (Parks et al., 1988b). Indeed, using serially passed cells as a model of developmental aging, we find that the marked drop in tropoelastin production with time in culture is not associated by any significant decrease in tropoelastin transcription (R.A. Pierce and W.C. Parks, unpublished observations). These observations indicate that posttranscriptional mechanisms may be important in controlling the cessation of elastin production.

The matrix-induction experiments demonstrate that the tissue milieu influences both the start and maintenance of elastin production. These studies also show that cell-matrix contact is required for differentiation, suggesting strongly that a cell surface molecule is involved and that the instructive signal for elastogenesis is a component of the ligament matrix. Unfortunately, the nature of matrix differentiation factor is unknown other than that it is not soluble and loses no bioactivity after treatment with agents that extract or denature many connective tissue macromolecules (Mecham, 1987a). The ligament matrix factor is probably not the sole extracellular component that affects tropoelastin expression and differentiation. Although serum does not induce tropoelastin expression in preelastogenic cells, it is absolutely required for maintenance of tropoelastin production in cultured cells (Mecham, 1987b). Depleting cells of serum results in nearly a complete repression of tropoelastin synthesis within 24 hours, indicating that circulating factors also participate in the regulation of elastin production. Indeed, as will be discussed, certain cytokines influence tropoelastin expression, and a factor in smooth muscle cell-conditioned medium has elastogenic properties.

B. Autoregulation

The limited developmental expression of elastin indicates that a finely tuned regulatory system controls production of this extracellular matrix protein. As discussed, the microenvironment influences differentiation to the elastin phenotype, and this induction involves signals which transverse the cell surface. It is unknown, however, if a similar extracellular signal mediates cessation of elastogenesis at maturity. As will be discussed, several physiologic modulators markedly downregulate tropoelastin expression, but their *in vivo* role has not yet been ascertained. One possible mechanism to downregulate production may involve a negative feedback signal from the matrix itself. In other words, once sufficient fiber mass has formed, the producing cells may receive an inhibitory signal to turn off tropoelastin expression. Conversely, in pathological conditions or in response to wounding, a loss of elastin may stimulate new production.

When the secretion of tropoelastin is blocked by exposure of fetal rat smooth muscle cells to monensin, tropoelastin synthesis and mRNA levels are dramatically reduced (Frisch et al., 1985). These findings indirectly suggest a negative feedback on tropoelastin gene expression that is mediated by an intracellular accumulation of tropoelastin, similar to that seen for tubulin (Cleveland and Kirschner, 1981) and α1(I) procollagen (Müller et al., 1986). Frisch et al. (1985) suggested several alternative explanations for their findings including monensin-mediated effects on intracellular pH, cytoskeletal proteins, and recycling of the elastin receptor, all of which may effect tropoelastin production independent of intracellular stores of the protein, but other models have as well suggested negative autoregulation of elastin production. Elastin deposition is decreased by cells grown in medium supplemented with ascorbic acid (Faris et al., 1983), and this probably occurs by overhydroxylation of tropoelastin which may hinder its incorporation into fibers. Apparently, this effect is associated with decreased production of tropoelastin (Davidson and Giro, 1986), but it is not clear if this diminished production is mediated by feedback from accumulated uncross-linked tropoelastin, or if ascorbate affects tropoelastin expression by a more direct mechanism. In another model, medium conditioned by an elastase-treated elastin matrix mediated a downregulation of tropoelastin expression in cultures of rat lung fibroblasts (Foster et al., 1990b). These findings suggest that elastin peptides generated by elastase treatment mediated a negative feedback control on the elastin-producing cells. A correlate of this predicts that a loss of elastin matrix may stimulate tropoelastin expression, and, indeed, the levels of tropoelastin protein and mRNA increased in the cultures that were subject to elastase treatment (Foster et al., 1990b).

Although these studies suggest that negative autoregulation affects tropoelastin expression, other observations indicate that this may not be an active regulatory mechanism. The results of the elastase-generated peptide experiments predict that tropoelastin expression would be elevated in conditions characterized by a loss of elastin fibers and downregulated in situations where elastin accumulates. In abdominal aneurysms, for example, the aortic media is devoid of elastic laminae, yet only minimal levels of tropoelastin mRNA are detected in the vessel wall (Mesh et al., 1992). Furthermore, diseases characterized by abnormal accumulation of elastin, such as pulmonary hypertension (Stenmark et al., 1988) and progeria (Sephel et al., 1988), are associated with elevated levels of tropoelastin expression. It is possible that regulation of elastin production may be aberrant in these disease models, thereby affecting any negative feedback; however, it would seem unlikely that these diverse conditions would be caused by a similar regulatory defect. In our studies, we found that prolonged exposure of bovine ear chondrocytes to high concentrations of α-elastin peptides did not influence tropoelastin synthesis or mRNA levels (R.P. Mecham and W.C. Parks, unpublished data). In addition, the α-elastin peptides used in these experiments were biologically active as demonstrated by their ability to induce chemotaxis (Senior et al., 1980). Since these peptides can also interact with the elastin receptor, our findings indicate that cell surface interactions with elastin do not transduce regulatory signals that downregulate tropoelastin production. Furthermore, tropoelastin may not affect its own production since fetal bovine ligament fibroblasts, which synthesize much tropoelastin, yet do not cross-link it, continue to express the protein at a constant rate regardless of its accumulation in the medium. Elastin peptides do, however, induce expression of elastonectin, a cell-surface elastin-binding protein (Hornebeck et al., 1986), and this interaction is associated with a stimulation of intracellular signaling pathways (Robert et al., 1989). There is, in fact, accumulating evidence that binding of elastin to the cell surface initiates a biological response within the cell. For example, extracellular elastin peptides have been reported to induce phosphatidylinositol breakdown with mobilization of intracellular calcium, to alter protein kinase C activity, and to provide chemotactic signals to several cell types (Senior et al., 1984; Jacob et al., 1987; Blood and Zetter, 1989; Robert et al., 1989; Varga et al., 1990). Although elastin binding proteins are capable of generating intracellular signals, it is not clear if the transduced signals influence tropoelastin metabolism.

C. Cytokines

Various cytokines stimulate tropoelastin production in cultured cells, and in general the effect is about two- to fourfold above control levels. Although this increase is seemingly small, a slight, yet sustained, increase in tropoelastin synthesis can result in a significant increase in the mass of deposited matrix because of the extreme durability of the insoluble fiber. On the hand, factors and conditions that downregulate tropoelastin synthesis have a profound effect on production, and with sufficient exposure can nearly completely repress tropoelastin expression.

Insulin-like growth factor 1 (IGF-1) stimulates tropoelastin synthesis and mRNA levels in neonatal arterial smooth muscle cells (Badesch et al., 1989; Foster et al., 1989; Foster et al., 1990a) and lung fibroblasts (Noguchi and Nelson, 1991). IGF-1 may be involved in embryonic production of elastin since the circulating concentration of and hepatic mRNA levels for this cytokine correlate with tropoelastin synthesis in the developing chick aorta (Foster et al., 1989). It is not clear, however, if hepatic production of IGF-1 is causally associated with vascular elastin production, or if factors intrinsic to the vessel are more important in modulating matrix deposition. This is a reasonable caveat since cytokines often act in a paracrine or autocrine fashion. Indeed, this study also showed that IGF-1 mRNA levels in the embryonic chick aorta correlated with the onset of tropoelastin expression in this tissue (Foster et al., 1989). If IGF-1 is responsible for regulating developmental elastogenesis, the control may be a paracrine effect specific to the aorta since the pattern of tropoelastin expression in other tissues did not parallel IGF-1 serum levels.

Transforming growth factor β1 (TGF-β_1) enhances tropoelastin expression in neonatal arterial smooth muscle cells (Liu and Davidson, 1988). TGF-β_1 also stimulates tropoelastin synthesis in neonatal rat lung fibroblasts (McGowan and McNamer, 1990), but this effect may be limited to a subpopulation of lung fibroblasts since tropoelastin expression by lipid-laden interstitial cells, a neonatal rat lung fibroblast subtype, is unaffected by TGF-β_1 (Berk et al., 1991). TGF-β_1 influences the production of various extracellular matrix proteins and has been proposed as being involved in regulating elastin production in vascular remodeling (Liu and Davidson, 1988; Perkett et al., 1990) and in emphysema (McGowan and McNamer, 1990). Although no *in vivo* experiments have been done to confirm these hypotheses, TGF-β_1 may not be a required factor for increased elastin deposition during vascular remodeling. For example, in hypoxic pulmonary hypertension, increased deposition and synthesis of elastin are associated with a marked decrease in TGF-β_1 mRNA levels in the vascular wall (Botney et al., 1991b). This model, however, is not associated with a prominent inflammatory response as is seen in other models of pulmonary hypertension, and, thus, the effects of this cytokine may be dependent on the etiology of the vascular remodeling. Although both TGF-β_1 and IGF-1 act pretranslationally, the precise regulatory mechanism controlling tropoelastin expression by theses cytokines is unknown. Since no NF-1 binding element has been located in the tropoelastin gene, TGF-β_1 may stimulate tropoelastin production at a posttranscriptional step (Chen et al., 1991).

During periods of elevated elastin deposition, increased expression of tropoelastin may be controlled by a soluble cytokine-like factor that may act in both an autocrine and paracrine fashion. This factor is called smooth muscle elastogenic factor, or SMEF, and is secreted by smooth muscle cells derived from intralobar arteries of calves with hypoxic pulmonary hypertension (Mecham et al., 1987). SMEF stimulates tropoelastin synthesis in actively producing cells and, interestingly, induces expression in preelastogenic ligament fibroblasts. SMEF may be a known cytokine, but cell culture studies demonstrate that the inductive and stimulatory effects of SMEF are not duplicated by acidic or basic fibroblast growth factor, or by platelet-derived growth factor (R.P. Mecham, unpublished observations). Furthermore, the low levels of TGF-β_1 mRNA in hypertensive pulmonary arteries (Botney et al., 1991b) suggest that SMEF is not this cytokine either. The biochemical properties of SMEF have not been characterized other than its approximate molecular mass of 25 kDa, as determined by the migration of its bioactivity through a gel filtration column, and the demonstration that its activity is susceptible to protease treatment and heat.

D. Steroids

The response of elastin production to glucocorticoids, which has been studied in various systems using cortisol or synthetic analogs such as dexamethasone, is particularly interesting in that it appears to be an age-specific mechanism. Elastogenic cells derived from fetal bovine nuchal ligament (Mecham et al., 1984b), fetal rat lung (Noguchi et al., 1990), or embryonic chick aorta (Burnett et al., 1980) demonstrate increased production of tropoelastin upon exposure to glucocorticoids, and, at least in rat lung, this effect is paralleled by a similar stimulation of tropoelastin mRNA levels. Glucocorticoids, however, do not induce tropoelastin expression in preelastogenic ligament fibroblast, and the stimulation in fetal fibroblasts is reversible indicating that glucocorticoids do not initiate a stable phenotypic change (Mecham et al., 1984b). In addition, the ability of ligament fibroblasts to respond to glucocorticoids is dependent upon growth rate and time in culture. In general, tropoelastin synthesis by cultured cells is stimulated by glucocorticoids during periods of maximal production, namely, during late log phase growth and early passage in culture. If these *in vitro* growth states reflect developmental changes in elastin production seen *in vivo*, then a comparative response to glucocorticoids may be seen in elastic tissues at different stages of development. Indeed, in chick aortic organ culture, glucocorticoids stimulate elastin production in embryonic tissues, but do not effect synthesis in newborn animals and actually inhibit production in aorta from 14-day-old chicks (Keeley and Johnson, 1987). Similarly, in rat lung fibroblasts, glucocorticoids mediate a stimulation of tropoelastin synthesis in cells isolated from lungs during periods of active alveolar elastogenesis (Noguchi et al., 1990).

A few reports, however, seem to contradict the age-specific pattern of glucocorticoid responsiveness. Distinct from the developing mammalian lung, dexamethasone does not affect elastin production in embryonic chick lung (Foster and Curtiss, 1990). Since glucocorticoids stimulate elastogenesis in embryonic chick aorta, the lack of effect in chick lung may indicate a tissue or species-specific regulation. In our studies, tropoelastin production by adult ligament fibroblasts was stimulated by exposure to dexamethasone (Parks et al., 1988b). Although these cells were derived from a tissue that has long completed its deposition of elastin, removal from the their *in vivo* environment may have affected their capacity to respond to mediators of elastin production.

The mechanism of glucocorticoid regulation of tropoelastin expression is not known. Glucocorticoids can alter transcription by direct binding of an activated glucocorticoid receptor to specific response elements within a gene promoter. Three consensus glucocorticoid responsive elements (GREs) are found between basepairs -1200 to -1420 of the human tropoelastin gene (Kähäri et al., 1990), but it is not known if these confer any biological activity. Since the response to glucocorticoids varies with age and the growth state of the treated cell, then transcriptional regulation mediated through GREs would require other, possibly more specific *trans*-acting factors.

E. Downregulation of Tropoelastin Expression

The temporally precise and nearly complete cessation of elastin production that occurs in most tissues presents an intriguing model for deciphering the control of a developmentally regulated gene. It is quite possible that distinct mechanisms regulate the initiation, maintenance, and repression of tropoelastin expression, and some *in vivo* observations support this idea. The response in vascular remodeling at different ages suggests that elastogenic cells do not retain the ability to recapitulate developmental regulation of elastin synthesis. For example, in neonates with pulmonary hypertension, increased elastin synthesis and deposition occurs in the media of elastic lobar arteries (Stenmark et al., 1988; Prosser et al., 1989), but in adults, new elastin deposition is confined to the neointimal expansion (Botney et al., 1991a).

Marked repression of tropoelastin synthesis and mRNA levels and elastin deposition in various cultured cells is mediated by epidermal growth factor achiro et al., 1990), recombinant interleukin-1β (rIL-1) (Berk et al., 1991), tumor necrosis factor-α (TNF-α), interferon-γ (Kähäri et al., 1992), 1,25 dihydroxyvitamin D_3 (1,25$[OH]_2$ D_3) (Hinek et al., 1991; Pierce et al., 1992), and 12-*o*-tetradecanol-phorbol-13-acetate (TPA) (Kähäri et al., 1992; Parks et al., 1992a). Recently, TNF-α and TPA were shown to downregulate the activity of a human tropoelastin promoter-driven CAT reporter gene (Kähäri et al., 1992). Deletion analysis demonstrated that this effect was dependent upon an intact TPA responsive element located about 200 basepairs 5′ of the transcription start site. This finding is particularly interesting in light of the fact that TNF-α and TPA stimulate the synthesis and formation of AP-1 heterodimers (cJun-cFos), and that AP-1 interacts with TPA responsive element (Chiu et al., 1988). This is the first demonstration of modulation of tropoelastin expression mediated at the level of the gene, but the decrease in CAT activity (about twofold) mediated by TNF-α or TPA did not correlate with the decrease in tropoelastin mRNA levels (about 10-fold). This discrepancy suggests that transcriptional regulation is not solely responsible for downregulation of tropoelastin. Indeed, results from our laboratory demonstrate that repression by TPA or 1,25$(OH)_2D_3$ is controlled primarily by a posttranscriptional mechanism that mediates an accelerated degradation of tropoelastin mRNA (Parks et al., 1992a; Pierce et al., 1992), and a similar mechanism occurs in serum-depleted cells (W.C. Parks, unpublished findings).

It is clear that mRNA stability contributes to the control of expression of many genes, including those coding for extracellular matrix proteins such as α2(I) procollagen (Dhawan et al., 1991) and α1(II) procollagen (Dozin et al., 1990). Changes in the turnover rate of tropoelastin mRNA suggests that this effect involves regulatory sequences within the transcript. Transcript stability may be affected by diverse mechanisms such as alternate 5′ processing (Rabbits et al., 1985), sequences and structures in the 3′ UTR (Shaw and Kamen, 1986; Müllner and Kühn, 1988; Lumelsky and Forget, 1991), the length of the poly(A) tail (Shaw et al., 1986; Green and Dove, 1988; Wilson and Treisman, 1988), and other mechanisms (Brawerman, 1987). Sequences or changes in the 5′ UTR of the tropoelastin transcript are not likely to be involved in the turnover of tropoelastin mRNA since multiple start sites are used during production (Yeh et al., 1989). The poly(A) tail of mRNA is critical to the stability of many mRNAs, and typically, poly(A) tail length correlates with transcript half-life (Sachs, 1991). The tropoelastin gene has two polyadenlyation signals (Figure 7), and thus, alternative usage of these sites may function in regulation. In many mRNAs, sequences in the 3′ UTR are required for the regulation of stability, and rapid degradation of the mRNA follows cleavage at specific sites in the 3′ UTR (Ratnasabapathy et al., 1990). The tropoelastin mRNA 3′ UTR is highly conserved among mammals, consistent with the hypothesis that this region may be important for regulating mRNA stability. The tropoelastin 3′ UTR does not contain any AU-rich regions which have been described to mediate rapid degradation of some transcripts (Brawerman, 1987).

Although both stimulators and repressors of tropoelastin production and elastin deposition have been identified, the actual physiological action of these modulators, if any, is not known. Deciphering the role of cytokines and other messengers in the process of elastogenesis is made difficult, in part, by the simplicity of *in vitro* models relative to the potential complexity of *in vivo* events which affect production. Typically, the effects of one or two modulators are studied in cells, but *in vivo*, multiple factors may be simultaneously interacting on elastogenic cells. The combination of positive and negative factors at different concentrations probably elicits a desired effect at the appropriate time. Furthermore, other conditions, such as developmental stage, composition of the extracellular matrix with which the cell interacts, proliferative potential of the producing cells, presence of migratory cells, and physical forces add to the influences affecting extracellular matrix gene-specific activity. Cell culture models, however, do provide a defined environment for identifying factors and conditions that mediate changes in protein production, and for characterizing the regulatory mechanisms involved in these processes. As models become more sophisticated, it may be possible to create a defined *in vitro* system which closely duplicates an *in vivo* environment. The tissue-specific, temporally regulated expression of tropoelastin *in vivo* is presumably subject to many controls, and regulation may be exerted at transcriptional, posttranscriptional, and translational levels. It is likely, given the tissue distribution of elastin, that multiple factors and diverse mechanisms will be important in regulating tropoelastin production in different cell types.

VII. TROPOELASTIN HETEROGENEITY

A. Alternative Splicing of Tropoelastin Pre-mRNA

One of the more interesting yet potentially complex areas of elastin biology is the extensive heterogeneity of tropoelastin. It is now well established that the primary transcript of tropoelastin undergoes extensive alternative splicing (Anwar, 1987; Indik et al., 1987b; Yeh et al., 1987; Baule and Foster, 1988; Fazio et al., 1988a,b; Yeh et al., 1989; Pierce et al., 1990) resulting in the translation of multiple heterogeneous protein isoforms (Davidson et al., 1982; Campagnone et al., 1987; Wrenn et al., 1987; Parks et al., 1988a). To date, at least 10 bovine and 6 human exons are known to be alternatively spliced individually or with adjacent or distant exons (Figure 7) (Yeh et al., 1989; Indik et al., 1990). In addition, analysis of rat mRNA has demonstrated alternative splicing of sequences homologous to bovine exons 13, 14, 15, and 33 (Pierce et al., 1990). Partial analysis of chick tropoelastin cDNA sequences show similar patterns of alternative splicing (Baule and Foster, 1988).

The structure of the human, bovine, and rat tropoelastin genes allows extensive alternative splicing without changing the reading frame, or introducing a premature stop codon in the mRNA. The number of nucleotides in exons 2 to 35 are multiples of three and have split codons at their boundaries which usually code for a glycine residue (Indik et al., 1987a; Yeh et al., 1987). Because exon 1 encodes the signal peptide and exon 36 encodes the conserved carboxy terminus and contains the only polyadenylation signals, these exons are presumably not subject to alternative splicing. Thus, deletion of these exons would result in either a protein that would not be translocated into the rough endoplasmic reticulum or an improperly processed pre-mRNA. Tropoelastin pre-mRNA is alternatively spliced in different patterns; these include casette alternative splicing resulting in exclusion of an entire exon from the processed transcript, and alternate donor or acceptor site usage resulting in the variable inclusion of parts of exons, such as the deletion of 26A.

In the human tropoelastin gene, three exons have unusual alternate splicing properties. The 3′ 18 nucleotides of exon 24, which encode the sequence ala-leu-leu-lys-leu-ala, are missing from essentially all human tropoelastin transcripts; in fact, the sequence has only been seen in one cDNA clone (Fazio et al., 1988a). Exon 22 is apparently deleted from all transcripts and was discovered only by analysis of genomic sequences (Indik et al., 1990). It is included as an exon because it is bounded by consensus splicing recognition sequences and, if translated, would encode a characteristic tropoelastin hydrophobic sequence that is very similar to its bovine homologue (see Table 2). The 3′ end of exon 26, which is referred to as 26A, has only been found in one tropoelastin cDNA clone from a human fetal aorta library (Indik et al., 1987a). Northern hybridization and analysis with a monospecific antibody support that this region of exon 26 is frequently alternatively spliced and, hence, usually lacking in human tropoelastin isoforms. Recently, however, low levels of 26A in mRNA and protein were detected in the neointima of pulmonary arteries from patients with primary pulmonary hypertension (Liptay et al., 1991). This disease is characterized by new elastin production and tropoelastin expression in the expanded neointima, and, thus, the expression of exon 26A may be a marker of new elastin production.

The frequency of alternative splicing of individual exons is developmentally regulated. In bovine nuchal ligament, the pattern of alternative splicing is qualitatively similar among fetal, neonatal, and adult samples in that exons 13, 14, 27, and 33 are deleted at all ages (Yeh et al., 1989). Whereas exon 33 is alternatively spliced at the same frequency for all ages, exons 13, 14, and 27 are deleted more often in late gestation mRNA than in neonatal mRNA. These developmental differences suggest that more of the tropoelastin protein isoforms produced in bovine ligament after birth would be of greater molecular mass than those synthesized *in utero*. This hypothesis is consistent with the relative amounts of tropoelastin isoforms produced by cell-free translation (Parks et al., 1988a). In contrast to the alternative splicing seen at earlier ages, more exons are deleted from adult ligament mRNA (see Figure 7). Whether all splice variants are translated and incorporated into elastic fibers is not yet clear. During active periods of elastogenesis, however, greater than half of the tropoelastin transcripts are lacking exons due to alternative splicing, and hence, it would be inefficient if these splice variants were not utilized for fibrillogenesis.

B. Role of Tropoelastin Heterogeneity

The multiplicity of tropoelastin isoforms suggests that tropoelastin heterogeneity may have functional significance. Beyond its role as a precursor for elastin fibrillogenesis, tropoelastin and elastic fibers may be involved in cell:matrix and matrix:matrix interactions, and tropoelastin heterogeneity may be pertinent to any of these processes. Of the many exons subject to alternative splicing in various species, exon 33 is most commonly alternatively spliced, and in rat, deletion of this exon is tissue-specific and developmentally regulated (Heim et al., 1991). In bovine tissue, exon 33 is alternatively spliced at the greatest frequency, with about 50% of mRNA lacking this exon (Yeh et al., 1989), suggesting a unique functional role for alternative splicing of this exon (Boyd et al., 1991).

The role of tropoelastin heterogeneity in fibrillogenesis is obscure because the process of fiber assembly is largely unknown. The ultrastructure of elastic tissues suggests that tropoelastin monomers associate with microfibrils during fiber assembly (see Figure 6); however, it is unknown if this interaction is influenced by the deletion of certain domains. In addition, alternative splicing of tropoelastin transcripts frequently results in abutting cross-linking domains (see Figure 7), but whether this rearrangement effects inter- and intramolecular cross-link formation is unclear.

The production of different isoforms may influence the distinct morphogenesis of elastic fibers among tissues. *In situ* hybridization studies with splice variant-specific probes, however, indicate that all elastogenic cells of bovine elastic tissues express all tropoelastin isoforms (Parks et al., 1992b), and, other than exon 33 in rat, unambiguous tissue-specific differences in the pattern of tropoelastin isoform production have not been detected (Parks et al., 1988a). If the pattern of tropoelastin isoforms produced is similar among tissues, then the architecture of elastic matrices may be directed by other components of the fiber. If alternative splicing is of minimal importance for fibrillogenesis, then variability in transcript processing may have an effect on other functions of elastic fibers such as the interaction with cells and other proteins.

VIII. ELASTIN BINDING PROTEINS

A. Mammalian Binding Proteins for Elastin

The first evidence for elastin binding proteins on mammalian cells came from studies documenting a chemotactic response by monocytes and fibroblasts to tropoelastin and peptide fragments of insoluble elastin (Senior et al., 1980; Senior et al., 1982; Senior et al., 1984). Subsequently, three cell-surface elastin-binding proteins have been described: (1) a 120-kDa protein (elastonectin) whose expression on the surface of smooth muscle cells is inducible by elastin peptides (Hornebeck et al., 1986); (2) a 59-kDa protein found on tumor cells that is coupled to protein kinase C (Blood et al., 1988; Blood and Zetter, 1989); and (3) a 67-kDa peripheral membrane protein found on most cell types that bind elastin (Hinek et al., 1988; Mecham et al., 1989a). These binding proteins may also be referred to as receptors since they are coupled with intracellular signalling pathways (Senior et al., 1984; Jacob et al., 1987; Blood and Zetter, 1989; Robert et al., 1989; Varga et al., 1990).

Elastin contains no arg-gly-asp (RGD) sequences, and there is no evidence that elastin or tropoelastin interacts with integrins. Instead, both the 67- and 59-kDa elastin-binding proteins recognize a hydrophobic hexapeptide, val-gly-val-ala-pro-gly (VGVAPG), which repeats several times in the human and bovine molecule (see Table 2, exon 24). Recent studies suggest that receptor recognition is not sequence specific but involves a conformational determinant that tolerates several different amino acids as long as the overall hydrophobic nature of the peptide is maintained. The binding properties of the 67-kDa protein are somewhat promiscuous in that laminin (Mecham et al., 1989b) and type IV collagen (Senior et al., 1989) compete with elastin for receptor binding.

The 67-kDa elastin-binding protein has properties of a peripheral, but not of an integral membrane protein. Biochemical studies have shown that the receptor is a galactoside-binding protein that has immunological similarity with a family of small molecular weight galactoside lectins (Barondes, 1988; Hinek et al., 1988). A striking feature of this receptor is that its affinity for elastin is greatly influenced by its lectin domain. In the absence of carbohydrate, the receptor binds to elastin with high affinity, whereas in the presence of galactoside sugar, little, if any, binding occurs (Hinek et al., 1988; Mecham et al., 1991). As will be discussed below, this property may be important for fiber assembly.

B. Bacterial Binding Protein for Elastin

Recent studies have suggested that several pathogenic bacteria specifically bind to elastin. Interestingly, a prominent bacterial-binding domain has been mapped to a 30-kDa region at the amino end of the elastin molecule (Park et al., 1991), obviously different from the mammalian receptor binding region which is located about one-third the distance in from the carboxy terminus. Agonists known to perturb elastin interaction with the mammalian receptor do not influence bacterial binding. Many bacteria, including *Staphylococcus aureus*, are known to infect and colonize elastin-rich organs like lung, skin, and blood vessels. Because invasive bacteria have a high probability of interacting with the elastin matrix of the vascular wall, the existence of a binding protein for elastin on the bacterial surface could facilitate bacterial invasion, in addition to adherence during colonization.

IX. ELASTIC FIBER FORMATION

Despite our knowledge of tropoelastin structure and expression and the cross-linking process, the conditions and components necessary for elastogenesis remain enigmatic. Multiple components are associated with the expression of elastin, and each of these may be required for fibrillogenesis (Table 4). The molecular interactions among these molecules, however, has yet to be described. Thus, the events and processes of elastin assembly, a regulated, ordered, and active process, presents a significant gap in our understanding of the extracellular matrix.

A. Posttranslational Modification and Secretion of Tropoelastin

Tropoelastin traffics through the cisternae of the endoplasmic reticulum, the Golgi stacks, and the *trans*-Golgi reticular network in accord with the classical secretory pathway for proteins. During posttranslational processing and secretion, tropoelastin undergoes little modification other than cleavage of the signal peptide and the hydroxylation of some proline residues. The functional significance of this hydroxylation may be minimal since inhibition of prolyl hydroxylase does not affect fiber formation (Uitto et al., 1976); in fact, overhydroxylation, which will influence the net hydrophobicity of the molecule, may hinder elastin assembly (Faris et al., 1983). Of interest is whether tropoelastin is cosecreted with other components of the elastic fiber such as microfibrillar proteins or lysyl oxidase.

Table 4. Products Associated with Elastic Fiber Production

- Tropoelastin
- Proline hydroxylase
- 67-kDa elastin receptor complex
- Microfibrils
  - -MAGP
  - -Fibrillin
    - -Fib 5
    - -Fib 15
  - -Others
    - -78-, 70-, and 25-kDa glycoproteins
- Lysyl oxidase
- Elastases and enzyme inhibitors

Cosecretion of components of the elastin fiber would temporally and spatially colocalize them in the extracellular space, and perhaps facilitate cross-linking. The 67-kDa elastin-binding protein colocalizes with tropoelastin in intracellular vesicles leading to the hypothesis that the receptor serves as a molecular chaperone (see receptor-mediated assembly below). In this capacity, the receptor would provide the trafficking signals necessary to direct tropoelastin to sites of active fibrillogenesis on the plasma membrane, while preventing intracellular aggregation.

B. Receptor-Mediated Assembly

With multiple binding domains, the 67-kDa elastin receptor is ideally suited to direct the specific association between tropoelastin and highly glycosylated microfibrillar components. Immunolocalization studies with elastin-producing cells reveal that tropoelastin and the 67-kDa protein colocalize intracellularly, suggesting that tropoelastin is bound to the 67-kDa protein in the secretory pathway. According to this model, tropoelastin remains bound to the 67-kDa protein at the plasma membrane until an interaction with a microfibril-associated galactoside sugar induces the transfer of tropoelastin to an acceptor site on the microfibril. Supporting a matrix-assembly function for the 67-kDa protein are studies showing inhibition of elastin fiber assembly by the addition of lactose or galactose sugars to the culture medium of elastin-producing cells (Hinek et al., 1988). As explained by the model in Figure 9, lactose competes with glycoconjugates on microfibrils for the sugar-binding site on the receptor, and causes premature release of tropoelastin. Biochemical studies show that, in the presence of lactose, the majority of newly synthesized tropoelastin was released directly into the medium with only low levels remaining associated with the cell layer. Electron microscopy confirmed the inhibitory effects of lactose on fiber assembly, and showed that the small amounts of elastin associated with the cell layer accumulated in the matrix as small, irregularly distributed globules that did not coalesce to form the large amorphous structures typical of mature fibers.

By facilitating the proper association of tropoelastin with microfibrils, the 67-kDa protein fits the definition of a molecular chaperone; a protein that ensures the folding and assembly of other proteins into oligomeric structures (Ellis, 1987). In addition to providing transport signals and facilitating fiber assembly, another function of the receptor may be to prevent intracellular coacervation of tropoelastin. While this may be an important aspect of fiber assembly in the extracellular space, aggregation within the secretory pathway could be detrimental to the cell. As a transporter protein, the receptor would prevent coacervation and insure that aggregation does not occur until tropoelastin is at the correct site on the cell surface.

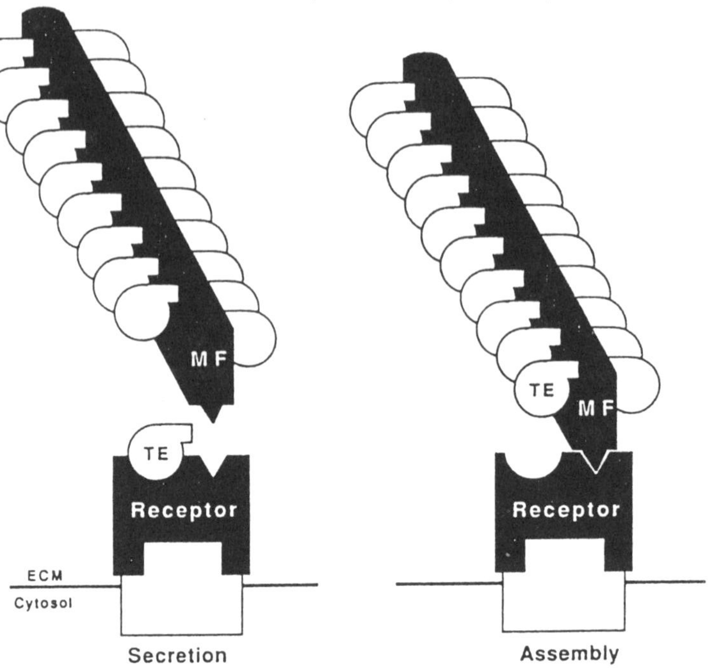

Figure 9. Receptor-mediated elastic fiber assembly. The 67-kDa elastin receptor is not an integral membrane protein, but rather is bound to a transmembrane protein. The receptor contains a second protein binding site that binds tropoelastin with high affinity. This interaction may occur intracellularly, and the two proteins may be secreted onto the cell surface as a complex. In the extracellular space, a third binding site, which recognizes galactoside sugars, may interact with carbohydrate moieties on microfibrils (MF). Binding of carbohydrate at the sugar binding site lowers the affinity at the protein binding site resulting in the transfer of tropoelastin from the cell to microfibrils. Once associated with microfibrils, the tropoelastin is rapidly incorporated into the growing elastic fiber.

C. Polymerization of Tropoelastin

Elastic fibers first appear in fetal development as aggregates of microfibrils arranged in parallel array, often occupying infoldings of the cell membrane. As development progresses, the microfibrils become coated with globular structures (probably tropoelastin monomers) that subsequently coalesce to form true elastic fibers (see Figure 6). The relative proportion of microfibrils to elastin declines with increasing age; adult elastic fibers having only a sparse peripheral mantle of microfibrillar material. Despite advances in understanding the structural complexity of microfibrils, the mechanisms by which they participate in fibrillogenesis remain speculative. The observation that microfibrillar aggregates appear in the extracellular matrix before elastin is deposited and take the form and orientation of presumptive elastic fibers led Ross et al. (1977) to suggest that these aggregates direct the morphogenesis of elastic fibers by acting as a scaffold on which elastin is deposited. It is likely that microfibrils serve to align tropoelastin molecules, the soluble secreted form of elastin, in precise register, so that cross-linking regions are juxtaposed prior to oxidation by lysyl oxidase. As yet, however, this hypothesis is supported by morphological observations alone since the biochemical events and structural interactions underlying fiber formation are not fully characterized. The VGVAPG repeating sequence encoded by exon 24 of bovine and human tropoelastin genes is believed to play an important role in fibrillogenesis by virtue of its interaction with the 6-kDa elastin receptor protein. Although this repeat is not found in either rat or chick tropoelastin, other hydrophobic repeats may have the same capacity in these molecules.

In addition to the receptor recognition sequences, other domains of tropoelastin must be involved in the initial steps of fiber assembly. These areas would most likely interact with microfibrils, but it is not known which domains are responsible for this interaction, or with what microfibrillar component tropoelastin is initially associated. Examination of the primary sequence of the tropoelastin does suggest that certain domains may be involved in fibrillogenesis. Of potential importance is the highly conserved carboxy terminus of tropoelastin encoded by exon 36 (see Table 2). Because of its invariant retention and unusual and highly conserved sequence, the carboxy terminal amino acids of tropoelastin may interact with other matrix molecules, such as microfibrils. The basic amino acids, which may protrude from the body of tropoelastin due to cysteine bonding (Brown et al., 1992), suggest that this region is biologically active, and possibly the C-terminus registers tropoelastin in an orderly fashion on microfibrils in preparation for cross-linking. Because of its net hydrophilicity, the C-terminal domain would repel from the mostly nonpolar tropoelastin and interact with the polar extracellular environment. Interestingly, rotary shadowing images of purified tropoelastin reveal a globular structure with a short protruding tail which may represent the C-terminal domain (Mecham and Heuser, 1991). This proposed tertiary structure of tropoelastin would facilitate an interaction of the C-terminus with microfibrils.

Some studies have suggested that the C-terminus is cleaved from tropoelastin before polymerization (Franzblau et al., 1989), but immunohistochemistry studies with a monospecific antibody demonstrated that this domain is a component of elastic fibers (Wrenn et al., 1987). It is not clear, however, if the C-terminal epitope detected by immunostaining is a component of the primary sequence of cross-linked tropoelastin monomers, or if it is indeed a cleavage product that is retained in the fiber. Indirect evidence, however, suggests that tropoelastin molecules which lack the C-terminal domain that may not be destined for fiber assembly. Tropoelastin produced by the ductus arteriosus after birth, at which time no new elastin is deposited in the tissue, is truncated at the C-terminal end, and it has been suggested that this provides a mechanism to prevent continued elastogenesis (Hinek and Rabinovitch, 1993).

X. TURNOVER OF ELASTIN

Physiological turnover of insoluble elastin appears to be very slow, with a half-life approaching that of the animal (Lefevre and Rucker, 1980; Shapiro et al., 1991). Thus, under normal conditions, very little remodeling of elastic fibers occurs in adult life. Some elastin-synthesizing cells, such as human skin fibroblasts and rat aortic smooth muscle cells, possess small amounts of elastase activity. While this activity is very low in comparison to that found in inflammatory cells, it is of potential importance in allowing cell migration in the developing extracellular matrix, and in modulating morphogenesis of newly synthesized elastic fibers. Although tropoelastin is sensitive to proteolytic degradation before cross-linking into insoluble elastin, there is no evidence that significant intracellular degradation occurs prior to secretion. Studies in cultured cells suggest that only about 1% of newly synthesized elastin is degraded per hour, in contrast to the situation observed in collagen synthesis where a larger percentage of newly synthesized interstitial collagen undergoes intracellular catabolism.

Increased elastin destruction does take place in certain pathological conditions, either as a result of the release of powerful elastinolytic serine proteases, or elastases, from inflammatory cells or bacteria, or because of a genetic deficiency in the naturally occurring elastase inhibitor α1-antitrypsin. Although inflammatory cell-derived elastases are unlikely to play a significant role in normal regulation of the elastic matrix, it is clear that they contribute substantially to the pathogenesis of some human diseases, including pulmonary emphysema, pollutional lung disease, atherosclerosis, and rheumatoid arthritis. Neutrophil and macrophage elastases are of particular importance in this regard, and have been the subject of much study. Of significant interest are the recent findings that human 92-kDa and 72-kDa type IV collagenases degrade elastin (Senior et al., 1991). These results extend the spectrum of human proteinases with elastolytic activity to metalloproteinases, and suggest a role for these enzymes in the pathogenesis of diseases involving damage to elastic fibers, such as pulmonary emphysema. Many mesenchymal cell types, including fibroblasts and smooth muscle cells, produce 72-kDa type IV collagenase as a major gene product (Stetler-Stevenson, 1990). The 92-kDa type IV collagenase is an abundant secretory product of human alveolar macrophages (Hibbs et al., 1987; Welgus et al., 1990) and some metastatic tumor cell lines (Liotta et al., 1988; Khokha and Denhardt, 1989). A full discussion of elastases and elastin turnover can be found elsewhere (Bieth, 1986; Robert and Hornebeck, 1989).

XI. GENETICS OF ELASTIN-RELATED DISEASES

Several human diseases involve a derangement of elastic fibers (Uitto and Ryhanen, 1987), and recently a disruption in the elastin gene was correlated to aortic supravalvular stenosis (Curren et al., 1993). Beyond this example, however, the tropoelastin has not been linked to any heritable disorder, but it remains a candidate for several heritable diseases of elastic tissue, including pseudoxanthoma elasticum and Buschke–Ollendorf syndrome. Numerous gene products, such as lysyl oxidase and the elastin receptor, are important for production of elastic fibers (see Table 4), and mutations in or aberrant expression of any of these genes may lead to an elastin disease. For example, certain variants of α-1-antitrypsin that do not efficiently inhibit elastase activity result in pulmonary emphysema. In patients with Marfan syndrome, an autosomal dominant connective tissue disease characterized by cardiovascular and musculoskeletal abnormalities, histologic studies reveal abnormal elastin fibers in affected tissues. Gene linkage studies, however, have excluded defects in certain collagens and the tropoelastin gene as a cause of Marfan syndrome (Kainulainen et al., 1990b). Recently, fibrillin has been linked to the occurrence of Marfan syndrome, and defects in its gene are associated with the disease (Dietz et al., 1991; Lee et al., 1991; Malsen et al., 1991).

Although the domain and exon organization, and the primary amino acid sequence of tropoelastin, is highly conserved among species, there is considerable variation in the coding sequence and little conservation of most intronic sequences. Many polymorphisms occur within noncoding DNA sequences (Sandell and Boyd, 1990), yet surprisingly, relatively few restriction fragment length polymorphisms (RFLP) have been described for the tropoelastin gene, especially in light of the large intron:exon ration (15:1). The human tropoelastin gene contains numerous *Alu*-like repeats in its 3′ end, and since such sequences may indicate genomic instability (Calabretta et al., 1982), it was suggested that variation among individual tropoelastin genes may be common (Indik et al., 1987b). However, the search for RFLPs has been difficult, and analysis with most restriction enzymes has revealed no variation among individuals. To date, only five RFLPs have been found for the tropoelastin gene. An A to G polymorphism, which results in a conversion of serine to glycine in the hydrophobic domain encoded by exon 20, is found in several populations, and the frequency of these two variants is very similar among alleles (Tromp et al., 1991). Two polymorphisms of multiple alleles are localized the 3′ end of the gene (Christiano and Boyd, 1991). Two additional RFLPs have been identified by cleavage with *Hind*III and *Eco*R1, and the frequency of the variants is not equally distributed within the test population (Kainulainen et al., 1990a).

The infrequency of genomic variation of the tropoelastin gene is unusual, but it may indicate an intolerance for change within certain domains. For example, mutation or deletion of key amino acids within the carboxyl terminus, the cell recognition domain, or cross-linking regions may severely hinder or completely prevent the ability of tropoelastin to be incorporated into fibers. A loss or partial decrease of the structural integrity of vascular elastic fibers may be lethal. Increased hemodynamic stress during vascular development would not be tolerated by the structurally compromised arterial walls. Thus, mutations may occur within certain areas of the tropoelastin gene, but because they may be lethal, are not found in the normal population.

On the other hand, identification of functionally important mutations in the tropoelastin gene will be complicated by an inherent tolerance of variability within homologous domains. Although the sequence of tropoelastin is similar with regard to overall hydrophobicity, comparison of human, bovine, chick, and rat tropoelastins indicates that extensive variability exists in the amino acids encoded by many hydrophobic exons (Indik et al., 1990; Parks and Deak, 1990; Boyd et al., 1991). Substitutions resulting from insertion or deletion of repetitive coding sequences are common among hydrophobic-encoding exons, suggesting that the net hydrophobicity is more critical to protein function than the primary amino acid sequence within these regions. The extent of variability among species suggests that functional constraints on tropoelastin hydrophobic domains are not rigorous with regard to size or sequence composition as compared to the highly conserved cross-linking domains and C-terminus.

XII. CONCLUSIONS

Our understanding of elastin structure and elastic fiber assembly and function is expanding rapidly because of advances in cell and molecular biology. The recent characterization of the elastin gene has provided the tools with which to delineate the cellular controls of elastogenesis. The precise developmental expression of tropoelastin makes this an appropriate model for studying tissue- and temporal-specific regulation at the molecular level. Progress should be made in coming years on describing these mechanisms, and on determining the contribution of different regulatory processes to the control of elastogenesis. In addition, new *in vitro* models are being developed by many laboratories that can be used to delineate the influence of various factors and conditions on elastin-producing cells. Such models will provide further insights into the mechanisms by which cells interact with, and are influenced by, the extracellular matrix. The future thus holds considerable promise for rapid advances in our understanding of normal elastin biosynthesis and its pathological alterations. Many important and interesting questions remain, however, such as determining the structure and composition of microfibrils, and resolving the pathway of tropoelastin secretion and fiber assembly.

ACKNOWLEDGMENTS

We thank our colleagues Drs. Charles Boyd and Mark Gibson for their permission to cite their unpublished data and Drs. Boyd and Edward Cleary for their contributions of figures. Our research is supported by NIH grants HL-48762, HL-41040, HL-26499, HL-41926, and HL-07317, and grants from the Dermatology Foundation and the American Heart Association. Dr. Parks is a Genetech/American Lung Association Career Investigator.

REFERENCES

Baccarani-Contri, M., Fornieri, C., & Pasquali Ronchetti, I. (1985). Elastin-proteoglycans association revealed by cytochemical methods. Connect. Tissue Res. 13, 237–249.

Baccarani-Contri, M., Vincenzi, D., Francesca, C., Mori, G., & Pasquali Ronchetti, I. (1990). Immunocytochemical localization of proteoglycans within normal elastin fibers. Eur. J. Cell Biol. 53, 305–312.

Badesch, D.H., Lee, P.K., Parks, W.C., & Stenmark, K.R. (1989). Insulin-like growth factor I stimulates elastin synthesis by bovine pulmonary artery smooth muscle cells. Biochem. Biophys. Res. Comm. 160, 382–387.

Barondes, S.H. (1988). Bifunctional properties of lectins: lectins redefined. Trends Biochem. Sci. 13, 480–482.

Bashir, M.M., Indik, A., Yeh, H., Ornstein-Goldstein, N., Rosenbloom, J.C., Abrams, W., Fazio, M., Uitto, J., & Rosenbloom, J. (1989). Characterization of the complete human elastin gene. Delineation of unusual features in the 5′-flanking region. J. Biol. Chem. 264, 8887–8891.

Baule, V.J. & Foster, J.A. (1988). Multiple chick tropoelastin mRNAs. Biochem. Biophys. Res. Commun. 154, 1054–1060.

Berk, J.L., Franzblau, C., & Goldstein, R.H. (1991). Recombinant interleukin-1β inhibits elastin formation by a neonatal rat lung fibroblast subtype. J. Biol. Chem. 266, 3192–3197.

Bieth, J.G. (1986). In: Regulation of Matrix Accumulation (Mecham, R.P., Ed.), pp. 217–320. Academic Press, New York.

Blood, C.H., Sasse, J., Brodt, P., & Zetter, B.R. (1988). Identification of a tumor cell receptor for VGVAPG, an elastin-derived chemotactic peptide. J. Cell Biol. 107, 1987–1993.

Blood, C.H. & Zetter, B.R. (1989). Membrane-bound protein kinase C modulates receptor affinity and chemotactic responsiveness of Lewis lung carcinoma sublines to an elastin-derived peptide. J. Biol. Chem. 264, 10614–10620.

Botney, M.D., Kaiser, L., Cooper, J., Mecham, R.P., Roby, J.D., & Parks, W.C. (1992a). Extracellular matrix protein gene expression in atherosclerotic pulmonary arteries. Am. J. Pathol. 140, 357–364.

Botney, M.D., Parks, W.C., Crouch, E.C., Stenmark, K.R., & Mecham, R.P. (1992b). TGF-β1 mRNA levels are decreased in remodeling hypertensive bovine pulmonary arteries. J. Clin. Invest. 89, 1629–1635.

Boyd, C.D., Christiano, A.M., R.A., P., Stolle, C.A., & Deak, S.B. (1991). Mammalian tropoelastin: multiple domains of the protein define an evolutionarily divergent amino acid sequence. Matrix 11, 235–241.

Brawerman, G. (1987). Determinants of messenger RNA stability. Cell 48, 5–6.

Breathnach, S.M., Melrose, S.M., Bhogal, B., de Beer, F.C., Dyck, R.F., Black, M.M., Tennent, G., & Pepys, M.B. (1981). Amyloid P component is located on elastic fibre microfibrils in normal human tissue. Nature 293, 652–654.

Bressan, G.M., Argos, P., & Stanley, K.K. (1987). Repeating structure of chick tropoelastin revealed by complementary DNA cloning. Biochemistry 26, 1497–1503.

Brody, J.S. & Vaccaro, C. (1979). Postnatal formation of alveoli: Interstitial events and physiological consequences. Fed. Proc. 38, 215–223.

Brown, P., Mecham, L., Tisdale, C., & Mecharn, R.P. (1992). The cysteine residues in the carboxy terminal domain of tropoelastin form an interchain disulfide bond creating a loop structure and positively charged pocket. Biochem. Biophys. Res. Com. 186, 549–555.

Burnett, W., Eichner, R., & Rosenbloom, J. (1980). Correlation of functional elastin messenger ribonucleic acid levels and rate of elastin synthesis in the developing chick aorta. Biochemistry 19, 1106–1111.

Burnett, W., Yoon, K., Finnigan-Bunick, A., & Rosenbloom, J. (1982). Control of elastin synthesis. J. Invest. Dermatol. 79, 138s–145s.

Calabretta, B., Robberson, D.L., Barrera-Saldana, H.A., Lambrou, T.P., & Saunders, G.F. (1982). Genomic instability in a region of human DNA enriched in Alu repeat sequences. Nature 296, 219–225.

Campagnone, R., Regan, J., Rich, C.B., Miller, M., Keene, D.R., Sakai, L., & Foster, J.A. (1987). Pulmonary fibroblasts: a model system for studying elastin synthesis. Lab. Invest. 56, 224–230.

Chen, Y.Q., Kähäri, V.-M., Olsen, D.R., Rhudy, R.W., Carillo, P.A., & Uitto, J. (1991). Transforming growth factor beta (TGF-beta) enhances elastin gene expression in human skin fibroblast cultures: evidence for posttranscriptional regulation. Clin. Res. 39, 510 (abstract).

Chiu, R., Boyle, W.J., Meek, J., Smeal, T., Hunter, T., & Karin, M. (1988). The c-Fos protein interacts with c-Jun/AP-1 to stimulate transcription of AP-1 responsive genes. Cell 54, 541–552.

Cleary, E.G. (1987). In: Connective Tissue Disease. Molecular Pathology of the Extracellular Matrix (Uitto, J. & Perejda, A.J., Eds.), pp. 55–81. Marcel Dekker, New York.

Cleary, E.G. & Gibson, M.A. (1983). Elastin-associated microfibrils and microfibrillar proteins. Int. Rev. Connect. Tissue Res. 10, 97–209.

Cleveland, D. & Kirschner, M. (1981). Autoregulatory control of the expression of α- and β-tubulins: implications for microtubule assembly. Cold Spring Harbor Symp. Quant. Biol. 46, 171–183.

Collett, A.J. & Des Biens, G. (1974). Fine structure of myogenesis and elastogenesis in the developing rat lung. Anat. Rec. 179, 343–360.

Colombatti, A., Bonaldo, P., Volpin, D., & Bressan, G.M. (1988). The elastin associated glycoprotein gp115 synthesis and secretion by chick cells in culture. J. Biol. Chem. 263, 17534–17540.

Cox, B.A., Starcher, B.C., & Urry, D.W. (1974). Coacervation of tropoelastin results in fiber formation. J. Biol. Chem. 249, 997–998.

Crissman, R.S. (1986). SEM observations of the elastic networks in canine femoral artery. Am. J. Anat. 175, 481–492.

Curren, M.E., Atkinson, D.L., Ewart, A.K., Morris, C.A., Leppart, M.F., & Keating, M.T. (1993). The elastin gene is disrupted by a translocation associated with supravalvular aortic stenosis. Cell 73, 159–168.

Dahlbäck, K., Ljungquist, A., Löfberg, H., Dahlbäck, B., Engvall, E., & Sakai, L.Y. (1990). Fibrillin immunoreactive fibers constitute a unique network in the human dermis. J. Invest. Dermat. 94, 284–291.

Dahlbäck, K., Löfberg, H., Alumets, J., & Dahlbäck, B. (1989). Immunohistochemical demonstration of age-related deposition of vitronectin (S-protein of complement) and terminal complement complex on dermal elastic fibers. J. Invest. Dermatol. 92, 727–733.

Damiano, V., Tsang, A.-L., Weinbaum, G., Christner, P., & Rosenbloom, J. (1984). Secretion of elastin in the embryonic chick aorta as visualized by immunoelectron microscopy. Collagen Rel. Res. 4, 153–164.

Davidson, J.M. & Giro, M.G. (1986). In: Regulation of Matrix Accumulation (Mecham, R.P., Ed.), pp. 177–216. Academic Press, New York.

Davidson, J.M., Hill, K.E., Mason, M.L., & Giro, M.G. (1985). Longitudinal gradients of collagen and elastin gene expression in the procine aorta. J. Biol. Chem. 260, 1901–1908.

Davidson, J.M., Shibahara, S., Boyd, C., Mason, M.L., Tolstoshev, P., & Crystal, R.G. (1984). Elastin mRNA levels during foetal development of sheep nuchal ligament and lung. Biochem. J. 220, 653–660.

Davidson, J.M., Smith, K., Shibahara, S., Tolstoshev, P., & Crystal, R.G. (1982). Regulation of elastin synthesis in developing sheep nuchal ligament by elastin mRNA levels. J. Biol. Chem. 257, 747–754.

Dhawan, J., Lichter, A.C., Rowe, D.W., & Farmer, S.R. (1991). Cell adhesion regulates pro-alpha1(I) collagen mRNA stability and transcription in mouse fibroblasts. J. Biol. Chem. 266, 8470–8475.

Dietz, H.C., Cutting, G.R., Pyeritz, R.E., C.L., M., Sakai, L.Y., Corson, G.M., Puffenberger, E.G., Hamosh, A., Nanthakumar, E.J., Curristin, S.M., Stetten, G., Meyers, D.A., & Francomano, C.A.

(1991). Marfan syndrome caused by a recurrent *de novo* missense mutation in the fibrillin gene. Nature 352, 337–339.

Dozin, B., Quarto, R., Rossi, F., & Cancedda, R. (1990). Stabilization of the mRNA follows transcriptional activation of type II collagen gene in differentiating chicken chondrocyte. J. Biol. Chem. 265, 7216–7220.

Ellis, J. (1987). Proteins as molecular chaperones. Nature (Lond.) 328, 378–379.

Fahrenbach, W.H., Sandberg, L.B., & Cleary, E.G. (1966). Ultrastructural studies on early elastogenesis. Anat.Rec. 155, 563–576.

Faris, B., Ferrera, R., Toselli, P., Nambu, J., Gonnerman, W.A., & Franzblau, C. (1983). Effect of varying amounts of ascorbate on collagen, elastin and lysyl oxidase synthesis in aortic smooth muscle cell cultures. Biochim. Biophys. Acta 797, 71–75.

Farquharson, C. & Robins, S.P. (1988). The distribution of elastin in developing and adult rat organs using immunocytochemical techniques. J. Anat. 165, 225–236.

Fazio, M.J., Kähäri, V.-M., Bashir, M.M., Saitta, B., Rosenbloom, J., & Uitto, J. (1990). Regulation of elastin gene expression: Evidence for functional promoter activity in the 5'-flanking region of the human gene. J. Invest. Dermatol. 94, 191–196.

Fazio, M.J., Mattei, M.-G., Passage, E., Chu, M.-L., Black, D., Solomon, E., Davidson, J.M., & Uitto, J. (1991). Human elastin gene: new evidence for localization to the long arm of chromosome 7. Am. J. Hum. Genet. 48, 696–703.

Fazio, M.J., Olsen, D.R., Kauh, E.A., Baldwin, C.T., Indik, Z., Ornstein-Goldstein, N., Yeh, H., Rosenbloom, J., & Uitto, J. (1988a). Cloning of full-length elastin cDNAs from a human skin fibroblast recombinant cDNA library: further elucidation of alternative splicing using exon-specific probes. J. Invest. Dermatol. 91, 458–464.

Fazio, M.J., Olsen, D.R., Kuivaniemi, H., Chu, M., Davidson, J.M., Rosenbloom, J., & Uitto, J. (1988b). Isolation and characterization of human elastin cDNAs, and age-associated variation in elastin gene expression in cultured skin fibroblasts. Lab. Invest. 58, 270–277.

Fleming, W.W., Sullivan, C.E., & Torchia, D.A. (1980). Characterization of molecular motions in ^{13}C-labelled elastin by ^{13}C-^{1}H magnetic double resonance. Biopolymers 19, 597–618.

Fornieri, C., Baccarani-Contri, M., Quaglino, D.J., & Pasquali-Ronchetti, I. (1987). Lysyl oxidase activity and elastin/glycosaminoglycans interactions in growing chick and rat aortas. J. Cell Biol. 105, 1463–1469.

Foster, J.A. (1982). Elastin structure and biosynthesis: an overview. Meth. Enzymol. 82, 559–570.

Foster, J.A. & Curtiss, S.W. (1990). The regulation of lung elastin synthesis. Am. J. Physiol. L13–23.

Foster, J.A., Miller, M.L., Benedict, M.R., Richmann, R.A., & Rich, C.B. (1989). Evidence for insulin-like growth factor-I regulation of chick aortic elastogenesis. Matrix 9, 328–335.

Foster, J.A., Rich, C.B., Fletcher, S., Karr, S.R., & Przybyla, A. (1980). Translation of chick aortic messenger ribonucleic acid. Comparison to elastin synthesis in chick aorta organ culture. Biochemistry 19, 857–864.

Foster, J.A., Rich, C.B., Miller, M., Benedict, M.R., Richman, R.A., & Florini, J.R. (1990a). Effect of age and IGF-I administration on elastin gene expression in rat aorta. J. Gerontol. 45, B113–118.

Foster, J.A., Rich, C.B., & Miller, M.F. (1990b). Pulmonary fibroblasts: an in vitro model of emphysema. Regulation of elastin gene expression. J. Biol. Chem. 265, 15544–15549.

Franzblau, C., Baris, B., Lent, R.W., Salcedo, L.L., Smith, B., Jaffe, R., & Crombie, G. (1969). In: Chemistry and Molecular Biology of the Intracellular Matrix. Volume 1 (Balazs, E.A., Ed.), pp. 617–639. Academic Press, New York.

Franzblau, C., Pratt, C.A., Faris, B., Colannino, N.M., Offner, G.D., Mogayzel, P.J., & Troxler, R.F. (1989). Role of tropoelastin fragmentation in elastogenesis in rat smooth muscle cells. J. Biol. Chem. 264, 15115–15119.

Frisch, S.M., Davidson, J.M., & Werb, Z. (1985). Blockage of tropoelastin secretion by monensin represses tropoelastin synthesis at a pretranslational level in rat smooth muscle cells. Mol. Cell. Biol. 5, 253–258.

Fukuda, Y., Ferrans, V.J., & Crystal, R.G. (1983). The development of alveolar septa in fetal sheep lung. An ultrastructural and immunohistochemical study. Am. J. Anat. 167, 405–439.

Gibson, M.A. & Cleary, E.G. (1987). The immunohistochemical localization of microfibril-associated glycoprotein (MAGP) in elastic and non-elastic tissues. Immunol. Cell Biol. 65, 345–356.

Gibson, M.A., Hughes, J.L., Fanning, J.C., & Cleary, E.G. (1986). The major antigen of elastin-associated microfibrils is a 31 kDa glycoprotein. J. Biol. Chem. 261, 11429–11436.

Gibson, M.A., Kumaratilake, J.S., & Cleary, E.G. (1989). The protein component of the 12-nanometer microfibrils of elastic and nonelastic tissues. J. Biol. Chem. 264, 4590–4598.

Gibson, M.A., Sandberg, L.B., Grosso, L.E., & Cleary, E.G. (1991). Complementary DNA cloning establishes microfibril-associated glycoprotein (MAGP) to be a discrete component of the elastin-associated microfibrils. J. Biol. Chem. 266, 7596–7601.

Gosline, J.M. (1976). The physical properties of elastic tissue. Int. Rev. Connect. Tissue Res. 7, 211–249.

Gosline, J.M. (1978). Hydrophobic interaction and a model for the elasticity of elastin. Biopolymers 17, 677–695.

Gosline, J.M. & Rosenbloom, J. (1984). In: Extracellular Matrix Biochemistry (Piez, K.A. & Reddi, A.H., Eds.), pp. 191–228. Elsevier, New York.

Gray, W.R. (1977). Some kinetic aspects of crosslink biosynthesis. Adv. Exp. Med. Biol. 79, 285–290.

Gray, W.R., Sandberg, L.B., & Foster, J.A. (1973). Molecular model for elastin structure and function. Nature (Lond.) 246, 461–466.

Green, L.L. & Dove, W.F. (1988). Correlation between tubluin mRNA stability and poly(A) tail length over the cell cycle of *Physarum polycephalum*. J. Mol. Biol. 200, 321–328.

Heim, R.A., Pierce, R.A., Deak, S.B., Riley, D.J., Boyd, C.D., & Stolle, C.A. (1991). Alternative splicing of rat tropoelastin mRNA is tissue-specific and developmentally regulated. Matrix 11, 359–366.

Hibbs, M.S., Hoidal, J.R., & Kang, A.H. (1987). Expression of a metalloproteinase that degrades native type V collagen and denatured collagens by cultured human alveolar macrophages. J. Clin. Invest. 80, 1644–1650.

Hinek, A., Botney, M.D., Mecham, R.P., & Parks, W.C. (1991). Inhibition of tropoelastin expression by 1,25 dihydroxyvitamin D_3. Connect. Tissue Res. 26, 155–166.

Hinek, A., & Rabinovitch, M. (1993). The ductus arteriosus migratory smooth muscle cell phenotype processes tropoelastin to a 52-kDa producted associated with impaired assembly of elastic laminae. J. Biol. Chem. 268, 1405–1413.

Hinek, A., Wrenn, D.S., Mecham, R.P., & Barondes, S.H. (1988). The elastin receptor is a galactoside binding protein. Science (Wash. DC) 239, 1539–1541.

Hoeve, C.A.J. (1974). The elastic properties of elastin. Biopolymers 13, 677–686.

Hornebeck, W., Tixier, J.M., & Robert, L. (1986). Inducible adhesion of mesenchymal cells to elastic fibers: Elastonectin. Proc. Natl. Acad. Sci. USA. 83, 5517–5520.

Horton, W., Miyashita, T., Kohno, K., Hassell, J.R., & Yamada, Y. (1987). Identification of a phenotype-specific enhancer in the first intron of the rat collagen II gene. Proc. Natl. Acad. Sci. USA 84, 8864–8868.

Ichiro, T., Tajima, S., & Nishikawa, T. (1990). Preferential inhibition of elastin synthesis by epidermal growth factor in chick aorta smooth muscle cells. Biochem. Biophys. Res. Comm. 168, 850–856.

Indik, Z., Yeh, H., Ornstein-Goldstein, N., & Rosenbloom, J. (1990). In: Extracellular Matrix Genes (Sandell, L. & Boyd, C.D., Eds.), pp. 221–250. Academic Press, New York.

Indik, Z., Yeh, H., Ornstein-Goldstein, N., Sheppard, P., Anderson, N., Rosenbloom, J.C., Peltonen, L., & Rosenbloom, J. (1987a). Alternative splicing of human elastin messenger-RNA indicated by sequence analysis of cloned genomic and complementary DNA. Proc. Natl. Acad. Sci. USA. 84, 5680–5684.

Indik, Z., Yoon, K., Morrow, S.D., Cicila, B., Rosenbloom, J.C., Rosenbloom, J., & Ornstein-Goldstein, N. (1987b). Structure of the 3′ region of the human elastin gene: great abundance of Alu repetitive sequences and few coding sequences. Connect. Tissue Res. 16, 197–211.

Jacob, M.P., Fulop Jr., T., Foris, G., & Robert, L. (1987). Effect of elastin peptides on ion fluxes in mononuclear cells, fibroblasts, and smooth muscle cells. Proc. Natl. Acad. Sci. USA. 84, 995–999.

Kagan, H.M. (1986). In: Regulation of Matrix Accumulation (Mecham, R.P., Ed.), pp. 322–399. Academic Press, New York.

Kagan, H.M. & Trackman, P.C. (1991). Properties and function of lysyl oxidase. Am. J. Respir. Cell Mol. Biol. 5, 206–210.

Kagan, H.M., Vaccaro, C.A., Bronson, R.E., Tang, S.S., & Brody, J.S. (1986). Ultrastructural immunolocalization of lysyl oxidase in vascular connective tissue. J. Cell Biol. 103, 1121–1128.

Kähäri, V.-M., Chen, Y.Q., Bashir, M.M., Rosenbloom, J., & Uitto, J. (1992). Tumor necrosis factor-α regulates human elastin gene expression: evidence for the role of AP-1 in the suppression of promoter activity. J. Biol. Chem. 267, 26134–26141.

Kähäri, V.-M., Fazio, M.J., Chen, Y.Q., Bushir, M.M., Rosenbloom, J., & Uitto, J. (1990). Deletion analysis of the 5′-flanking region of the human elastin gene. Delineation of functional promoter and regulatory cis-elements. J. Biol. Chem. 265, 9485–9490.

Kainulainen, K., Rosenbloom, J., & Peltonen, L. (1990a). Two polymorphisms for the human elastin gene. Nucl. Acids Res. 18, 3114.

Kainulainen, K., Savolainen, A., Palotie, A., Kaitila, I., Rosenbloom, J., & Peltonen, L. (1990b). Marfan's syndrome: exclusion of genetic linkage to five genes coding for connective tissue components in the long arm of chromosome 2. Hum. Genet. 84, 233–236.

Kao, R.T., Hall, J., & Stern, R. (1986). Collagen and elastin synthesis in human stroma and breast carcinoma cell lines: modulation by the extracellular matrix. Con. Tis. Res. 14, 245–255.

Keeley, F.W. & Johnson, D.J. (1987). Age differences in the effect of hydrocortisone on the synthesis of insoluble elastin in aortic tissue of growing chicks. Connect. Tissue Res. 16, 259–268.

Khokha, R. & Denhardt, D.T. (1989). Matrix metalloproteinases and tissue inhibitor of metalloproteinases: a review of their role in tumorigenesis and tissue invasion. Invasion Metastasis 9, 391–405.

Krishnan, R. & Cleary, E.G. (1990). Elastin gene expression in elastotic human breast cancers and epithelial cell lines. Cancer Res. 50, 2164–2171.

Lee, B., Godfrey, M., Vitale, E., Hori, H., Mattei, M.-G., Sarfarazi, M., Tsipouras, P., Ramirez, F., & Hollister, D.W. (1991). Linkage of Marfan syndrome and a phenotypically related disorder to two different fibrillin genes. Nature 352, 330–334.

Lefevre, M. & Rucker, R.B. (1980). Aortic elastin turnover in normal and hypercholesterolemic Japanese quail. Biochim. Biophys. Acta 630, 519–529.

Liotta, L.A., Wewer, U., Rao, N.C., Schiffmann, E., Stracke, M., Guirguis, R., Thorgeirsson, U., Muschel, R., & Sobel, M. (1988). Biochemical mechanisms of tumor invasion and metastases. Adv. Exp. Med. Biol. 233, 161–169.

Liptay, M.J., Botney, M.D., Mecham, R.P., Cooper, J.D., & Kaiser, L.R. (1991). Detection of a unique elastin isoform in human hypertensive pulmonary arteries. Surg. Forum (In press).

Liu, J.-M. & Davidson, J.M. (1988). The elastogenic effect of recombinant transforming growth factor-beta on porcine aortic smooth muscle cells. Biophys. Biochem. Res. Commun. 154, 895–901.

Lumelsky, N.L. & Forget, B.G. (1991). Negative regulation of globin gene expression during megakaryocytic differentiation of a human erythroleukemic cell line. Mol. Cell. Biol. 11, 3528–3536.

Maddox, B.K., Sakai, L.Y., Keene, D.R., & Glanville, R.W. (1989). Connective tissue microfibrils; isolation and characterization of three large pepsin-resistant domains of fibrillin. J. Biol. Chem. 264, 21381–21385.

Malsen, C.L., Corson, G.M., Maddox, B.K., Glanville, R.W., & Sakai, L.Y. (1991). Partial sequence of a candidate gene for the Marfan syndrome. Nature 352, 334–337.

McGowan, S.E. & McNamer, R. (1990). Transforming growth factor-β increases elastin production by neonatal rat lung fibroblasts. Am. J. Respir. Cell Mol. Biol. 3, 369–376.

Mecham, R.P. (1987a). The biology of connective tissue: Regulation of mesenchymal cell differentiation by extracellular matrix. Bone Min. Res. 5, 185–208.

Mecham, R.P. (1987b). Modulation of elastin synthesis: In vitro models. Meth. Enzymol. 144, 232–246.

Mecham, R.P. & Heuser, J. (1990). Three-dimensional organization of extracellular matrix in elastic cartilage as viewed by quick freeze, deep etch electron microscopy. Connect. Tissue Res. 24, 83–93.

Mecham, R.P. & Heuser, J. (1991). In: Cell Biology of Extracellular Matrix (Hay, E.D., Ed.), pp. 79–109, Plenum Press, New York.

Mecham, R.P., Hinek, A., Entwistle, R., Wrenn, D.S., Griffin, G.L., & Senior, R.M. (1989a). Elastin binds to a multifunctional 67-kDa peripheral membrane protein. Biochemistry 28, 3716–3722.

Mecham, R.P., Hinek, A., Griffin, G.L., Senior, R.M., & Liotta, L. (1989b). The elastin receptor shows structural and functional similarities to the 67-kDa tumor cell laminin receptor. J. Biol. Chem. 264, 16652–16657.

Mecham, R.P., Lange, G., Madaras, J.G., & Starcher, B.C. (1981). Elastin production by ligamentum nuchae fibroblasts: effects of culture conditions and extracellular matrix on elastin production. J. Cell Biol. 90, 332–338.

Mecham, R.P., Madaras, J.G., & Senior, R.M. (1984a). Extracellular matrix-specific induction of elastogenic differentiation and maintenance of phenotypic stability in bovine ligament fibroblasts. J. Cell Biol. 98, 1804–1812.

Mecham, R.P., Morris, S.L., Levy, B.D., & Wrenn, D.S. (1984b). Glucocorticoids stimulate elastin production in differentiated bovine ligament fibroblasts but do not induce elastin synthesis in undifferentiated cells. J. Biol. Chem. 259, 12414–12418.

Mecham, R.P., Whitehouse, L., Hay, M., Hinek, A., & Sheetz, M. (1991). Ligand affinity of the 67-kDa elastin/laminin binding protein is modulated by the protein's lectin domain: Visualization of elastin/laminin receptor complexes with gold-tagged ligands. J. Cell Biol. 113, 187–194.

Mecham, R.P., Whitehouse, L.A., Wrenn, D.S., Parks, W.C., Griffin, G.L., Senior, R.M., Crouch, E.C., Stenmark, K.R., & Voelkel, N.F. (1987). Smooth muscle-mediated connective tissue remodeling in pulmonary hypertension. Science (Wash. DC) 237, 423–426.

Mesh, C.L., Baxter, T., Pearce, W.H., Chisholm, R.L., McGee, G.S., & Yao, J.S.T. (1992). Collagen and elastin gene expression in aortic aneuysms. Surgery 112, 256–262.

Morocutti, M., Raspanti, M., Govoni, P., Kadar, A., & Ruggeri, A. (1988). Ultrastructural aspects of freeze-fractured and etched elastin. Connect. Tissue Res. 18, 55–64.

Müller, P.K., Nerlich, A.G., Böhm, J., Phan-Than, L., & Krieg, T. (1986). In: Regulation of Matrix Accumulation (Mecham, R.P., Ed.), pp. 100–114. Academic Press, New York.

Müllner, E.W. & Kühn, L.C. (1988). A stem-loop in the 3′ untranslated region mediates iron-dependent regulation of transferrin receptor mRNA stability in the cytoplasm. Cell 53, 815–825.

Niculescu, F., Rus, H.G., & Vlaicu, R. (1987). Immunohistochemical localization of C5b-9, S-protein, C3d and apolipoprotein B in human arterial tissues with atherosclerosis. Atherosclerosis 65, 1–11.

Noguchi, A., Firsching, K., Kursar, J.D., & Reddy, R. (1990). Developmental changes of tropoelastin synthesis by rat pulmonary fibroblasts and effects of dexamethasone. Pediatr. Res. 28, 379–382.

Noguchi, A. & Nelson, T. (1991). IGF-I stimulates tropoelastin synthesis in neonatal rat pulmonary fibroblasts. Pediatr. Res. 30, 248–251.

Noguchi, A., Reddy, R., Kursar, J.D., Parks, W.C., & Mecham, R.P. (1989). Smooth muscle isoactin and elastin in fetal bovine lung. Exp. Lung Res. 15, 537–552.

Parfey, N.A. & Doyle, C.T. (1985). Elastosis in benign and malignant breast disease. Hum. Pathol. 16, 674–676.

Park, P.W., Roberts, D.D., Grosso, L.E., Parks, W.C., Rosenbloom, J., Abrams, W.R., & Mecham, R.P. (1991). Binding of tropoelastin to *Staphylococcus aureus*. J. Biol. Chem. (In press).

Parks, W.C. & Deak, S.B. (1990). Tropoelastin heterogeneity: Implications for protein function and disease. Am. J. Respir. Cell Mol. Biol. 2, 399–406.

Parks, W.C., Pierce, R.A., & Kolodziej, M.E. (1992a). Phorbol ester repress tropoelastin synthesis by a posttranscriptional mechanism. Biochemistry 31, 6639–6645.

Parks, W.C., Roby, J.D., Wu, L.C., & Grosso, L.E. (1992b). Cellular expression of tropoelastin mRNA splice variants. Matrix 12, 156–162.

Parks, W.C., Secrist, H., Wu, L.C., & Mecham, R.P. (1988a). Developmental regulation of tropoelastin isoforms. J. Biol. Chem. 263, 4416–4423.

Parks, W.C., Whitehouse, L.A., Wu, L.C., & Mecham, R.P. (1988b). Terminal differentiation of nuchal ligament fibroblasts: characterization of synthetic properties and responsiveness to external stimuli. Devel. Biol. 129, 555–564.

Partridge, M.S. (1969). In: Chemistry and Molecular Biology of the Intracellular Matrix (Balazs, E.A., Ed.), Vol. 1, pp. 593–616. Academic Press, New York.

Pasquali-Ronchetti, I. & Fornieri, C. (1984). In: Ultrastructure of the Connective Tissue Matrix (Ruggeri, A., & Motta, P.M., Eds.), pp. 126–139. Nijhoff, The Hague.

Pasquali-Ronchetti, I., Fornieri, C., Baccarani-Contri, M., & Volpin, D. (1979). The ultrastructure of elastin revealed by freeze-fracture electron microscopy. Micron 10, 89–99.

Perkett, E.A., Lyons, R.M., Moses, H.L., Brigham, K.L., & Meyrick, B. (1990). Transforming growth factor-beta activity in sheep lung lymph during the development of pulmonary hypertension. J. Clin. Invest. 86, 1459–1464.

Pierce, R.A., Deak, S.B., Stolle, C.A., & Boyd, C.D. (1990). Heterogeneity of rat tropoelastin mRNA revealed by cDNA cloning. Biochemistry 29, 9677–9683.

Pierce, R.A., Kolodziej, M.E., & Parks, W.C. (1992). 1,25 dihydroxyvitamin D_3 represses tropoelastin expression by a posttranscriptional mechanism. J. Biol. Chem. 267, 11593–11599.

Prosser, I.W., Stenmark, K.R., Suthar, M., Crouch, E.C., Mecham, R.P., & Parks, W.C. (1989). Regional heterogeneity of elastin and collagen gene expression in intralobar arteries in response to hypoxic pulmonary hypertension as demonstrated by *in situ* hybridization. Am. J. Pathol. 135, 1073–1088.

Rabbits, P.H., Forster, A., Stinson, M.A., & Rabbits, T.H. (1985). Truncation of exon 1 from the c-*myc* gene results in prolonged c-*myc* mRNA stability. EMBO J. 4, 3727–3733.

Raju, K. & Anwar, R.A. (1987). Primary structure of bovine elastin a, b, and c deduced from the sequences of cDNA clones. J. Biol. Chem. 262, 5755–5762.

Ratnasabapathy, R., Hwang, S.-P.L., & Williams, D.L. (1990). The 3′-untranslated region of apolipoprotein II mRNA contains two independent domains that bind distinct cytosolic factors. J. Biol. Chem. 265, 14050–14055.

Robert, L. & Hornebeck, W. (1989) Elastin and Elastases. CRC Press, Boca Raton, FL.

Robert, L., Jacob, M.P., Fülöp, T., Timar, J., & Hornebeck, W. (1989). Elastonectin and the elastin receptor. Path.Biol.37, 736–741.

Rosenquist, T.H., Beall, A.C., Modis, L., & Fishman, R. (1990). Impaired elastic matrix development in the great arteries after ablation of the cardiac neural crest. Anat. Rec. 226, 347–359.

Ross, R., Fialkow, P.J., & Altman, K. (1977). The morphogenesis of elastic fibers. Adv. Exp. Med. Biol. 79, 7–17.

Sachs, A. (1991). The role of poly(A) in the translation and stability of mRNA. Cur. Opinion Cell Biol. 2, 1092–1098.

Sakai, L.Y., Keene, D.R., & Engvall, E. (1986). Fibrillin, a new 350-kD glycoprotein, is a component of extracellular microfibrils. J. Cell Biol. 103, 2499–2509.

Sandberg, L.B. (1976). Elastin structure in health and disease. Internat. Rev. Conn. Tis. Res. 7, 159–210.

Sandberg, L.B. & Davidson, J.M. (1984). Elastin and its gene. Peptide Protein Rev. 3, 169–193.

Sandberg, L.B., Weissman, N., & Smith, D.W. (1969). The purification and partial characterization of a soluble elastin-like protein from copper-deficient porcine aorta. Biochemistry 7, 2940–2945.

Sandell, L.J. & Boyd, C.D. (1990). In: Extracellular Matrix Genes (Sandell, L.J. & Boyd, C.D., Eds.), pp. 1–56, Academic Press, New York.

Senior, R.M., Griffin, G.L., Fliszar, C.J., Shapiro, S.D., Goldberg, G.I., & Welgus, H.G.. (1991). Human 92-kDa and 72-kDa type IV collagenases are elastases. J. Biol. Chem. 266, 7870–7875.

Senior, R.M., Griffin, G.L., & Mecham, R.P. (1980). Chemotactic activity of elastin-derived peptides. J. Clin. Invest. 66, 859–862.

Senior, R.M., Griffin, G.L., & Mecham, R.P. (1982). Chemotactic responses of fibroblasts to tropoelastin and elastin-derived peptides. J. Clin. Invest. 70, 614–618.

Senior, R.M., Griffin, G.L., Mecham, R.P., Wrenn, D.S., Prasad, K.U., & Urry, D.W. (1984). Val-Gly-Val-Ala-Pro-Gly, a repeating peptide in elastin, is chemotactic for fibroblasts and monocytes. J. Cell Biol. 99, 870–874.

Senior, R.M., Hinek, A., Griffin, G.L., Pipoly, D.J., Crouch, E.C., & Mecham, R.P. (1989). Neutrophils show chemotaxis to type IV collagen and its 7S domain and contain a 67 kD type IV collagen binding protein with lectin properties. Am. J. Respir. Cell Mol. Biol. 1, 479–487.

Sephel, G.C. & Davidson, J.M. (1986). Elastin production in human skin fibroblasts and its decline with age. J. Invest. Dermatol. 86, 279–285.

Sephel, G.C., Sturrock, A., Giro, M.G., & Davidson, J.M. (1988). Increased elastin production by progeria skin fibroblasts is controlled by the steady-state levels of elastin mRNA. J. Invest. Dermatol. 90, 643–647.

Serafini-Fracassini, A., Ventrella, G., Field, M.J., Hinnie, J., Onyezili, N.I., & Griffiths, R. (1981). Characterization of a structural glycoprotein from bovine ligamentum nuchae exhibiting dual amine oxidase activity. Biochemistry 20, 5424–5429.

Shapiro, S.D., Endicott, S.K., Province, M.A., Pierce, J.A., & Campbell, E.J. (1991). Marked longevity of human lung parenchymal elastic fibers deduced from prevalence of D-aspartate and nuclear weapons-related radiocarbon. J. Clin. Invest. 87, 1828–1834.

Sharrow, L., Tinker, D., Davidson, J.M., & Rucker, R.B. (1989). Accumulation and regulation of elastin in the rat uterus. Proc. Soc. Exp. Biol. Med. 192, 121–126.

Shaw, G. & Kamen, R. (1986). A conserved AU sequence from the 3′ untranslated region of GM-CSF mRNA mediates selective mRNA degradation. Cell 46, 659–667.

Shaw, P., Sordat, B., & Schibler, U. (1986). Developmental coordination of α-amylase and *psp* gene expression during mouse parotid gland differentiation is controlled posttranscriptionally. Cell 47, 107–112.

Shibahara, S., Davidson, J.M., Smith, K., & Crystal, R.G. (1981). Modulation of tropoelastin production and elastin messenger ribonucleic acid activity in developing sheep lung. Biochemistry 20, 6577–6584.

Sjölund, M., Madsen, K., von der Mark, K., & Thyberg, J. (1986). Phenotypic modulation of primary cultures of smooth-muscle cells from rat aorta. Synthesis of collagen and elastin. Differentiation 32, 173–180.

Smith, L.T., Holbrook, K.A., & Byers, P.H. (1982). Structure of the dermal matrix during development and in the adult. J. Invest. Dermatol. 79, 93s–104s.

Stenmark, K.R., Orton, E.C., Reeves, J.T., Voelkel, N.F., Crouch, E.C., Parks, W.C., & Mecham, R.P. (1988). Vascular remodeling in neonatal pulmonary hypertension. Role of the smooth muscle cell. Chest 93, 127S–133S.

Stetler-Stevenson, W.G. (1990). Type IV collagenases in tumor invasion and metastasis. Cancer Metastasis Rev. 9, 289–303.

Streeten, B.W. & Licari, P.A. (1983). The zonules and the elastic microfibrillar system in the ciliary body. Invest. Opthal. Visual Sci. 21, 130–135.

Suyama, K. & Nakamura, F. (1990). Isolation and characterization of new cross-linking amino acid "allodesmosine" from hydrolysate of elastin. Biochem. Biophys. Res. Comm. 170, 713–718.

Tokimitsu, I., Tajima, S., Nishikawa, T., Tajima, M., & Fukasawa, T. (1987). Sequence analysis of elastin cDNA from chick aorta and tissue-specific transcription of the elastin gene in developing chick embryo. Arch. Biochem. Biophys. 256, 455–461.

Tromp, G., Christiano, A.M., Goldstein, N., Indik, Z., Rosenbloom, J., Boyd, C.D., Deak, S.B., Prockop, D.J., & Kuivaniemi, H. (1991). A to G polymorphism in the elastin gene. Nucl. Acids Res. 19, 4314.

Uitto, J. (1987). In: Dermatology in General Medicine, third edition (Fitzpatrick, T.B., Eisen, A.Z., Wolff, K., Freedberg, I.M., & Austen, K.F., Eds.), pp. 303–314. McGraw-Hill, New York.

Uitto, J., Hoffman, H.P., & Prockop, D.J. (1976). Synthesis of elastin and procollagen by cells from embryonic aorta. Differences in the role of hydroxyproline and the effects of proline analogs on the secretion of the two proteins. Arch. Biochem. Biophys. 173, 187–200.

Uitto, J., Ryhanen, L., Abraham, P.A., & Perejda, A.J. (1982). Elastin in diseases. J. Invest. Dermatol. 79, 160s–168s.

Uitto, J. & Ryhanen, L.J. (1987). In: Connective Tissue Diseases (Uitto, J. & Peredja, A.J., Eds.), pp. 394–422. Marcel Dekker, New York.

Urry, D.W. (1974). Studies on the conformation and interactions of elastin. Adv. Exp. Med. Biol. 43, 211–243.

Urry, D.W. (1983). What is elastin; what is not. Ultrastruct. Pathol. 4, 227–251.

Urry, D.W. & Long, M.M. (1977). On the conformation, coacervation and function of polymeric models of elastin. Adv. Exp. Biol. Med. 79, 685–714.

Vaccaro, C. & Brody, J.S. (1978). Ultrastructure of developing alveoli. I. The role of the interstitial fibroblast. Anat. Rec. 192, 457–480.

Varga, Z., Jacob, M.-P., Csongor, J., Robert, L., Leövey, A., & Fülöp, T.J. (1990). Altered phosphatidylinositol breakdown after k-elastin stimulation in PMNLs of elderly. Mechan. Age. Develop. 52, 61–70.

Venkatachalam, C.M. & Urry, D.W. (1981). Development of a linear helical conformation from its cyclic correlate. β-spiral model of the elastin poly(pentapeptide), (VPGVG)n. Macromolecules 14, 1225.

Weis-Fogh, T. & Andersen, S.O. (1970). New molecular model for the long-range elasticity of elastin. Nature (Lond) 227, 718–721.

Welgus, H.G., Campbell, E.J., Cury, J.D., Eisen, A.Z., Senior, R.M., Wilhelm, S.M., & Goldberg, G.I. (1990). Neutral metalloproteinases produced by human mononuclear phagocytes. Enzyme profile, regulation, and expression during cellular development. J. Clin. Invest. 86, 1496–1502.

Werth, V.P., Ivanov, I.E., & Nussenzweig, V. (1988). Decay-accelerating factor in human skin is associated with elastic fibers. J. Invest. Dermatol. 91, 511–516.

Wilson, T. & Treisman, R. (1988). Removal of poly(A) and consequent degradation of c-*fos* mRNA facilitated by 3′ AU-rich sequences. Nature 336, 396–399.

Wrenn, D.S. & Mecham, R.P. (1987). Immunology of elastin. Methods Enzymol. 144, 246–259.

Wrenn, D.S., Parks, W.C., Whitehouse, L.A., Crouch, E.C., Kucich, U., Rosenbloom, J., & Mecham, R.P. (1987). Identification of multiple tropoelastins secreted by bovine cells. J. Biol. Chem. 262, 2244–2249.

Yeh, H., Anderson, N., Ornstein-Goldstein, N., Bashir, M.M., Rosenbloom, J.C., Abrams, W., Indik, Z., Yoon, K., Parks, W.C., Mecham, R.P., & Rosenbloom, J. (1989). Structure of the bovine elastin gene and S1 nuclease analysis of alternative splicing of elastin mRNA in the bovine nuchal ligament. Biochemistry 28, 2365–2370.

Yeh, H., Ornstein-Goldstein, N., Indik, Z., Shepard, P., Anderson, N., Rosenbloom, J.C., Cicila, G., Yoon, K., & Rosenbloom, J. (1987). Sequence variation of bovine elastin mRNA due to alternative splicing. Collagen Rel. Res. 7, 235–247.

STRUCTURE AND FUNCTION OF BASEMENT MEMBRANE COMPONENTS:

LAMININ, NIDOGEN, COLLAGEN IV, AND BM-40

Monique Aumailley

Advances in Molecular and Cell Biology
Volume 6, pages 183–206

ISBN: 1-55938-515-4

I. INTRODUCTION

Basement membranes are complex anatomical structures forming thin boundaries between different tissues or organs and separating epithelial, endothelial, and nerve cells from the mesenchyme. Their physiological role is to serve as a barrier and filter for cells and molecules. They also act as a structural support within tissues. However, a major and more recently discovered function is that, in specifically interacting with cells, basement membranes control several cellular activities like adhesion, spreading, migration, differentiation, proliferation, and protein synthesis and secretion. During development, as well as during tissue repair, they guide and then organize cells and interstitial extracellular matrix into different compartments (for review see Timpl and Dziadek, 1986; Timpl, 1989).

At a biochemical level, basement membranes are composite extracellular matrices composed of several different and rather insoluble constituents. The discovery that the Engelbreth–Holm–Swarm (EHS) tumor transplantable into mouse (Orkin et al., 1977) was producing basement membranes in large quantities led to a breakthrough in their chemical and structural characterization. The constitutive molecules extracted from the tumor represent the archetypes for basement membrane components. Immunological and molecular biology reagents developed first to study the EHS tumor components later helped to investigate normal basement membranes and led to the discovery of isoforms. This has now opened new insights in the field of basement membrane biology and pathology in terms of structural and functional polymorphism for basement membranes of different anatomical location.

In the first part of this chapter, the structure and mode of assembly of basement membrane components will be described. The second part will present the current concepts on the molecular mechanisms involved in the cellular interactions with these components.

II. STRUCTURE OF BASEMENT MEMBRANE COMPONENTS: ARCHETYPE MOLECULES AND ISOFORMS

Basement membrane are complex and polymorphic extracellular matrices. They are composed of large multidomain and multifunctional molecules like collagen IV, laminin, nidogen, and proteoglycans, and of smaller components such as BM-40/SPARC and BM-90 (Table 1). Whereas all these are ubiquitous components of basement membranes, additional molecules have been described with a localization restricted to certain basement membranes.

A. Collagen IV

Collage IV represents the major collagenous structural component and the first molecule which has been identified in basement membranes (Kefalides, 1973). It is formed by three polypeptide α chains (~180 kDa), the most common being $\alpha 1$(IV) and $\alpha 2$(IV). The association of two $\alpha 1$(IV) and one $\alpha 2$(IV) chains [$\alpha 1$ (IV)$_2$ $\alpha 2$(IV)], represents the archetype molecule. The main characteristic features of collagen IV molecules are that at the amino terminus the three chains are folded together to form a short, triple-helical segment of 30 nm, called 7S, followed by a 360-nm long triple-helical structure, and a globular noncollagenous domain, NC1, at the carboxy-terminus (Kühn et al., 1981; Timpl et al., 1981). The central triple helix is interrupted several times by short noncollagenous sequences (Brazel et al., 1987), giving flexibility to the molecule. More detailed information on the structure of collagen IV will be found in a chapter on collagens in this volume.

Less frequent chains of collagen IV have been described; namely $\alpha 3$ (Saus et al., 1988), $\alpha 4$ (Gunwar et al., 1990), and $\alpha 5$ (Hostikka et al., 1990). Combinations between the different α chains give rise to collagen IV isoforms, yet only partially characterized (Johansson et al., 1992). Expression of these different collagen IV α chains varies with anatomical localization as shown by detection of specific mRNAs (Boot-Handford et al., 1987), and indirect immunofluorescence staining of tissues with chain specific antibodies (Butkowski et al., 1987). In particular,

Table 1. Components of Basement Membranes

Components	*Molecular Composition*
Collagen IV	three α chains of 170–185 kDa
Laminin	one A chain of ~400 kDa, two B chains of ~200 kDa
Nidogen (Entactin)	single chain of ~150 kDa
Heparan sulfate proteoglycan	protein core of 500 kDa, 3 heparan sulfate chains
BM-40/SPARC	single chain of 38–43 kDa
BM-90	single chain of 90 kDa

positive staining with antibodies directed against α3 or α4 is preferentially seen in synaptic and muscle basement membranes (Sanes et al., 1990).

B. Laminin

Laminin is the most abundant noncollagenous protein specific of basement membranes. It is a large glycoprotein (~800 kDa) composed of three genetically distinct chains: one A chain (~400 kDa) and two different B chains (~200 kDa) assembled to form a cross-shaped molecule as shown in electron microscopy after rotary shadowing (Timpl et al., 1979; Beck et al., 1990). The cross is composed of one long arm of 77 nm terminating with a large globular domain, and three short arms, two with a length of 34 nm and each bearing two globular domains separated by rod-like segments, and one of 48 nm with three globular domains.

By cDNA and peptide sequencing, the entire primary structure of these chains has been determined, first for the mouse EHS tumor laminin (Sasaki et al., 1987, 1988; Sasaki and Yamada, 1987), and then for several other species including human (for review see Paulsson, 1991). The human A (Haaparanta et al., 1991; Nissinen et al., 1991) and B (Pikkarainen et al., 1987, 1988) chains show 76–78% and 90–95% homology, respectively, to the mouse EHS tumor laminin counterparts. All three chains have homologous sequences and contain several distinct structural domains (Figure 1). The short arms are constituted by cysteine-poor sequences (domains IV and VI) folded into globular domains and connected by cysteine-rich repeats (domains III and V) forming rod-like elements. These repeats, composed of about 50 amino acids, have a certain degree of homology with motifs found in epidermal growth factor (EGF), although they contain 8 cysteines each instead of 6 as in EGF. These motifs are present in several other basement membrane components and might represent structural spacing elements facilitating protein-protein interactions, as well as providing exposed loops with biological functions (Beck et al., 1990; Davis, 1990).

The long arm (domains I and II) is formed by the three chains folded together into a coiled-coil α helix (Paulsson et al., 1985). The amino acids contributing to this region are arranged in characteristic repetitive heptad sequences (Sasaki et al., 1988). The A chain is longer than the B chains, and its carboxy-terminus is folded into the large globular domain at the end of the long arm. The amino acid sequence in this part of the A chain can be divided into five successive homologous repeats; the G domains (Sasaki et al., 1988) likely to represent five small globular structures seen in electron microscopy (Beck et al., 1990). The chains are bound together by disulfide bridges (Timpl et al., 1979) at the center of the cross, and between the carboxy-termini of the B chains (Paulsson et al., 1985). Laminin sequences contain numerous potential glycosylation sites (Sasaki et al., 1987, 1988; Sasaki and Yamada, 1987), particularly in the long arm portion of the A chain. Carbohydrates comprise 15–25% of the laminin molecular mass (Aumailley et al., 1983; Fujiwara et al., 1988).

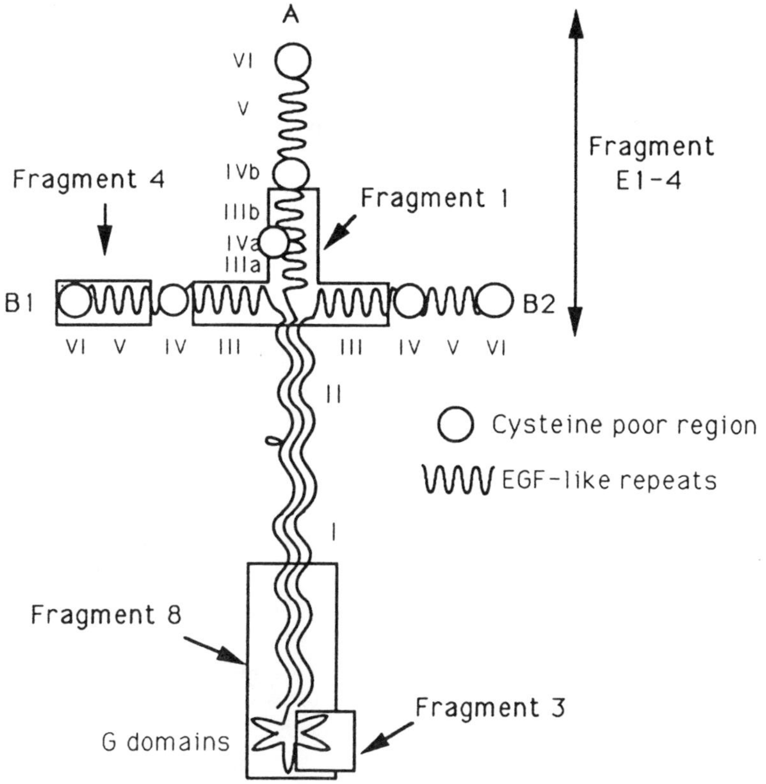

Figure 1. Schematic representation of laminin. The roman numerals represent the different domains individualized by cDNA sequencing (Sasaki et al., 1988) and seen in electron microscopy after rotary shadowing. The major proteolytic fragments of laminin are indicated.

Based on cDNA sequencing A chains[1], A1 and A2 (merosin; Engvall et al., 1990), and four B chains, B1, B2, B3 (S-laminin; Hunter et al., 1989b), and B4 (B2t; Kallunki et al., 1992), have been described, and the various associations of one A chain and two B chains determine laminin isoforms (Engvall et al., 1990). The archetype molecule isolated from the EHS tumor corresponds to the trimer A1B1B2. Isoforms, extracted from mouse heart (Paulsson and Saladin, 1989) or from human placenta (Ehrig et al., 1990), appear in electron microscopy similar to the EHS laminin. However, these molecules do not contain A1 chain but instead a 300-kDa component, initially called merosin (Ehrig et al., 1990) and later designated as the A2 or Am chain forming a trimer of molecular composition A2B1B2 (Paulsson et al., 1991). The amino acid sequence deduced from cDNA sequencing of the carboxy-terminal part of the A2 chain or merosin indicates 40% homology to the corresponding A1 chain sequence (Ehrig et al., 1990). The B3 chain or S-laminin is homologous to the B1 chain and is involved in the formation of trimers whose composition is A1B3B2 or A2B3B2 (Engvall et al., 1990). A laminin-like molecule, lacking the A chain, has been identified in cultures of rat Schwannoma cells (Edgar et al., 1988), and laminin A chain variants have been found to be synthesized by endothelial cells *in vitro* (Tokida et al., 1990). Most recently a new isoform, K-laminin, has been isolated from skin and contains the classical B1 and B2 chains associated with a variant A chain of 190 kDa (Marinkovich et al., 1992). In addition, other molecules localizing to the anchoring filaments of the dermal-epidermal junction were identified in different laboratories and assigned the names of kalinin (Rousselle et al., 1991), epiligrin (Carter et al., 1991), and BM-600/Nicein (Verrando et al., 1988). They were recently found to represent a unique laminin isoform with shorter chains (Domloge-Hultsch et al., 1992; Verrando et al., in press).

Expression of the different laminin chains varies with development and tissue localization. Immunological staining with chain-specific antibodies showed that some, but not all, basement membranes were reactive, while antibodies against whole laminin stained all basement membranes (Wewer et al., 1987). Measurements of mRNA levels specific for the A1, B1, and B2 chains demonstrated an uncoordinated expression in various tissues (Boot-Handford et al., 1987; Kleinman et al., 1987). Particularly, the A1 chain is transiently and locally expressed at the time of induction of epithelial cell polarity during kidney organogenesis (Ekblom et al., 1990; Klein et al., 1990). Also in mouse sciatic nerve, A1 chain-specific mRNA is barely detected, while mRNA for both B chains are expressed (Kücherer-Ehret et al., 1990). However, at this time it was not determined whether or not the missing A1 chain was replaced by another A chain, such as the A2 chain. Merosin (A2 chain) is expressed late in development and preferentially in nerve and striated muscle basement membranes (Leivo et al., 1988). The B3 chain (S-laminin) is remarkably concentrated in basement membranes apposed on both sides by cell layers, such as in the synaptic cleft of the neuromuscular junction and glomerular endothelium, while basement membranes separating cells from underlying interstitial connective tissue are poor in S-laminin (Hunter et al., 1989b). B4 chain (B2t) expression was localized to specific epithelial cells in skin, lung, and kidney (Kallunki et al., 1992). Expression of the different laminin chains appears therefore to be spatio-temporally regulated, which determines a molecular heterogeneity of basement membranes (Sanes et al., 1990).

C. Nidogen/Entactin

Nidogen isolated from the mouse EHS tumor (Timpl et al., 1983; Dziadek et al., 1985), is identical to the sulfated 158-kDa glycoprotein identified in cultured cells, called entactin (Carlin et al., 1981). It is composed of a single polypeptide chain of which the sequence is entirely known for mouse (Durkin et al., 1988; Mann et al., 1989) and human (Nagayoshi et al., 1989). The central portion of the polypeptide chain contains five consecutive EGF-like motifs forming a 16-nm long rod

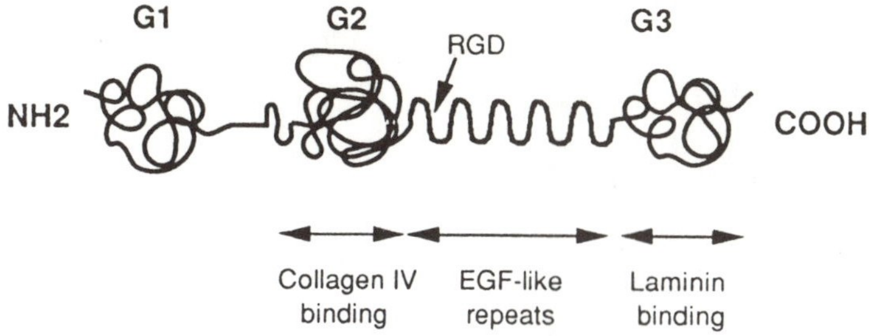

Figure 2. Schematic model of nidogen deduced from cDNA sequencing (Mann et al., 1989) and electron microscopy of purified recombinant nidogen (Fox et al., 1991). The biologically active sites are indicated.

(Paulsson et al., 1986; Fox et al., 1991). This region is flanked by large sequences interrupted on the amino terminal side by one EGF-like repeat (Figure 2). Nidogen purified from the EHS tumor was visualized in electron microscopy as a rod with a globular domain at both ends. More recently, electron microscopy observation of recombinant nidogen, likely to have a more native conformation, revealed that two globes, G1 and G2, separated by a short string, are present at the amino terminus (Fox et al., 1991). Nidogen contains two consensus sequences for tyrosine sulfation (Durkin et al., 1988; Mann et al., 1989) localized in the internal globular domain G2 (Fox et al., 1991); by *in vitro* labeling, sulfate is incorporated in the corresponding recombinant domain (Fox et al., 1991). There are also consensus sequences for glycosylation (oligosaccharides represent 5% of the molecular mass), and an RGD sequence in one of the central EGF-like motifs.

D. BM-40 / SPARC

BM-40, the small calcium-binding protein of ~ 40 kDa isolated from the EHS tumor (Dziadek et al., 1986), is identical to SPARC (secreted protein acid and rich in cysteine) isolated from cultures of endothelial (Sage et al., 1984) and parietal endoderm cells (Mason et al., 1986). By sequencing the corresponding cDNA (Mason et al., 1986; Lankat-Buttgereit et al., 1988), it was found to be identical to a major protein isolated from bone (Bolander et al., 1988), called osteonectin (Termine et al., 1981). BM-40/SPARC is, in fact, not restricted to basement membranes, and has a widespread distribution in extracellular matrices. On the basis of the amino acid sequence, a structural model was proposed (Figure 3) with four different domains (Engel et al., 1987). Both domain 1, rich in acidic residues and located at the amino-terminus, and domain III have a high degree of α helicity. Domain II is rich in cysteines involved in disulfide bonds, and has homology to the ovomucoid inhibitor. Domain IV is characterized by the presence of an EF-hand consensus sequence known to bind calcium in other proteins. Calcium-binding studies have shown that indeed BM-40 contains one high affinity site and other sites of lower affinity (Engel et al., 1987). Although the sequence of BM-40/SPARC appears to be very much conserved through species and tissues and no isoform has

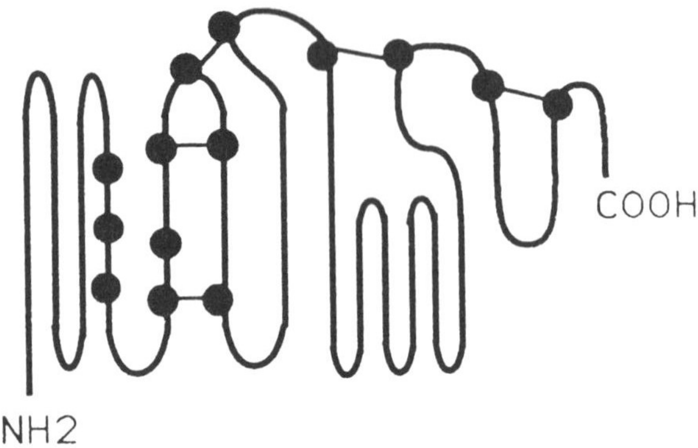

Figure 3. Schematic model of the calcium binding protein BM-40/SPARC. Known disulfide bridges are indicated between cysteine represented by black dots (modified from Engel et al., 1987).

been yet identified, a polymorphism might be given by the extent of posttranslational modifications including glycosylation (Kelm et al., 1990) and serine phosphorylation (Engel et al., 1987). Also immunological studies indicate a certain degree of polymorphism since only low immunological cross-reactions were observed between mouse and human BM-40 (Nischt et al., 1991), and between rat and bovine osteonectin (Malavel et al., 1987).

E. Proteoglycans and Additional Components

Proteoglycans with heparan or chondroitin sulfate side chains are present in basement membranes (Kanwar et al., 1979; Hassell et al., 1980). The low density heparan sulfate proteoglycan extracted from the mouse EHS tumor is formed by a large protein core and three heparan sulfate chains (Paulsson et al., 1987b). The protein core consists of a single polypeptide chain of 500 kDa folded into six globular domains aligned in a 80-nm row with the heparan sulfate chains branched at one pole. The amino acid sequence of the protein core has been determined by cDNA sequencing, and presents several typical domains including EGF-like motifs and domains homologous to the G domains of the laminin A chain (Noonan et al., 1988). Details on proteoglycans will be found in the corresponding chapter on proteoglycans in this volume.

Additional components are present in basement membranes. Some of them are less well characterized and others have a restricted occurrence. Bamin (Robinson et al., 1989) and BM-90 (Kluge et al., 1990)—both isolated from the mouse EHS tumor—and fibulin (Argraves et al., 1990) could be related molecules. Collagen VII is a component of the anchoring fibrils in basement membranes underlining stratified squamous epithelium (for details see the chapter on collagens).

III. BINDING SITES FOR ASSEMBLY OF BASEMENT MEMBRANE COMPONENTS

Basement membrane components are associated to form orderly structures which appear as meshworks in electron microscopy (for review see Yurchenco and Schittny, 1990). The supramolecular organization is likely to provide basement membranes with their physiological and biological functions. Homotypic and heterotypic associations participate in the supramolecular assembly of basement membranes (Table 2).

A. Homotypic Associations

One of the best defined homotypic association is that of collagen IV where the molecules have the property to self assembly (Yurchenco and Furthmayr, 1984) by several interactions between specific domains (Timpl et al., 1981; Kühn et al., 1981). The assembly involves dimeric associations between the NC1 domains, formation of tetramers between the so-called 7S domains, and lateral aggregations of the triple helices (Yurchenco and Schittny, 1990). (Details on these associations will be found in the chapter on collagens.) Assembly of the collagen IV molecules gives rise to polymers forming a large network which gives basement membranes tensile strength and a scaffold on which other components can associate.

Laminin is also a self-aggregating protein (Yurchenco et al., 1985; Hunter et al., 1990). Formation of laminin networks is dependent on the presence of divalent cations (Paulsson, 1988), and is mediated by interactions essentially between the terminal globular domains of the short arms corresponding to fragment 4 (Schittny and Yurchenco, 1990). A third type of homotypic association has been suggested for heparan sulfate proteoglycans (reviewed in a chapter in this volume).

Table 2. Supramolecular Assembly of Basement Membranes

Homotypic associations:
Collagen IV networks
Laminin networks
HS proteoglycan oligomers
Heterotypic associations:
Laminin-nidogen complex
Laminin-HS proteoglycan
Collagen IV-nidogen
Collagen IV-BM-40
Nidogen-HS proteoglycan
Collagen IV-HS proteoglycan

B. Heterotypic Interactions

Nidogen is a key element for at least three heterotypic associations. The two best characterized are between laminin and nidogen which was the first described, and between collagen IV and nidogen. The third is between nidogen and heparan sulfate proteoglycan. Other heterotypic associations occur between collagen IV and BM-40, and between laminin and heparan sulfate proteoglycan.

Nidogen and Laminin

Nidogen and laminin have the property to form stable (Kd ~ 10–20 nM) equimolar complexes which can be extracted from tissues by neutral buffers containing EDTA (Paulsson et al., 1987a), and were first characterized within the archetype molecules of the EHS tumor. Later such complexes were seen for mouse normal heart tissue showing that laminin isoform also binds nidogen (Paulsson and Saladin, 1989). The complexes are formed by an interaction between the carboxy-terminal globe G3 of nidogen and the inner rod-like segment of the laminin B2 chain (Mann et al., 1988; Fox et al., 1991; Gerl et al., 1991). Until now, the B2 chain has been found to participate in all laminin isoforms (Engvall et al., 1990) which indicates that nidogen can be complexed to all of the laminin chains.

Nidogen and Collagen IV

Nidogen binds to the major triple helix of collage IV at a preferential site located ~80 nm away from NC1, and to a less frequent site at ~180 nm (Aumailley et al., 1989). As shown by radio-ligand binding and competition assays using recombinant nidogen, collagen IV and laminin interact independently with nidogen. While the carboxy-terminal globe G3 binds to laminin, the collagen IV binding site is located on the amino-terminal region of nidogen close to or on the G2 globe (Fox et al., 1991; Aumailley et al., 1993). Studies with recombinant nidogen indicate that the interactions with collagen IV and laminin are strong enough (Kd < 1 nM) to allow *in vitro* reconstitution of ternary complexes (Fox et al., 1991) (Figure 4). This suggests that *in vivo* nidogen could link together the large networks of collagen IV and laminin.

Collagen IV and BM-40

EHS tumor extracted and recombinant human BM-40 bind specifically to collagen IV and not to other collagens or known basement membrane components (Mayer et al., 1991; Nischt et al., 1991). This interaction is calcium-dependent and requires the triple-helical conformation of collagen IV, as well as a correct folding of BM-40. Binding occurs preferentially on a site located about the middle length of the major triple helix of collagen IV (Mayer et al., 1991). In contrast, SPARC/osteonectin has been shown to bind preferentially to interstitial collagens (Sage et al., 1989). The apparent discrepancies between these observations could be due to a polymorphism of BM-40/SPARC in terms of glycosylation, as shown for bone and platelet osteonectin which are differently glycosylated and exhibit different collagen binding properties (Kelm et al., 1991).

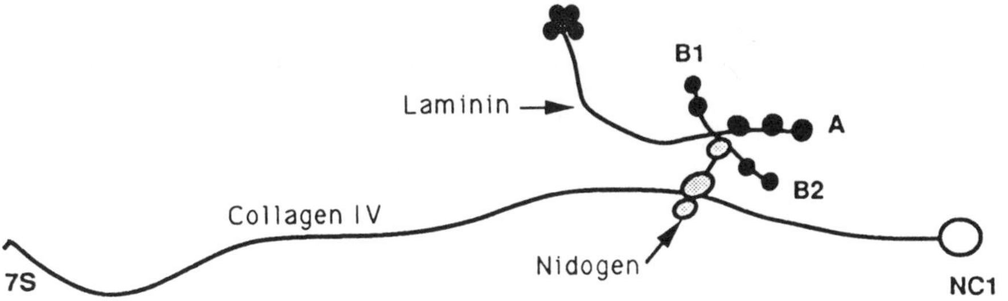

Figure 4. Schematic model representing *in vitro* reconstituted ternary complexes between collagen IV, nidogen and laminin as described in Fox et al., 1991).

Interactions with Heparan Sulfate Proteoglycans

The carboxy terminal end of the A chain, fragment E3, binds heparin (Ott et al., 1982) and heparan sulfate chains of proteoglycans (Battaglia et al., 1992). Observation by electron microscopy after rotary shadowing of mixtures made from purified basement membrane heparan sulfate proteoglycan and collagen IV showed that the heparan sulfate proteoglycan binds preferentially to a site on the collagen IV helix located at about 200-nm from NC1 domain (Laurie et al., 1986). Binding to collagen IV was also demonstrated for heparan sulfate proteoglycan synthesized by hepatocytes using a solid phase assay. Binding of collagen IV was reduced by about 50% when free GAG chains were used as a ligand in comparison to the whole proteoglycan, suggesting that both the protein core and the GAG chains are involved in binding to collagen IV (Babu and Sudhakaran, 1991). However, this heterotypic association between heparan sulfate proteoglycan and collagen IV is probably of much lower affinity than the others described above. Recently, an interaction between nidogen and proteoglycan protein core has been observed which could be involved in bridging collagen IV and proteoglycan (Battaglia et al., 1992).

IV. CELLULAR INTERACTIONS OF BASEMENT MEMBRANE COMPONENTS

Extracellular matrices play an important role in controlling cellular behavior, and much work has been devoted to determine the molecular basis of these effects, especially for basement membranes since they represent critical barriers, mechanical supports, and control elements for cells.

Collagen IV, laminin, nidogen, BM-40/SPARC, and heparan sulfate proteoglycan have been shown to develop *in vitro* specific interactions with cells, and to induce distinct cellular responses. These interactions are complex and involve several types of molecular mechanisms which are still not clearly related to the different cellular responses observed. Only in some cases, *in vivo* studies have corroborated the biological significance of these interactions. As for other extracellular matrices, cellular receptors of the integrin family are implicated in the recognition of domains or peptide sequences of basement membrane components. Other interactions involve carbohydrate chains of the basement membrane components, and lectins associated with the cell surfaces.

A. Cell-Binding Sites on Laminin

Laminin induces diverse cellular responses, such as adhesion, proliferation, migration, and differentiation (Kleinman et al., 1985; Beck et al., 1990). This multidomain molecule contains several independent sites for cellular interactions with at least two being of high affinity (Aumailley et al., 1987; Nurcombe et al., 1989). Four sites have been mapped to distinct structures of the molecule.

Two different sites, inducing cell adhesion, are located on the short arms and are contained in fragment P1 (see Figure 1) corresponding to the center of the cross (Aumailley et al., 1987; Goodman et al., 1987, 1991). One is cryptic in the intact mouse molecule (Nurcombe et al., 1989), and its activity is due to the RGD sequence from domain IIIb of the A chain (Aumailley et al., 1990a; Goodman et al., 1991). In intact laminin, this sequence is not available to cells and is presumably masked by folding of the globular domain IVa over it (Aumailley et al., 1990a). The RGD sequence also induces cell migration, and some neural cell lines produce neurites (Tashiro et al., 1991). A second site on fragment 1 (Table 3) is available on intact laminin. It is recognized only by few different cell types through an apparently RGD-independent mechanism (Goodman et al., 1987, 1991; Hall et al., 1990). Chemotactic activity has been assigned to a sequence of the B1 chain, YIGSR, contained in fragment P1 (Graf et al., 1987). Participation of this sequence has been reported for more complex phenomena, such as endothelial tube formation (Kubota et al., 1988; Grant et al., 1989), and organization of intestinal epithelial cells into multicellular structures (Olson et al., 1991). Another cellular interaction with this fragment, not yet assigned to a precise sequence and exposed in the intact molecule, induces mitogenesis (Terranova et al., 1986; Panayotou et al., 1989).

Other cell interaction sites are located on the terminal portion of the long arm and have been mapped to fragments 3 and 8 (Table 3). Fragment 3, corresponding to the most distal portion of the A chain contains a minor cell adhesion site (Sonnenberg et al., 1990; Taraboletti et al., 1990; Gehlsen et al., 1992; Sorokin et al., 1992). Fragment 8 initiates various cellular responses. It represents the major site on native laminin for adhesion and spreading of many different cell types (Aumailley et al., 1987; Goodman et al., 1987). Neurite outgrowth promoting activity of laminin (Baron van Evercooren et al., 1982) is localized on this fragment (Edgar et al., 1984; Dillner et al., 1988). It activates cell locomotion (Goodman et al., 1989b) and, *in situ*, participates in kidney tubulogenesis by inducing epithelial cell polarity and differentiation (Klein et al., 1988; Ekblom et al., 1990). Fragment 8 has a complex structure and is formed by the carboxy-termini of the B chains; part of the rod-like portion and proximal part of the globular domain contributed by the A chain (see Figure 1). Different short amino acid sequences could be

Table 3. Laminin Domains and Receptors Involved in Cellular Interactions

Domains	*Cellular Responses*	*Receptors*
Fragment 1		
	RGD-dependent adhesion (cryptic)	β3αv
	RGD-independent adhesion (exposed)	β1α1
	spreading	?
	mitogenic signal	?
	chemotactic signal, multicellular organization (YIGSR)	67 Kda
Fragment 3		
	adhesion	?
Fragment 8		
	adhesion	β1α6, β1α3
	spreading	?
	induction of neurite outgrowth	?
	polarization signal	β1α6
	locomotion signal	β1α7 ?

implicated in the determination of the biological activities of fragment 8. A synthetic peptide representing a sequence from the A chain, IKVAV, induces cell adhesion and neurite outgrowth; however, depending on the cell line, adhesion is not always followed by spreading (Tashiro et al., 1989). The sequence LRE found on the B3 chain (S-laminin) was active in adhesion of motor neurons (Hunter et al., 1989a). However, induction by fragment 8 of more specific cellular responses, including adhesion plus spreading or neurite outgrowth from neurons, requires more than primary amino acid sequences, and secondary and tertiary structures are necessary to elicit these responses (Deutzmann et al., 1990).

B. Cellular Receptors for Laminin

Integrins participate in the recognition of laminin by cells. This has been demonstrated mainly by the isolation of various integrins by affinity chromatography on laminin or laminin fragments, or by using function-inhibiting anti-integrin antibodies, namely cell adhesion inhibiting antibodies. With these approaches, several integrins from the β1 and β3 families have been shown to mediate interactions with laminin. For some of them, their reactive sites on laminin have been precisely mapped.

Integrin α1β1 binds to laminin (Forsberg et al., 1990; Hall et al., 1990; Ignatus et al., 1990) and recognizes more specifically the noncryptic RGD-independent site on the short arms (Goodman et al., 1991). The cryptic RGD-dependent site on fragment P1 is a ligand for integrins of the β3 family, including αvβ3 (Sonnenberg et al., 1990, Goodman et al., 1991). Integrin α6β1 mediates cell adhesion to intact laminin (Sonnenberg et al., 1988) via the fragment 8 site (Sonnenberg et al., 1990; Aumailley et al., 1990b; Hall et al., 1990). This integrin also plays a role in the induction of epithelial cell polarity in kidney tubulogenesis (Sorokin et al., 1990). Up to now, fragment 8 is the only ligand for β1α6. The recently discovered α7 integrin subunit associating with β1 is an alternative receptor for fragment 8 on myoblasts (von der Mark et al., 1991), and on melanoma cells (Kramer et al., 1991). α2β1 and α3β1, two promiscuous integrins, function also as laminin receptors on endothelial (Languino et al., 1989) and glioma cells (Gehlsen et al., 1988), respectively. Another subunit, α8, predominantly expressed during development along major axon tracts and on ectodermal tissue, colocalizes with basal lamina suggesting that the ligand for this integrin is in the basal lamina, and that it could be specific for axon-basement membrane interactions (Bossy et al., 1991).

Besides integrins, other cell surface-associated components can interact with laminin and could be involved in laminin-induced cellular responses (Mercurio, 1990). Several galactoside-binding lectins have affinity for laminin, and much work has been devoted to one of them, the 67-kDa laminin-binding protein (for review see von der Mark and Kühl, 1985 and Mecham, 1991). However, the role of lectins in laminin-induced cellular responses is still elusive. Lectin interactions with laminin carbohydrates could be necessary for cellular responses, such as neurite outgrowth by certain cell lines or cell spreading (Chandrasekaran et al., 1991). Unglycosylated laminin allows cell adhesion, but does not support cell spreading or neurite outgrowth (Dean et al., 1990). A 72 Kda cell-surface concanavalin A-binding glycoprotein was shown to play a role in the spreading of chick embryo fibroblasts onto laminin (Moutsita et al., 1991), and different lectins are implicated in the spreading of B16F1 murine melanoma cells on laminin fragments 1 and 8 (Bouzon et al., 1990). The neurite outgrowth promoting activity of laminin fragment 8 on PC12 cells involves an interaction between cell surface β1,4 galactosyltransferase, and terminal N-linked oligosaccharides of laminin (Begovac et al., 1991). These and other cell surface-associated components might act in concert with integrins (Brown et al., 1990; Mecham et al., 1991).

Most of the observations described above were made with EHS laminin, and there is still little known on the cellular interactions of laminin isoforms and the relevant molecular mechanisms involved. Schwannoma laminin, which does not contain the A1 chain, can promote neurite outgrowth (Edgar et al., 1988), and S-laminin and merosin have adhesive properties (Hunter et al., 1989). Pepsin-extracted human placenta laminin induces cell adhesion (Aumailley et al., 1990a); however, the RGD sequence on the A1 chain of EHS laminin is present as RAD on the human A1 chain (Haaparanta et al., 1991; Nissinen et al., 1991). Nevertheless, as on EHS laminin fragment 1, this adhesion can be blocked by addition of RGD synthetic peptides (Aumailley et al., 1990a). Conversely, RAD synthetic peptides maintained in a constrained conformation inhibit adhesion of some cell lines to pepsin-treated mouse laminin (Aumailley et al., 1991). In contrast to this apparent conservation of the molecular mechanism involved in the activity of pepsin-extracted laminin, different integrin receptors could be implicated in the recognition of native laminin isoforms. Antibodies against the integrin α6 subunit prevent cell adhesion to EHS laminin (Aumailley et al., 1990b) while they do not block adhesion of the same cells to human placenta laminin (Brown and Goodman, 1991). Since some laminin isoforms, such as A2 chain-containing laminin, appear late in development, it could be that there is a parallel variation of the corresponding receptors. It is already known that functional receptors for EHS laminin are lost during myoblast differentiation (Goodman et al., 1989a), and in the late developmental stage of retinal ganglion cells (Cohen et al., 1986, 1989). In the latter case, it could be shown that the α6 integrin subunit is specifically downregulated (deCurtis et al., 1991). Integrin expression might also vary according to the tissue localization of cells of the same lineage (Defilippi et al., 1991).

Cell surface components, especially several integrins, have the ability to bind laminin and to transmit signals from laminin to the intracellular compartment. However, definitive conclusions on the specificity and the significance of these various integrin-laminin interactions are difficult to make at this point. According to their integrin repertoire, cells could use one or the other integrin that they possess to bind laminin. Some integrins, like the α6β1, could be highly specific for laminin fragment 8; however, a cell devoided of that integrin could use a promiscuous receptor such as α1β1, α2β1, or α3β1 (Gehlsen et al., 1988; Elices and Hemler, 1989; Forsberg et al., 1990). It therefore remains to be elucidated whether the signals transmitted through these various integrin-laminin interactions are the same or not (Gimond and Aumailley, 1992).

C. Cellular Interactions with Collagen IV

Collagen IV has been shown to induce adhesion and spreading of many different type of cells either by itself (Aumailley and Timpl, 1986), or in concert with laminin (Terranova et al., 1980), and to behave as a migratory support (Herbst et al., 1988; Chelberg et al., 1989). Adhesion to collagen IV requires the triple-helical conformation (Aumailley and Timpl, 1986), and one cell-binding site has been localized to the CB3 peptide corresponding to the middle part of major triple helix (Vandenberg et al., 1991). More specialized cell lines, such as PC12 and murine melanoma cells, may use the NC1 domain as a substrate (Chelberg et al., 1989; Lein et al., 1991). Collagen IV is preferentially recognized by the α1β1 integrin (Syfrig et al., 1991) which has been implicated in the interactions with the helicoidal CB3 peptide (Vandenberg et al., 1991), as well as with the NC1 domain for the growth of axons (but not dendrites) of peripheral neurons (Lein et al., 1991). β1α2 and β1α3, two promiscuous integrins, can also participate in cell binding to collagen IV.

D. Cellular Interactions with Other Basement Membrane Components

BM-40/SPARC could have a "negative" interaction with cells. It reverses spreading of endothelial cells plated onto collagens (Sage et al., 1989), and two distinct sites of SPARC could be involved in cell shape modulation (Lane and Sage, 1990). Similar effects have been observed already for tenascin and thrombospondin, and they are still poorly understood (Sage and Bornstein, 1991).

Nidogen prepared from the mouse EHS tumor (Mann et al., 1989) was found to be able to induce adhesion of a few cell types. Also recombinant nidogen made in a bacculovirus system could trigger cell adhesion (Chakravarti et al., 1990). However, information on nidogen adhesive properties is still too limited to be conclusive.

The core protein of the basement membrane heparan sulfate proteoglycan serves as anchoring support for hepatocytes by interacting with a 80-kDa cell surface component (Clément et al., 1989). Other types of cells adhere to heparan sulfate proteoglycan by a mechanism involving β1 integrins (Battaglia et al., 1993).

V. CONCLUSIONS

Mechanical, physiological, and biological functions of basement membranes are likely to be determined by their many components and their assembly within the basement membrane. Several specific homotypic and heterotypic associative patterns are already determined for the archetype molecules. However, since major components such as collagen IV and laminin exist as isoforms whose expression is spatio-temporally regulated, it could well be that the associative patterns vary with time and with the anatomical location providing mechanical and functional diversity to basement membranes. These structures offer multiple and different recognition sites for cellular receptors which initiate diverse cellular functions. Some of the basement membrane-induced cellular responses have been shown to be triggered by specific peptide sites of the constitutive molecules, and corresponding receptors, essentially integrins, have been precisely identified. However, it is still largely unknown which biochemical signals are transduced across the cell membrane in order to modulate cellular behavior. It remains to be elucidated whether interactions with different integrins, or with other cell surface molecules, activate the same or different signalling pathways leading a cell to proliferate, to adhere, to differentiate, or to change its gene expression.

ACKNOWLEDGMENTS

I wish to warmly thank Dr. Klaus Kühn and Dr. Rupert Timpl for their support and guidance during the many years that I have spent at the Max-Planck Institut für Biochemie in Martinsried, Germany, where my contribution to the work referenced in this review has been made.

NOTES

1. With the terminology used here A1, B1, and B2 refer to the mouse EHS tumor laminin chains; alternatively the A1 chain is called Ae referring to its EHS tumor origin. A2 refers to merosin, the A chain variant isolated from human placenta and from normal mouse and bovine heart tissues, and alternatively called Am. B3 refers to a B1 chain homologue, initially named S-laminin because it was identified in the synaptic cleft of the neuromuscular junction; this variant is also designated by B1s. B4 refers to the truncated B2 chain homolog B2t.

REFERENCES

Argraves, W.S., Tran, H., Burgess, W.H., & Dickerson, K. (1990). Fibulin is an extracellular matrix and plasma glycoprotein with repeated domain structure. J. Cell Biol. 111, 3155–3164.

Aumailley, M, Battaglia, C, Mayer, U, Reinhardt, D., Nischt, R., Timpl, R., & Fox J.W. (1993). Nidogen mediates the formation of ternary complexes of basement membrane components. Kidney Int. 43, 7–12.

Aumailley, M., Gerl, M., Sonnenberg, A., Deutzmann, R., & Timpl, R. (1990a). Identification of the Arg-Gly-Asp sequence in laminin A chain as a latent cell-binding site being exposed in fragment P1. FEBS Lett. 262, 82–86.

Aumailley, M., Gurrath, M., Müller, G., Calvete, J., Timpl, R., & Kessler, H. (1991). Arg-Gly-Asp constrained within cyclic pentapeptides: strong and selective inhibitors of cell adhesion to vitronectin and laminin fragment P1. FEBS Lett. 291, 50–54.

Aumailley, M., Nowack, H., & Timpl, R. Laminin: structure and cell binding activity. (1983). In: Structural Carbohydrates of the Liver (Popper, H., Reutter, W., Gudat, W., Kottgen, E., Eds.), pp. 375–384. MTP Press, Lancaster, England.

Aumailley, M., Nurcombe, V., Edgar, D., Paulsson, M., & Timpl, R. (1987). The cellular interactions of laminin fragments. Cell adhesion correlates with two fragment-specific high affinity binding sites. J. Biol. Chem. 262, 11532–11538.

Aumailley, M., Timpl, R., & Sonnenberg, A. (1990b). Antibody to integrin α6 subunit specifically inhibits cell-binding to laminin fragment 8. Exp. Cell Res. 188, 55–60.

Aumailley, M. & Timpl, R. (1986). Attachment of cells to basement membrane collagen type IV. J. Cell Biol. 103, 1569–1575.

Aumailley, M., Wiedemann, H., Mann, K., & Timpl, R. (1989). Binding of nidogen and the laminin-nidogen complex to basement membrane collagen type IV. Eur. J. Biochem. 184, 241–248.

Babu, S.P.B. & Sudhakaran, P.R. (1991). Isolation of heparan sulfate proteoglycan from beneath the monolayers of rat hepatocytes and its binding to type IV collagen. J. Cell. Biochem. 46, 48–53.

Baron von Evercooren, A., Kleinman, H.K., Ohno, S., Marangos, P., Schwartz, J.P., & Dubois-Dalcq, M.E. (1982). Nerve growth factor, laminin and fibronectin promote neurite outgrowth in human fetal sensory ganglia cultures. J. Neurosci. Res. 8, 179–193.

Battaglia, C., Aumailley, M., Mann K., Mayer, U., & Timpl, R. (1993). Cell adhesion to a large heparan sulfate proteoglycan from basement membranes mediated by β1 integrins. Eur. J. Cell Biol. (In press).

Battaglia, C., Mayer, U., Aumailley, M., & Timpl, R. (1992). Basement-membrane heparan sulfate proteoglycan binds to laminin by its heparan sulfate chains and to nidogen by sites in the protein core. Eur. J. Biochem. 203, 359–366.

Beck, K., Hunter, I., & Engel, J. (1990). Structure and function of laminin: anatomy of a multidomain glycoprotein. FASEB J. 4, 148–160.

Begovac, P.C., Hall, D.E., & Shur, B.D. (1991). Laminin fragment E8 mediates PC12 cell neurite outgrowth by binding to cell surface β1,4 galactosyltransferase. J. Cell Biol. 113, 637–644.

Bolander, M.E., Young, M.F., Fisher, L.W., Yamada, Y., & Termine, J.D. (1988). Osteonectin cDNA sequence reveals potential binding regions for calcium and hydroxyapatite and shows homologies with both a basement membrane protein (SPARC) and a serine protease inhibitor (ovomucoid). Proc. Natl. Acad. Sci. USA. 85, 2919–2923.

Boot-Handford, R.P, Kurkinen, M., & Prockop, D.J. (1987). Steady-state levels of mRNAs coding for the type IV collagen and laminin polypeptide chains of basement membranes exhibit marked tissue-specific stoichiometric variations in the rat. J. Biol. Chem. 262, 12475–22478.

Bossy, B., Bossy-Wetzel, E., & Reichardt, L.F. (1991). Characterization of the integrin α8 subunit: a new integrin β1-associated subunit, which is prominently expressed on axons and on cells in contact with basal laminae in chick embryos. EMBO J. 10, 2375–2385.

Bouzon, M., Dussert, C., Lissitzky, J.-C., & Martin, P.-M. (1990). Spreading of B16F1 cells on laminin and its proteolytic fragments P1 and E8: involvement of laminin carbohydrate chains. Exp. Cell Res. 190, 47–56.

Brazel, D., Oberbäumer, I., Dieringer, H., Babel, W., Glanville, R.W., Deutzmann, R., & Kühn, K. (1987). Completion of the α1 chain of human basement membrane collagen (type IV) reveals 21 non-triplet interruptions located within the collagenous domain. Eur. J. Biochem. 168, 529–536.

Brown, J. & Goodman, S.L. (1991). Different cellular receptors for human placental laminin and murine EHS laminin. FEBS Lett. 282, 5–8.

Brown, E., Hooper, L., Ho, T., & Gresham, H. (1990). Integrin-associated protein: a 50-kD plasma membrane antigen physically and functionally associated with integrins. J. Cell Biol. 111, 2785–2794.

Butkowski, R.J., Langeveld, J.P.M., Wieslander, J., Hamilton, J., & Hudson, B.G. (1987). Localization of the goodpasture epitope to a novel chain of basement-membrane collagen. J. Biol. Chem. 262, 7874–7877.

Carlin, B., Jaffe, R., Bender, B., & Chung, A.E. (1981). Entactin, a novel basal lamina associated sulfated glycoprotein. J. Biol. Chem. 256, 5209–5214.

Carter, W.G., Ryan, M.C., & Gahr, P.J. (1991). Epiligrin, a new cell adhesion ligand for integrin α3β1 in epithelial basement membranes. Cell 65, 599–610.

Chakravarti, S., Tam, M.F., & Chung, A.E. (1990). The basement membrane glycoprotein entactin promotes cell attachment and binds calcium ions. J. Biol. Chem. 265,10597–10603.

Chandrasekaran, S., Dean, J.W., Giniger, M.S., & Tanzer, M.L. (1991). Laminin carbohydrates are implicated in cell signaling. J. Cell Biochem. 46, 115–124.

Chelberg, M.K., Tsilibary, E.C., Hauser, A.R., & McCarthy, J.B. (1989). Type IV collagen-mediated melanoma cell adhesion and migration: involvement of multiple, distinct domains of the collagen molecule. Cancer Res. 49, 4796–4802.

Clément, B., Segui-Real, B., Hassell, J.R., Martin, G.R., & Yamada, Y. (1989). Identification of a cell surface-binding protein for the core protein of the basement membrane proteoglycan. J. Biol. Chem. 264, 12467–12471.

Cohen, J., Burne, J.F., Winter, J., & Bartlett, P.F. (1986). Retinal ganglion cells lose response to laminin with maturation. Nature (London) 322, 465–467.

Cohen, J., Nurcombe, V., Jeffrey, P., & Edgar, D. (1989). Developmental loss of functional laminin receptors on retinal ganglion cells is regulated by their target tissue, the optic tectum. Development 107, 381–387.

de Curtis, I., Quaranta, V., Tamura, R.N., & Reichardt, L.F. (1991). Laminin receptors in the retina: sequence analysis of the chick integrin α6 subunit. J. Cell Biol. 113, 405–416.

Davis, C.G. (1990). The many faces of epidermal growth factor repeats. New Biol. 2, 410–419.

Dean, J.W., Chandrasekaran, S., & Tanzer, M.L. (1990). A biological role of the carbohydrate moieties of laminin. J. Biol. Chem. 265, 12553–12562.

Defilippi, P., van Hinsberg, V., Bertoletto, A., Rossino, P., Silengo, L., & Tarone, G. (1991). Differential distribution and modulation of expression of alpha1/beta1 integrin on human endothelial cells. J. Cell Biol. 114, 855–863.

Deutzmann, R., Aumailley, M., Wiedemann, H., Pysny, W., Timpl, R., & Edgar, D. (1990). Cell adhesion, spreading and neurite stimulation by laminin fragment E8 depends on maintenance of secondary and tertiary structure in its rod and globular domain. Eur. J. Biochem. 191, 513–522.

Dillner, L., Dickerson, K., Manthorpe, M., Ruoslahti, E., Engvall, E. (1988). The neurite-promoting domain of human laminin promotes attachment and induces characteristics morphology in non-neuronal cells. Exp. Cell Res. 177, 186–198.

Domloge-Hultsch, N, Gammon, W.R, Briggaman, R.A., Gil, S.G., Carter, W.G., & Yancaey, K.B. (1992). Epiligrin, the major human keratinocyte integrin ligand, is a target in both an inherited subepidermal blistering skin disease. J. Invest. Dermatol. 90, 1628–1633.

Durkin, M.E., Chakravarti, S., Bartos, B.B., Liu, S.–H., Friedman, R.L., & Chung, A.E. (1988). Amino acid sequence and domain structure of entactin. Homology with epidermal growth factor precursor and low density lipoprotein receptor. J. Cell Biol. 107, 2749–2756.

Dziadek, M., Paulsson, M., & Timpl, R. (1985). Identification and interaction repertoire of large forms of the basement membrane protein nidogen. EMBO J. 4, 2513–2518.

Dziadek, M., Paulsson, M., Aumailley, M., & Timpl, R. (1986). Purification and tissue distribution of a small protein (BM-40) extracted from a basement membrane tumor. Eur. J. Biochem. 161, 455–464.

Edgar, D., Timpl, R., & Thoenen, H. (1984). The heparin-binding domain of laminin is responsible for its effects on neurite outgrowth and neuronal survival. EMBO J. 3, 1463–1468.

Edgar, D., Timpl, R., & Thoenen, H. (1988). Structural requirements for the stimulation of neurite outgrowth by two variants of laminin and their inhibition by antibodies. J. Cell Biol. 106,1299–1306.

Ehrig, K., Leivo, I., Argraves, W.S., Ruoslahti, E., & Engvall, E. (1990). Merosin, a tissue-specific basement membrane protein, is a laminin-like protein. Proc. Natl. Acad. Sci. USA. 87, 3264–3268.

Ekblom, M., Klein, G., Mugrauer, G., Fecker, L., Deutzmann, R., Timpl, R., & Ekblom, P. (1990). Transient and locally restricted expression of laminin A chain mRNA by developing epithelial cells during kidney organogenesis. Cell 60, 337–346.

Elices, M.J. & Hemler, M.E. (1989). The human integrin VLA-2 is a collagen receptor on some cells and a collagen/laminin receptor on others. Proc. Natl. Acad. Sci. USA. 86, 9906–9910.

Engel, J., Taylor, M., Paulsson, M., Sage, H., & Hogan, B.L.M. (1987). Calcium binding domains and calcium-induced conformational transition of SPARC/BM-40/osteonectin, an extracellular glycoprotein expressed in mineralized and nonmineralized tissues. Biochemistry 26, 6958–6965.

Engvall, E., Earwicker, D., Haaparanta, T., Ruoslahti, E., & Sanes, J.R. (1990). Distribution and isolation of four laminin variants; tissue restricted distribution of heterotrimers assembled from five different subunits. Cell Regul. 1, 731–740.

Forsberg, E., Paulsson, M., Timpl, R., & Johansson, S. (1990). Characterization of a laminin receptor on rat hepatocytes. J. Biol. Chem. 265, 6376–6381.

Fox, J., Mayer, U., Nischt, R., Aumailley, M., Reinhardt, D., Wiedemann, H., Mann, K., Timpl, R., Krieg, T., Engel, J., & Chu, M.-L. (1991). Recombinant nidogen consists of three globular domains and mediates binding of laminin to collagen IV. EMBO J. 10, 3137–3146.

Fujiwara, S., Shinkai, H., Deutzmann, R., Paulsson, M., & Timpl, R. (1988). Structure and distribution of N-linked oligosaccharide chains on various domains of mouse tumor laminin. Biochem. J. 252 453–461.

Gehlsen, K.R., Dillner, L., Engvall, E., & Ruoslahti, E. (1988). The human laminin receptor is a member of the integrin family of cell adhesion receptors. Science 241, 1228–1229.

Gehlsen, K.R., Sriramarao, P., Furcht, L.T., & Skubitz, P.N. (1992). A synthetic peptide derived from the carboxy terminus of the laminin A chain represents a binding site for the $\alpha3\beta1$ integrin. J. Cell Biol. 117, 449–459.

Gerl, M., Mann, K., Aumailley, M., & Timpl, R. (1991). Localization of a major nidogen binding site to domain lll of laminin B2 chain. Eur. J. Biochem. 202, 167–174.

Gimond, C., & Aumailley, M. (1992). Cellular interactions with the extracellular matrix are coupled to diverse transmembrane signaling pathways Exp. Cell Res. 203, 365–373.

Goodman, S.L., Aumailley, M., & von der Mark, H. (1991a). Multiple cell surface receptors for the short arms of laminin: $\alpha1\beta1$ integrin and RGD-dependent proteins mediate cell attachment only to domains III in murine tumor laminin. J. Cell Biol. 113, 931–941.

Goodman, S.L., Deutzmann, R., & von der Mark, K. (1987). Two distinct cell-binding domains in laminin can independently promote nonneuronal cell adhesion and spreading. J. Cell Biol. 105, 589–598.
Goodman, S.L., Deutzmann, R., & Nurcombe, V. (1989a). Locomotory competence and laminin-specific cell surface binding-sites are lost during myoblast differentiation. Development 106, 795–802.
Goodman, S.L., Risse, G., & von der Mark, K. (1989b). The E8 subfragment of laminin promotes locomotion of myoblasts over extracellular matrix. J. Cell Biol. 109, 799–809.
Graf, J., Iwamoto, Y., Sasaki, M., Martin, G.R., Kleinman, H.K., Robey, F.A., & Yamada, Y. (1987). Identification of an amino acid sequence in laminin mediating cell attachment, chemotaxis and receptor binding. Cell 48, 989–996.
Grant, D.S., Tashiro, K.I., Segui-Real, B., Yamada, Y., Martin, G.R., & Kleinman, H.K. (1989). Two different laminin domains mediate the differentiation of human endothelial cells into capillary-like structures *in vivo*. Cell 58, 933–943.
Gunwar, S., Saus, J., Noelken, M.E., & Hudson, B.G. (1990). Glomerular basement membrane. Identification of a fourth chain, α4, of type IV collagen. J. Biol. Chem. 265, 5466–5469.
Haaparanta, T., Uitto, J., Ruoslahti, E., & Engvall, E. (1991). Molecular cloning of the cDNA encoding human laminin A chain. Matrix 11, 151–160.
Hall, D.E., Reichardt, L.F., Crowley, E., Holley, B., Moezzi, H., Sonnenberg, A., & Damsky, C.H. (1990). The α1/β1 and α6/β1 integrin heterodimers mediate cell attachment to distinct sites on laminin. J. Cell Biol. 110, 2175–2184.
Hassel, J.R., Gehron Robey, P., Barrach, H.-J., Wilczek, J., Rennard, S.I., & Martin, G.R. (1980). Isolation of a heparan sulfate-containing proteoglycan from basement membrane. Proc. Natl. Acad. Sci. USA. 77, 4494–4498.
Herbst, T.J., McCarthy, J.B., Tsilibary, E.C., & Furcht, L.T. (1988). Differential effects of laminin, intact type IV collagen, and specific domains of type IV collagen on endothelial cell adhesion and migration. J. Cell Biol. 106, 1365–1373.
Hostikka, S.L., Eddy, R.L., Byers, M.G., Höyhtyä, M., Shows, T.B., & Tryggvason, K. (1990). Identification of a distinct type IV collagen α chain with restricted kidney distribution and assignment of its gene to the locus of X chromosome-linked Alport syndrome. Proc. Natl. Acad. Sci. USA. 87, 1606–1610.
Hunter, I., Schulthess, T., Bruch, M., Beck, K., & Engel, J. (1990). Evidence for a specific mechanism of laminin assembly. Eur. J. Biochem. 188, 205–211.
Hunter, D.D., Porter, B.E., Bulock, J.W., Merlie, J.P., & Sanes, J.R. (1989a). Primary sequence of a motor neuron-selective adhesive site in the synaptic basal lamina protein S-laminin. Cell 59, 905–913.
Hunter, D.D., Shah, V., Merlie, J.P., & Sanes, J.R. (1989b). A laminin-like adhesive protein concentrated in the synaptic cleft of the neuromuscular junction. Nature (London) 338, 229–234.
Ignatus, M.J., Large, T.H., Houde, M., Tawil, J.W., Barton, A., Esch, F., Carbonetto, S., & Reichardt, L.F. (1990). Molecular cloning of the rat integrin α1-subunit: a receptor for laminin and collagen. J. Cell Biol. 111, 709–720.
Johansson, C., Butkowski, R., & Wieslander, J. (1992). The structural organization of type IV collagen. Identification of three NC1 populations in the glomerular basement membrane. J. Biol. Chem. 267, 24533–24537.
Kallunki, P., Sainio, K., Eddy, R., Byers, M., Kallunki, T., Beck, K., Hirvonene, H., Shows, T.B., & Tryggvason, K. (1992). A truncated laminin chain homologous to the B2 chain: structure, spatial expression, and chromosomal assignments. J. Cell Biol. 119, 679–693.
Kanwar Y.S. & Farquhar M.G. (1979). Presence of heparan sulfate in the glomerular basement membrane. Proc. Natl. Acad. Sci. USA. 76, 1303–1307.
Kefalides, N.A. (1973). Structure and biosynthesis of basement membrane. Int. Rev. Connect. Tissue Res. 6, 63–104.
Kelm, R.J. & Mann, K.G. (1991). The collagen binding specificity of bone and platelet osteonectin is related to differences in glycosylation. J. Biol. Chem. 266, 9632–9639.

Klein, G., Ekblom, M., Fecker, L., Timpl, R., & Ekblom, P. (1990). Differential expression of laminin A and B chains during development of embryonic mouse organs. Development 110, 823–837.

Klein, G., Langegger, M., Timpl, R., & Ekblom, P. (1988). Role of laminin A chain in the development of epithelial cell polarity. Cell, 55, 331–341.

Kleinman, H.K., Cannon, F.B., Laurie, G.W., Hassell, J.R., Aumailley, M., Terranova, V.P., Martin, G.R., & Dubois-Dalcq, M. (1985). Biological activities of laminin. J. Cell. Biochem. 27, 317–325.

Kleinman, H.K., Ebihara, I., Killen, P.D., Sasaki, M., Cannon, F.B., Yamada, Y., & Martin, G.R. (1987). Genes for basement membrane proteins are coordinately expressed in differentiating F9 cells but not in normal adult murine tissues. Dev. Biol. 122, 373–378.

Kluge, M., Mann, K., Dziadek, M., & Timpl, R. (1990). Characterization of a novel calcium-binding 90-kDa glycoprotein (BM-40) shared by basement membranes and serum. Eur. J. Biochem. 193, 651–659.

Kramer, R.H., McDonald, K.A., & Vu, M.P. (1989). Human melanoma cells express a novel integrin receptor for laminin. J. Biol. Chem 264, 15642–15649.

Kubota, Y., Kleinman, H.K., Martin, G.R., & Lawley, T.J. (1988). Role of laminin and basement membrane in the morphological differentiation of human endothelial cells in capillary-like structures. J. Cell Biol. 107, 1589–1598.

Kücherer-Ehret, A., Pottgiesser, J., Kreutzberg, G.W., Thoenen, H., & Edgar, D. (1990). Developmental loss of laminin from the interstitial extracellular matrix correlates with decreased laminin gene expression. Development 110, 1285–1293.

Kühn, K., Wiedemann, H., Timpl, R., Risteli, J., Dieringer, H., Voss, T., & Glanville, R.W. (1981). Macromolecular structure of basement membrane collagens. Identification of 7S collagen as a crosslinking domain of type IV collagen. FEBS Lett. 125, 123–128.

Lane, T.F. & Sage, E.H. (1990). Functional mapping of SPARC: peptides from two distinct Ca^{++}-binding sites modulate cell shape. J. Cell Biol. 111, 3065–3076.

Languino, L.R., Gehlsen, K.R., Wayner, E.A., Carter, W.G., Engvall, E., & Ruoslahti, E. (1989). Endothelial cells use $\alpha 2\beta 1$ integrin as a laminin receptor. J. Cell Biol. 109, 2455–2462.

Lankat-Buttgereit, B., Mann, K., Deutzmann, R., Timpl, R., & Krieg, T. (1988). Cloning and complete amino acid sequences of human and murine basement membrane protein BM-40 (SPARC, osteonectin). FEBS Lett. 236, 352–356.

Laurie, G.W., Bing, J.T., Kleinman, H.K., Hassell, J.R., Aumailley, M., Martin, G.R., & Feldmann, R.J. (1986). Localization of binding sites for laminin, heparan sulfate proteoglycan and fibronectin on basement membrane (type IV) collagen. J. Mol. Biol. 188, 205–216.

Lein, P.J., Higgins, D., Turner, D.C., Flier, L.A., & Terranova, V.P. (1991). The NC1 domain of type IV collagen promotes axonal growth in sympathetic neurons through interaction with the $\alpha 1\beta 1$ integrin. J. Cell Biol. 113, 417–428.

Leivo, I. & Engvall, E. (1988). Merosin, a protein specific for basement membranes of Schwann cells, striated muscle, and trophoblast, is expressed late in nerve and muscle development. Proc. Natl. Acad. Sci. USA 85, 1544–1548.

Malavel, L., Fournier, B., & Delmas, P.D. (1987). Circulating levels of osteonectin in normal subjects and in patients with thrombocytopenia. J. Bone Mineral Res. 2, 457–465.

Mann, K., Deutzmann, R., Aumailley, M., Timpl, R., Raimondi, L., Yamada, Y., Pan, T.-C., Conway, D., & Chu, M.-L. (1989). Amino acid sequence of mouse nidogen, a multidomain basement membrane protein with binding activity for laminin, collagen IV and cells. EMBO J. 8, 65–72.

Mann, K., Deutzmann, R., & Timpl, R. (1988). Characterization of proteolytic fragments of the laminin-nidogen complex and their activity in ligand-binding assays. Eur. J. Biochem. 178, 71–76.

Marinkovich, M.P., Lunstrum, G.R., Keene, D.R., & Burgeson, R.E. (1992). The dermal-epidermal junction of human skin contains a novel laminin variant. J. Cell Biol. 119, 695–703.

Mason, I.J., Taylor, A., Williams, J.G., Sage, H., & Hogan, B.L.M. (1986). Evidence from molecular cloning that SPARC, a major product of mouse embryo parietal endoderm is related to an endothelial cell "culture shock" glycoprotein of M_r 43,000. EMBO J. 5, 1465–1472.

Mayer, U., Aumailley, M., Mann, K., Timpl, R., & Engel, J. (1991). Calcium-dependent binding of BM-40 (osteonectin, SPARC) to basement membrane collagen type IV. Eur. J. Biochem. 198,141–150.

Mecham, R.P. (1991). Receptors for laminin on mammalian cells. FASEB J. 5, 2538–2546.

Mecham, R.P., Whitehouse, L., Hay, M., Hinek, A., & Sheetz, M.P. (1991). Ligand affinity of the 67-kDa elastin/laminin binding protein is modulated by the protein's lectin domain: visualization of elastin/laminin-receptor complexes with gold-tagged ligands. J. Cell Biol. 113, 187–194.

Mercurio, A.M. (1990). Laminin: multiple forms, multiple receptors. Current Opinion in Cell Biology 2, 845–849.

Moutsita, R., Aubery, M., & Codogno, P. (1991). A M_r 72K cell surface concanavalin A binding glycoprotein is specifically involved in the spreading of chick embryo fibroblasts onto laminin substrate. Exp. Cell Res. 192, 236–242.

Nagayoshi, T., Sanborn, D., Hickok, N.J., Olsen, D.R., Fazio, M.J., Chu, M.-L., Knowlton, R., Mann, K., Deutzmann, R., Timpl, R., & Uitto, J. (1989). Human nidogen: complete amino acid sequence and structural domains deduced from cDNAs and evidence for a polymorphism of the gene. DNA 8, 581–594.

Nischt, R., Pottgiesser, J., Krieg, T., Mayer, U., Aumailley, M., & Timpl, R. (1991). Recombinant expression and properties of the human calcium-binding extracellular matrix protein BM-40 (osteonectin, SPARC). Eur. J. Biochem. 200, 529–536.

Nissinen, M., Vuolteenaho, R., Boot-Handford, R., Kallunki, P., & Tryggvason, K. (1991). Primary structure of the human laminin A chain. Limited expression in human tissues. Biochem. J. 276, 369–378.

Noonan, D.M., Horigan, E.A., Ledbetter, S.R., Vogeli, G., Sasaki, M., Yamada, Y., & Hassell, J.R. (1988). Identification of cDNA clones encoding different domains of the basement membrane heparan sulfate proteoglycan. J. Biol. Chem. 263, 16379–16387.

Nurcombe, V., Aumailley, M., Timpl, R., & Edgar, D. (1989). The high affinity binding of laminin to cells. Assignation of a major cell-binding site to the long arm of laminin and a latent cell-binding site to its short arms. Eur. J. Biochem. 180, 9–14.

Olson, A., Pysher, T., & Bienkowski, R. (1991). Organization of intestinal cells into multicellular structures requires laminin and functional actin microfilaments. Exp. Cell Res. 192, 543–549.

Orkin, R.W., Gehron, P., McGoodwin, E.B., Martin, G.R., Valentine, T., & Swarm, R. (1977). A murine tumor producing a matrix of basement membrane. J. Exp. Med. 145, 204–220.

Ott, U., Odermatt, E., Engel, J., Furthmayr, H., & Timpl, R. (1982). Protease resistance and conformation of laminin. Eur. J. Biochem. 123, 63–72.

Panayotou, G., End, P., Aumailley, M., Timpl, R., & Engel, J. (1989). Domains of laminin with growth-factor activity. Cell 56, 93–101.

Paulsson, M. (1988). The role of Ca^{2+} binding in the self-aggregation of laminin-nidogen complexes. J. Biol. Chem. 263, 5425–5430.

Paulsson, M. (1991). Basement membrane proteins: structure, assembly, and cellular interactions. (1991). Crit. Rev. Biochem. Mol. Biol. (In press).

Paulsson, M., Aumailley, M., Deutzmann, R., Timpl, R., Beck, K., & Engel, J. (1987a). Laminin-nidogen complex. Extraction with chelating agents and structural characterization. Eur. J. Biochem. 166, 11–19.

Paulsson, M., Deutzmann, R., Dziadek, M., Nowack, H., Timpl, R., Weber, S., & Engel, J. (1986). Purification and structural characterization of intact and fragmented nidogen obtained from a basement membrane tumor. Eur. J. Biochem. 156, 467–478.

Paulsson, M., Deutzmann, R., Timpl, R., Dalzoppo, D., Odermatt, E., & Engel, J. (1985). Evidence for coiled-coil α-helical regions in the long arm of laminin. EMBO J. 4, 309–316.

Paulsson, M. & Saladin, K. (1989). Mouse heart laminin. Purification of the native protein and structural comparison with Engelbreth-Holm-Swarm tumor laminin. J. Biol. Chem. 264, 18726–18732.

Paulsson, M., Saladin, K., & Engvall, E. (1991). Structure of laminin variants. The 300 kDa chains of murine and bovine heart laminins are related to the human placenta merosin heavy chain and replace the A chain in some laminin variants. J. Biol. Chem. 266,17545–17551.

Paulsson, M., Yurchenco, P.D., Ruben, G.C., Engel, J., & Timpl, R. (1987b). Structure of low density heparan sulfate proteoglycan isolated from a mouse tumor basement membrane. J. Mol. Biol. 197, 297–313.

Pikkarainen, T., Eddy, R., Fukushima, Y., Byers, M., Shows, T., Pihlajaniemi, T., Saraste, M., & Tryggvasson, K. (1987). Human laminin B1 chain. A multidomain protein with gene (LAMB1) locus in the q22 region of chromosome 7. J. Biol. Chem. 262, 10454–10462.

Pikkarainen, T., Kallunki, T., & Tryggvasson, K. (1988). Human laminin B2 chain. Comparison of the complete amino acid sequence with the B1 chain reveals variability in sequence homology between different structural domains. J. Biol. Chem. 263, 6751–6758.

Robinson, L.K., Murrah, V.A., Mayer, M.P., & Rohrbach, D.H. (1989). Characterization of a novel glycoprotein isolated from the basement membrane matrix of the Engelbreth-Holm-Swarm tumor. J. Biol. Chem. 264, 5141–5147.

Rousselle, P., Lunstrum, G., Keene, D.R., & Burgeson, R.E. (1991). Kalinin: an epithelium-specific basement membrane adhesion molecule that is a component of anchoring filaments. J. Cell Biol. 114, 567–576.

Sage, E.H. & Bornstein, P. (1991). Extracellular proteins that modulate cell-matrix interactions. SPARC, tenascin, and thrombospondin. J. Biol. Chem. 266, 14831–14834.

Sage, E. H., Johnson, C., & Bornstein, P. (1984). Characterization of a novel serum albumin-binding glycoprotein secreted by endothelial cells in culture. J. Biol. Chem. 259, 3993–4007.

Sage, H., Vernon, R.B., Funk, S., Everitt, E.A., & Angello, J. (1989). SPARC a secreted protein associated with cellular proliferation, inhibits cell spreading *in vitro* and exhibits Ca^{2+}-dependent binding to the extracellular matrix. J. Cell Biol. 109, 341–346.

Sanes, J.R., Engvall, E., Butkowski, R., & Hunter, D.D. (1990). Molecular heterogeneity of basal laminae: isoforms of laminin and collagen IV at the neuromuscular junction and elsewhere. J. Cell Biol. 111, 1685–1699.

Sasaki, M., Kato, S., Kohno, K., Martin, G.R., & Yamada, Y. (1987). Sequence of the cDNA encoding the laminin B1 chain reveals a multidomain protein containing cysteine-rich repeats. Proc. Natl. Acad. Sci. USA 84, 935–939.

Sasaki, M., Kleinman, H.K., Huber, H., Deutzmann, R., & Yamada, Y. (1988). Laminin, a multidomain protein. The A chain has a unique globular domain and homology with the basement membrane proteoglycan and the laminin B chains. J. Biol. Chem. 263, 16536–16544.

Sasaki, M. & Yamada, Y. (1987). The laminin B2 chain has a multidomain structure homologous to the B1 chain. J. Biol. Chem. 262, 17111–17117.

Saus, J., Wieslander, J., Langeveld, J.P., Quinones, S., & Hudson, B.G. (1988). Identification of the Goodpasture antigen as the α3(IV) chain of collagen IV. J. Biol. Chem. 263, 13374–13380.

Schittny, J.C. & Yurchenco, P.D. (1990). Terminal short arm domains of basement membrane laminin are critical for its self-assembly. J. Cell Biol. 110, 825–832.

Sonnenberg, A., Modderman, P.W., & Hogervorst, F. (1988). Laminin receptor on platelets is the integrin VLA-6. Nature 336, 487–489.

Sonnenberg, A., Linders, C.J.T., Modderman, P.W., Damsky, C.H., Aumailley, M., & Timpl R. (1990). Integrin recognition of different cell-binding fragments of laminin (P1, E3, E8) and evidence that α6β1 but not α6β4 functions as a major receptor for fragment E8. J. Cell Biol. 110, 2145–2155.

Sorokin, L.M., Conzelmann, S., Ekblom, P., Battaglia, C., Aumailley, M., & Timpl, R. (1992). Monoclonal antibodies against laminin A chain fragment E3 and their effects on binding to cells and proteoglycan and on kidney development. Exp. Cell Res. 201, 137–144.

Sorokin, L., Sonnenberg, A., Aumailley, M., Timpl, R., & Ekblom, P. (1990). Recognition of the laminin E8 cell-binding site by an integrin possessing the α6 subunit is essential for epithelial polarization in developing kidney tubules. J. Cell Biol. 111, 1265–1273.

Syfrig, J., Mann, K., & Paulsson, M. (1991). An abundant chick gizzard integrin is the avian α1β1 integrin heterodimer and functions as a divalent cation-dependent collagen IV receptor. Exp. Cell Res 194, 165–173.

Taraboletti, D., Rao, C.N., Krutzsch, H.C., Liotta, L.A., & Roberts, D.D. (1990). Sulfatide-binding domain of the laminin A chain. J. Biol. Chem. 265, 12253–12258.

Tashiro, K.-I., Sephel, G.C., Greatorex, D., Sasaki, M., Shirashi, N., Martin, G.R., Kleinman, H.K., & Yamada, Y. (1991). The RGD containing site of the mouse laminin A chain is active for cell attachment, spreading, migration and neurite outgrowth. J. Cell. Physiol. 146, 451–459.

Tashiro, K.-I., Sephel, G.C., Weeks, B., Sasaki, M., Martin, G.R., Kleinman, H.K., & Yamada, Y. (1989). A synthetic peptide containing the IKVAV sequence from the A chain of laminin mediates cell attachment, migration, and neurite outgrowth. J. Biol. Chem. 264, 16174–16182.

Termine, J.D., Kleinman, H.K., Whitson, S.W., Conn, K.M., McGarvey, M.L., & Martin, G.R. (1981). Osteonectin, a bone-specific protein linking mineral to collagens. Cell 26, 99–105.

Terranova, V.P., Aumailley, M., Sultan, L.H., Martin, G.R., & Kleinman, H.K. (1986). Regulation of cell attachment and cell number by fibronectin and laminin. J. Cell. Physiol. 127, 47479.

Terranova, V.P., Rohrbach, D.H., & Martin, G.R. (1980). Role of laminin in attachment of PAM 212 (epithelial) cells to basement membrane collagen. Cell 22, 719–726.

Timpl, R. (1989). Structure and biological activity of basement membrane proteins. Eur. J. Biochem. 180, 487–502.

Timpl, R. & Dziadek, M. (1986). Structure, development and molecular pathology of basement membranes. Int. Rev. Exp. Pathol. 29, 1–112.

Timpl, R., Dziadek, M., Fujiwara, S., Nowack, H., & Wick, G. (1983). Nidogen: a new, self-aggregating basement membrane protein. Eur. J. Biochem. 137, 455–465.

Timpl R., Rhodes, H., Robey, P., Rennard, S., Foidart, J.-M., & Martin, G.R. (1979). Laminin—a glycoprotein from basement membranes. J. Biol. Chem. 254, 9933–9937.

Timpl, R., Wiedemann, H., van Delden, V., Furthmayr, H., & Kühn, K. (1981). A network model for the organization of type IV collagen molecules in basement membranes. Eur. J. Biochem. 120, 203–211.

Tokida, Y., Aratani, Y., Morita, A., & Kitagawa, Y. (1990). Production of two variant laminin forms by endothelial cells and shift of their relative levels by angiostatic steroids. J. Biol. Chem. 265, 18123–18129.

Vandenberg, P., Kern, A., Ries, A., Luckenbill-Edds, L., Mann, K., & Kühn, K. (1991). Characterization of a type IV collagen major cell binding site with affinity to the $\alpha1\beta1$ and the $\alpha2\beta1$ integrins. J. Cell Biol. 113, 1475–1483.

Verrando, P., Pisani, A., & Ortonne, J.P. (1988). The new basement membrane antigen recognized by the monoclonal antibody GB3 is a large size glycoprotein: modulation of its expression by retinoic acid. Biochim. Biophys. Acta 942, 45–56.

Verrando, P., Vailly, J., Champliaud, M.-F., Gerecke, D., Wagman, W., Baudoin, C., Aberdam, D., Burgeson, R.E., Bauer, E., & Ortonne, J.-P. (1993). A new laminin variant defined by the cDNA sequence of the 100 kDa subunit of the basement membrane protein nicein(BM-600)/kalinin (In press).

von der Mark, H., Dürr, J., Sonnenberg, A., von der Mark, K., Deutzmann, R., & Goodman, S.L. (1991). Skeletal myoblasts utilize a novel b1 serie integrin and not a6b1 for binding to the E8 and T8 fragments of laminin. J. Biol. Chem. 266, 23593–23601.

von der Mark, K. & Kühl, U. (1985). Laminin and its receptor. Biochim. Biophys. Acta 823, 147–160.

Wewer, U.M., Tichy, D., Damjanov, A., Paulsson, M., & Damjanov, I. (1987). Distinct antigenic characteristics of murine parietal yolk sac laminin. Dev. Biol. 121, 397–407.

Yurchenco, P.D. & Furthmayr, H. (1984). Self-assembly of basement membrane collagen. Biochemistry 23, 1839–1850.

Yurchenco, P.D. & Schittny, J.C. (1990). Molecular architecture of basement membranes. FASEB J. 4, 1577–1590.

Yurchenco, P.D., Tsilibary, E.C., Charonis, A.S., & Furthmayr, H. (1985). Laminin polymerization *in vitro*. Evidence for a two-step assembly with domain specificity. J. Biol. Chem. 260, 7636–7644.

EXTRACELLULAR MATRIX AND BONE MORPHOGENETIC PROTEINS IN CARTILAGE AND BONE DEVELOPMENT AND REPAIR

Slobodan Vukicevic, Vishwas M. Paralkar, and A.H. Reddi

Advances in Molecular and Cell Biology
Volume 6, pages 207–224

ISBN: 1-55938-515-4

I. INTRODUCTION

The origin and evolution of multicellular eukaryotes is accompanied with the appearance of the extracellular matrix (ECM). In certain skeletal tissues, such as bone and cartilage, the ECM is prominent and accounts for much of the mass and volume. The biosynthesis and supramolecular assembly of ECM in bone and cartilage is intimately associated with the development and morphogenesis of these tissues. The ECM provides the framework for the tissues of the skeletal system and, in addition, plays an important physiological role in sequestration of growth and differentiation factors involved in repair, regeneration, and remodeling.

The development and differentiation of bone occurs by one of two processes. First, there is a direct development of bone from mesenchyme as in the bones of the skull and facial skeleton. This process is referred to as intramembranous bone formation. Second, a transient cartilage model precedes differentiation of bone and occurs in a majority of bones. This process is known as endochondral ossification. It is noteworthy that these two methods of ossification signify only the environment of bone formation, and there appears to be no distinct differences in the bone formed. Angiogenesis and vascular invasion are prerequisites for both intramembranous and endochondral ossification. However, the underlying mechanistic interactions between vascular invasion and osteogenesis are not clear. The aim of this chapter is to describe recent progress in the interface of extracellular matrix components and growth and differentiation factors with special reference to bone and cartilage development and repair.

II. CARTILAGE AND BONE DIFFERENTIATION

The bone has a remarkable potential for repair and regeneration. Regeneration processes, as fracture repair, recapitulate events that occur in the course of embryonic bone development. The experimental means of inducing new bone formation at non-bony sites have been known for years. In early 1930, Huggins (1931a,b) discovered that surgical transplantation of living urinary bladder epithelium into abdominal muscles in dogs resulted in new cartilage and bone formation. The phenomenon was termed "epithelial osteogenesis." Urist (1965) and Reddi and Huggins (1972) have shown that implantation of demineralized bone matrix into intramuscular or subcutaneous sites induced a sequence of cellular events leading to the formation of cartilage, bone, and bone marrow. The events were termed "matrix-induced endochondral bone formation." Inductive properties of demineralized bone matrix implantation at heterotopic sites are also reminiscent of embryonic bone development. It is believed that the putative molecules responsible for epithelial and matrix-induced osteogenesis may have common structural properties, although it has not yet been demonstrated.

A. Demineralized Bone Matrix-Induced Osteogenesis

The cascade of matrix-induced endochondral bone formation consists of chemotaxis, attachment of mesenchymal stem cells to the matrix, proliferation of progenitor cells, and differentiation of cartilage. This is followed by cartilage resorption, bone formation, and finally culminating in the formation of an ossicle filled with hematopoietic marrow (Reddi, 1981). On day 1, fibronectin is deposited on the matrix facilitating cell attachment. On days 2 and 3, proliferation of mesenchymal cells takes place. This cell-matrix interaction results in the differentiation of chondroblasts at day 5, with appearance of cartilaginous matrix as shown by an increased ^{35}S (sulfate) incorporation into proteoglycans, and the appearance of hypertrophic chondrocytes at day 7. Angiogenesis with cartilage resorption starts on day 9, and new bone formation in association with ^{45}Ca incorporation into new bone mineral occurs at day 11. Extensive bone remodeling with hematopoietic marrow formation, as evidenced by ^{59}Fe incorporation, is maximal on day 21. This developmental program is a single cycle of cartilage and bone formation, and recapitulates the embryonic development of limbs. It could be a useful model for fracture healing. Moreover, this system enables molecular studies at the different stages of endochondral bone formation, and a better understanding of the role of endogenous and exogenous growth and differentiation factors involved.

B. Osteogenin and Bone Morphogenetic Proteins

Identification of osteogenic proteins in mammalian bone matrix responsible for this cascade of cartilage and bone differentiation has been a difficult task due to the low abundance of osteogenin and related bone morphogenetic proteins (BMPs) tightly bound to the bone extracellular matrix. Progress in recent years in the study of BMPs has been aided by four important technical developments, namely: (1) the development of a functional bioassay in a subcutaneous site in the rat to monitor the specific biology of osteogenic activity (Sampath and Reddi, 1981, 1983); (2) the development of a specific purification procedure involving heparin affinity chromatography (Sampath et al., 1987); (3) the use of electroendosmotic elution techniques after preparative sodium dodecyl sulfate gel electrophoresis to achieve final purification homogeneity (Luyten et al., 1989); and (4) the use of recombinant DNA methodologies for the cloning and expression of several members of the BMP family (Wozney et al., 1988; Celeste et al., 1990; Ozkaynak et al., 1990; Hammonds et al., 1991).

A breakthrough in the isolation of osteogenic activity has been the dissociative extraction of demineralized bone matrix by 4-M guanidinium hydrochloride, or 8-M urea (Sampath and Reddi, 1981). Neither the soluble extract nor the residual bone powder alone had detectable activity when implanted subcutaneously in the rat. However, if the extract was reconstituted with the inactive residue, the biological activity of bone and cartilage formation was recovered. This reconstitution assay has permitted the further purification of osteogenic components. Additional experiments revealed that irrespective of the species, the lower molecular weight fractions of a 4-M guanidinium hydrochloride extract following molecular sieve chromatography induced new cartilage and bone formation in rats (Sampath and Reddi, 1983). This discovery set the stage for the large scale purification of bone- and cartilage-inducing proteins.

A group of related bone morphogenetic factors has been discovered that include osteogenin, bone morphogenetic protein-2A (BMP-2A), BMP-2B (now BMP-4), BMP-3, BMPs 5-7, and osteogenic protein-l (OP-1) (Wozney et al., 1988; Luyten et al., 1989; Celeste et al., 1990; Ozkaynak et al., 1990; Wang et al., 1990). Osteogenin was purified to homogeneity from bovine bone matrix, and the sequences of several tryptic peptides were determined (Luyten et al., 1989). The sequences were similar to portions of the amino acid sequence deduced from the cDNA clone of bone morphogenetic protein-3 (BMP-3) (Wozney et al., 1988). Osteogenin shows 49 and 48% sequence identity in its carboxyl-terminal quarter to BMP-2A and BMP-2B, respectively (Wozney et al., 1988). BMP-7 was found to be identical to OP-l (Wang et al., 1990). Based on structural and amino acid sequence homologies, these osteogenic proteins are grouped among members of the transforming growth factor beta (TGF-β) superfamily, all of which share a striking homology in the C-terminal domain. As in the case with TGF-β, members of TGF-β superfamily are synthesized as a large precursor, four times greater than the mature form of the molecule. The mature proteins are processed from the precursor at the C-terminal as disulfide-linked homodimers containing a region that includes the highly conserved seven cystein residues (Massague, 1990).

Osteogenin and related BMPs initiate cartilage and bone formation *in vivo*. BMP-3 is most abundant in the highly purified preparations, but other members, especially BMP-2 and BMP-7, are also present in the bioactive highly purified bone-derived fractions. BMPs 2-5 and BMP-7 have been reported to induce cartilage and bone as single recombinant proteins in conjunction with residue (Wozney et al., 1988; Ozkaynak et al., 1990; Wang et al., 1990; Hammonds et al., 1991). The availability of all the recombinant proteins will allow researchers to address the question of particular functions of different members in endochondral bone formation more accurately. The studies of the mechanism of action of the bone morphogenetic proteins will add considerable new information on the molecular signals controlling cartilage and bone formation, as well as cartilage and bone regeneration.

C. *In Vitro* Activities of Cartilage and Bone Differentiation Factors

Recently, experimental studies have been performed to define the role of bone morphogenetic proteins in development, interaction with extracellular matrix, cellular binding and localization, as well as *in vitro* and *in vivo* activities. There is accumulating evidence that the TGF-β superfamily members are important regulators of many morphogenetic events during early vertebrate embryogenesis. Using iodinated osteogenin in sections of developing rat embryo, maximal binding of the protein was observed in mesodermal tissues such as cartilage, bone, perichondrium, and periosteum (Vukicevic et al., 1990a) (Figure 1). Applying *in situ* hybridization techniques, investigators have localized BMP-2, BMP-4, and Vgr-1 (BMP-6)

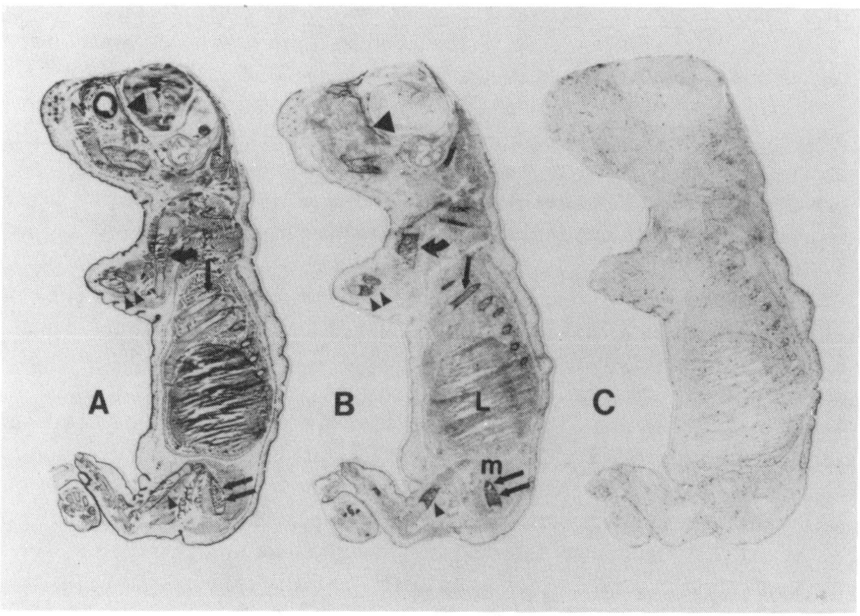

Figure 1. Autoradiographic images of osteogenin (BMP-3) binding to longitudinal sections of an 18-day embryo. Magnification, 9×. (**A**) Section stained with hematoxylin and eosin. Vascular invasion into the cartilaginous centers of the long bones is seen. Arrows and arrowheads correspond to the osteogenin binding sites of section B. (**B**) Total binding after incubation of an adjacent section with 2.5-ng/ml ^{125}I-osteogenin. Note the intensive binding to areas of endochondral ossification: ribs (arrow), femur (double arrow), tibia (small arrowheads), radius and ulna (double small arrowheads), humerus (thick arrow), as well as areas on intramembranous ossification; calvarial bones (big arrowheads). L stands for liver. (**C**) Nonspecific binding by incubation of an adjacent section with 2.5-ng/ml ^{125}I-osteogenin and 260-ng/ml unlabeled osteogenin. Reprinted with permission: Vukicevic et al., Dev. Biol. 140:209, 1990.

transcripts in several specialized organ systems during murine embryonic development. They include: the early development of the limb; myogenic layer of the atrioventricular cushions of the developing heart; developing hair and whisker follicles; tooth buds and palate; developing central nervous system; and craniofacial tissues (Lyons et al., 1989, 1990).

Recent OP-1 (BMP-7) localization data during human development suggested that OP-1 is present during critical morphogenetic stages in early human embryonic development, particularly in tissues of mesodermal origin, as well as in basement membranes separating epithelium from the underlying stroma (Vukicevic et al., in preparation). These data, together with the specific affinity of osteogenin and BMP-2B for the type IV collagen of the basement membrane matrix (Paralkar et al., 1990; see later), underscores the potential critical role of the extracellular matrix in sequestering differentiation factors, and in regulating their bioavailability. Moreover, it also sheds light on the possible role of basement membrane components of blood vessel walls (as laminin and type IV collagen) in both bone formation and regeneration (Vukicevic et al., 1990c; Ripamonti and Reddi, 1992).

In vitro studies allowed us to explore the effects of osteogenin and related bone morphogenetic proteins on both cartilage and bone cells. Osteogenin stimulated alkaline phosphatase activity and collagen synthesis in rat periosteal cells and calvarial osteoblasts (Vukicevic et al., 1989). There was also an increase in the formation of alkaline phosphatase-positive colonies in rat bone marrow stromal cell cultures (Vukicevic et al., 1989). The promotion of the osteogenic phenotype was confirmed in experiments with MC3T3-E1 osteoblastic cells; osteogenin inhibited growth and stimulated alkaline phosphatase activity within 72 hours (Vukicevic et al., 1990b), while usually about 12 days are needed for their spontaneous differentiation *in vitro* (Kodama et al., 1981).

In the same cells, high-affinity receptors for BMP-2B have been identified and partially characterized (Paralkar et al., 1991) (Figure 2, Table 1). Several studies using either BMP-2, 3, or 4 have confirmed and extended these original data on BMPs promoting the osteoblastic phenotype (Chen et al., 1991a; Hiraki et al., 1991; Takuwa et al., 1991; Yamaguchi et al., 1991). A profound stimulation of proteoglycan synthesis in fetal rat chondroblasts and rabbit articular chondrocytes (Vukicevic et al., 1989), together with increased cartilage matrix synthesis in chick limb bud cell cultures (Carrington et al., 1991; Chen et al., 1991b), demonstrated the role of osteogenin in the promotion of the chondrogenic phenotype. Osteogenin and BMP-2B (BMP-4) were equipotent in the maintenance of proteoglycan metabolism in articular cartilage explant cultures of young, adolescent and adult tissues by increasing proteoglycan synthesis and decreasing proteoglycan catabolism (Luyten et al., 1992). Analysis of the size of the newly synthesized proteoglycans, the glycosaminoglycan chain size, and the glycosaminoglycan type of explants treated with osteogenin or BMP-4 indicated that they were very comparable to each other. Thus, both osteogenin and BMP-4 alone are capable of stimulating and maintaining the chondrocyte phenotype *in vitro* (Luyten et al., 1992). Taken together, these *in*

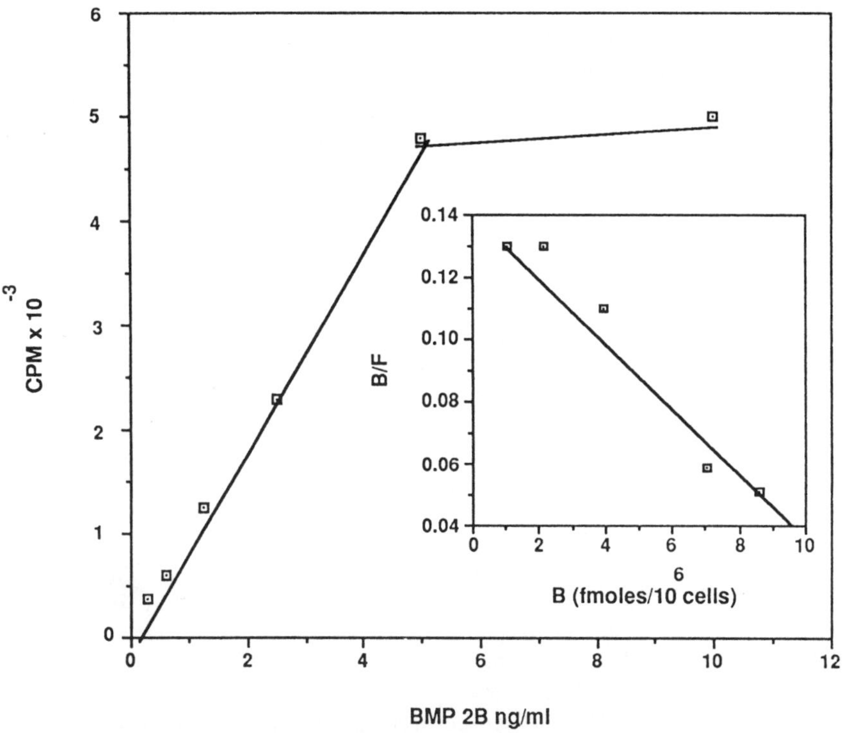

Figure 2. Binding of radiolabeled BMP 2B to MC3T3-E1 cells. Increasing concentrations of radiolabeled BMP-2B were incubated with MC3T3-E1 cells at room temperature for 1 hr and the cell-associated radioactivity was determined. Points represent specifically bound radioactivity determined by subtracting nonspecific binding (10–40% of total binding) from total radioactivity bound. (*Inset*) Scatchard analysis of the binding data. B,bound; F, free. Reprinted with permission: Paralkar et al., Proc. Natl. Acad. Sci. USA 88:3397, 1991.

vitro data suggest that osteogenin and related bone morphogenetic proteins do not only play a role in the initiation of cartilage and bone formation, but clearly promote the expression and maintenance of the chondrogenic and osteogenic phenotype.

D. *In Vivo* Role of Cartilage and Bone Differentiation Factors

The availability of bone morphogenetic proteins made it possible to extend the knowledge of their *in vivo* morphogenetic potential in rodents to primates, as a prerequisite for the exploration of potential therapeutic applications for the regeneration of bone in man. The healing potential of osteogenin on calvarial defects in a series of adult primates of genus *Papio* (baboon) has been studied. A critical size

Table 1. Competition With Growth Factors For the Binding of Iodinated BMP-2B to MC3T3-E1 and NIH 3T3 Cells

Growth Factors	*CPM Bound*	
	MC3T3-E1	*NIH 3T3*
^{125}I-BMP 2B	9,600 ± 800	10,480 ± 680
BMP 2B	4,000 ± 230	2,300 ± 265
PDGF	11,720 ± 1000	10,050 ± 202
FGF	8,210 ± 217	9,510 ± 230
IGF-1	8,930 ± 110	9,670 ± 275
EGF	9,806 ± 600	9,070 ± 660
TGF β1	9,260 ± 460	9,570 ± 330

Note: Each factor was in a 100-fold excess of the iodinated BMP-2B. The values are mean ± standard deviation of four observations. Reprinted with permission: Paralkar et al. Proc. Natl. Acad. Sci. U.S.A. 88, 3397, 1991.

defect (CSD)-dependent nonunion of the baboon calvaria model—that is, a defect that does not repair spontaneously with bone, requiring a graft of viable bone, or alternative substitutes to heal—has been established (Ripamonti et al., 1991). In 48 adult male baboons, calvarial defects were implanted with a graft of autogenous bone harvested from the iliac crest, and with different osteoconductive and osteoinductive substrata. Before calvarial implantation, osteogenin, isolated from both baboon and bovine bone matrix (and with biological activity in rats), was tested for biological activity in the rectus abdominis of an additional 16 baboons (Ripamonti and Reddi, 1992). The extraskeletal implantation permitted the unequivocal histological investigation of bone formation by induction, avoiding possible ambiguities of the orthotopic site. Reconstitution of calvarial defects with baboon osteogenin induced huge amounts of new bone as early as 30 days. At 3 months, bone formation was extensive, culminating in complete regeneration of the craniotomy defects (Ripamonti et al., 1992).

Osteogenin was recently tested for its regenerating effects in human periodontal intrabony defects (Bowers et al., 1991). Combined with demineralized freeze-dried bone allografts, osteogenin significantly enhanced regeneration of a new attachment apparatus and component tissues in a submerged environments (Bowers et al., 1991). Using transformation of rat muscle flaps, it was possible to generate *in vivo*, autogenous, well-perfused bones in the shapes of femoral heads and mandibles (Khouri et al., 1991).

Significant progress has been made in the characterization of the cartilage and bone differentiating proteins. A family of unique proteins has been described, and there is ample evidence that they are directly responsible for *de novo* cartilage and bone formation *in vivo*. Extensive research is underway to develop appropriate and optimal delivery systems based on extracellular matrix components. It is likely that bone morphogenetic proteins will play a crucial role in bone regeneration and repair.

III. INTERACTION OF GROWTH FACTORS WITH EXTRACELLULAR MATRIX

Renewed attention has recently been focused on extracellular matrix macromolecules with the discovery of the interactions of these macromolecules with growth and differentiation factors. The interaction of heparan sulfate proteoglycans with fibroblast growth factors (FGFs) has been known for some time (for review see Burgess and Maciag, 1989). Recently, it has been demonstrated that basic fibroblast growth factor (bFGF) binds to heparan sulfate prior to the binding to its receptor (Yayon et al., 1991).

Many other growth and differentiation factors are known to bind heparin and these include granulocyte-macrophage colony-stimulating factor, neurite-promoting factor, pleitropin, and platelet factor 4 (Roberts et al., 1988; Li et al., 1990; Merenmies and Rauvala, 1990; Ruoslahti and Yamaguchi, 1991). Transforming growth factor beta (TGF-β) has been shown to interact with the protein core of proteoglycan, named betaglycan, which is present on the cell surface of many cell types (Andres et al., 1989). TGF-β has also been shown to bind to extracellular matrix proteoglycan, decorin (Yamaguchi et al., 1990).

During studies on cartilage and bone development, we have shown that osteogenin/BMP-3, isolated from bovine bone, binds to collagen type IV which is a major component of the basement membrane (Paralkar et al., 1990). Using radioactive osteogenin, we examined the affinity of osteogenin for various components of extracellular matrix. Osteogenin binds with highest affinity to collagen type IV, but also binds to collagen types I and IX, and to heparin (Table 2). Osteogenin binds to both chains of collagen type IV, and this binding is pH- and time-dependent. The binding is also reversible with an apparent Kd of 4×10^{-11} M. Since osteogenin is also a member of TGF-β supergene family, we also tested the ability of TGF-β to bind extracellular matrix macromolecules. TGF-β binds to collagen type IV with high affinity (Paralkar et al., 1991; Tables 3 and 4). As both osteogenin and TGF-β bind collagen type IV, we tested some of the other growth factors for their ability to bind collagen type IV. As can be seen from Table 3, basic FGF and platelet-derived growth factor (PDGF) also bind collagen type IV. The ability of aFGF to bind collagen type IV has also been shown (Thompson et al., 1988). Thus, FGFs, TGF-β, and osteogenin all bind to proteoglycans as well as to collagen type IV of the extracellular matrix.

Table 2. Binding of Osteogenin (Bone Morphogenetic Protein-3) to Various Extracellular Matrix Macromolecules Immobilized on Nitrocellulose

Macromolecules	*cpm/µg Macromolecule*[a]
Heparin	910 ± 70
Collagen type I	990 ± 55
Collagen type II	ND[b]
Collagen type III	ND
Collagen type IV	4540 ± 100
Collagen type V	ND
Collagen type IX	770 ± 20
Collagen type X	ND
Laminin	ND
Fibronectin	ND
Chondroitin sulfate	ND
Keratin sulfate	ND
Dermatan sulfate	ND
Heparan sulfate	ND
Hyaluronic acid	ND

Notes: [a]Each macromolecule (2 µg/spot) was spotted in triplicate and the experiment was repeated three times. Each point represents mean ± S.D. of triplicate experiments with nine observations.

[b]ND, not detectable.

Reprinted with permission: Paralkar et al., J. Biol. Chem. 265, 17281, 1990

Table 3. Specificity of Binding of Various Growth Factors to Type IV Collagen

Growth Factors	*Growth factor binding to Type IV Collagen m mole of Growth Factor/Mole of Collagen*
Transforming growth factor β_1	1
b Fibroblast growth factor	0.07
Platelet-derived growth factor	0.09
Epidermal growth factor	ND*
Insulin-like growth factor-1	ND

Notes: *ND = not detectable. Values shown are the mean of triplicate determinations. The various radioiodinated growth factors were incubated with type IV collagen in the solid state on nitrocellulose paper as described in Materials and Methods. The specific activities were : TGF β_1 (80 µCi/µg); bFGF (52 µCi/µg); Platelet-derived growth factor (70 µCi/µg); Epidermal growth factor (174 µCi/µg); and Insulin-like growth factor-1 (188 µCi/µg).

Reprinted with permission: Paralkar et al., Dev. Biol. 143, 303, 1991.

Table 4. Binding of ^{125}I-TGF-β_1 to type IV collagen in solution. Effect of TGF-β_2, fibronectin and laminin

Addition	*CPM*
Experiment 1	
1 ng ^{125}I-TGF-β_1	6575 ± 1526
1 ng ^{125}I-TGF-β_1 + 200 ng TGF-β_2	29 ± 4
Experiment 2	
Collagen alone	26160 ± 2760
Collagen + fibronectin	30560 ± 510
Collagen + laminin	32890 ± 7260

Notes: Experiment 1. Collagen type IV (0.5 μg) was incubated with 1 ng radiolabeled TGF-β_1 (specific activity 100 μCi/μg) in the presence or absence of TGF-β_2 (200 ng) in a final volume of 500 μl of PBS plus 0.05% Tween 20 (TPBS) for 12 hr at 4°C. Type IV collagen precipitates under these conditions. After centrifugation at 12,000 × g for 15 min the precipitated collagen pellet was washed 2 times with TPBS. The radioactivity incorporated into the pellet was determined in a gamma counter. Each point represents mean ± S.D. of triplicate observations. *Experiment 2.* ^{125}I-TGF-β_1 (100 μCi/μg) was incubated with 5 μg of type IV collagen in the presence or absence of fibronectin and laminin (5 μg each) and the amount of radioactivity bound to collagen was determined by a gamma counter. Each point represents mean ± S.D. of triplicate observations. Reprinted with permission: Paralkar et al., Dev. Biol. 143, 303, 1991.

These observations could have far reaching implications for the regulatory role of growth and differentiation factors in the solid state. It is plausible that the extracellular matrix might protect these factors from proteolytic degradation, or alter the activity of these factors by their interactions. The latter was seen to be the case in our studies (Paralkar et al., 1991). In this work, MC3T3-El cells were grown in 24 well-tissue culture plates precoated with laminin, collagen I, and collagen type IV, or on plastic without any precoating. TGF beta inhibits the proliferation of these cells grown on plastic. When TGF beta type 1 was added to these cells grown on various matrix macromolecules, a transient inhibition of cell proliferation (40 to 50%) for up to 18 hours was seen in cells grown on all substrata. These cells were seen to recover from the inhibitory effects of TGF-β after 36 hours. However, in the case of cells grown on collagen type IV, the inhibitory effects on proliferation were sustained for up to 54 hours (Figure 3). Thus, *in vitro* collagen type IV seems to be able to modulate the activity of TGF beta 1. Similarly, heparin and heparan sulfate have been shown to be able to alter the activity of FGF's *in vitro* (Damon et al., 1986). Heparin has also been shown to stimulate the EGF receptor mediated phosphorylation of tyrosine and threonine residues (Gupta et al., 1991). Also

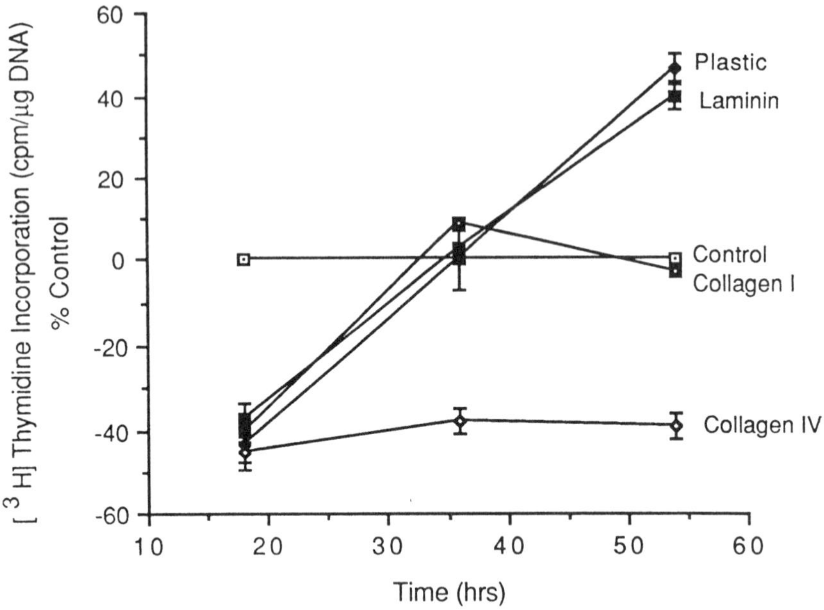

Figure 3. Sustained inhibition of [^{3}H]thymidine incorporation into acid-precitable DNA in osteoblastic cells by TGF-β_1 in the presence of type IV collagen. The inhibition was transient (only at 18 hr) when cultured on plastic, collagen type I, and laminin. Osteoblastic cells (MC3T3-E1) were plated at 10,000 cells per well in 24-well plates precoated with the indicated matrix macromolecules (50 µg/well), and half of the medium was exchanged daily. Cells were incubated in the presence or absence of TGF-β_1 (50 pg/ml) added at time 0. [^{3}H]Thymidine (5 µCi/ml) was added at the indicated time points in medium containing 0.5% fetal bovine serum for the last 4 hr of culture. Incorporation of [^{3}H]thymidine into DNA was determined as described. The results are presented as a percentage of control for each matrix substratum at 18, 36, and 54 hr. A control curve for each experimental substratum (presence of collagen type IV, laminin, collagen type I, and plastic) was determined in the absence of TGF-β. Each point is the mean ±SD of eight observations. Reprinted with permission: Paralkar et al., Dev. Biol. 143:303, 1991.

extracellular matrix-associated macromolecules act synergistically with ciliary neurotrophic factor to induce type two astrocyte development (Lillien et al., 1990).

The interactions of growth factors with extracellular matrix macromolecules could result in tight physiological regulation. These interactions could localize a potent bioactive molecule to the appropriate site. Indeed, the *Wnt* genes present in vertebrates code for a protein that acts over short distances because the protein is bound tightly to the extracellular matrix (Tabin, 1991). It is also possible that, depending upon the developmental stage of the embryo, the growth factors could be localized by the extracellular matrix (thus providing a storage of active molecules at the appropriate site), or they may be in the soluble form, thus being able to act at sites distant from the site of synthesis. With the discovery that differentiation inhibiting activity (DIA) is synthesized both as a diffusible protein as well as one that is matrix bound, this hypothesis seems more plausible (Rathjen et al., 1990). The diffusible and matrix bound forms of DIA seem to be transcribed from different promoters, thus raising the possibility that the two proteins could be differentially regulated depending upon the need for DIA at a local or distant site during embryonic development. It is also possible that soluble factors can have differential activity on the cells, depending on the composition of the extracellular matrix surrounding the cells.

A typical example of the *in vivo* role of extracellular matrix is endochondral bone induction by demineralized bone matrix (Urist, 1965; Reddi and Huggins, 1972; Luyten et al., 1989). Bone induction is due to the combined action of osteogenin/BMPs and the collagenous matrix (Reddi, 1981). It is known that vascular invasion is a prerequisite for new bone formation (Foidart and Reddi, 1980; Reddi and Kuettner, 1981). It is therefore possible that basement membrane, especially collagen type IV and heparan sulfate proteoglycan around the endothelial cells of the invading capillaries, may function as an affinity matrix to present soluble growth and differentiation factors in the solid state, and to responding mesenchymal cells and osteoprogenitors to initiate osteogenesis. We also propose that collagen type IV functions as a physiological delivery system by sequestering factors involved in embryonic induction, morphogenesis, and epithelial-mesenchymal interactions. These interactions may help explain the general role of basement membrane in embryonic development. The interaction of bone-inductive proteins with collagen type IV also helps to link demineralized bone matrix-induced bone formation to epithelium-induced bone formation.

IV. DIFFERENTIATION OF OSTEOBLAST-LIKE CELLS ON RECONSTITUTED BASEMENT MEMBRANE: ANGIOGENESIS AND OSTEOGENESIS

Vascular invasion is a prerequisite for bone formation (Trueta, 1963; Foidart and Reddi, 1980). Since developing osteoprogenitor cells and osteoblasts are in contact with the basement membrane of the invading vascular system, it is possible that the endothelial cell matrix might function during critical phases of capillary invasion in endochondral bone formation (Rhodin, 1974), growth (Scott and Pease, 1956; Kimmel and Jee, 1980), remodelling (Jaworski, 1976), and fracture repair (Simmons, 1980). In the matrix-dependent bone induction model, angiogenesis is correlated with chondrolysis and concomitant osteogenesis. Basement membrane components including laminin, collagen IV, and factor VIII antigen are localized around invading capillaries prior to and during osteogenesis (Foidart and Reddi, 1980). These observations, together with data on TGF-β and osteogenin binding to type IV collagen (Paralkar et al., 1990, 1991), prompted us to examine whether basement membrane extracellular matrix components might influence bone cell differentiation *in vitro*.

A. Osteoblastic Cells Recognize Basement Membrane Components

Primary calvarial cells and an osteoblastic cell line (MC3T3-E1) recognized components of basement membrane and underwent profound morphological changes when cultured on the surface of basement membrane. Colonies and canalicular cell processes that required both collagen IV and laminin were observed (Vukicevic et al., 1990c). Antibodies to laminin and to a laminin receptor, as well as certain biologically active laminin-derived synthetic peptides, blocked the cellular differentiation on the basement membrane (Matrigel). Direct interaction of bone cells with the laminin-derived peptides was demonstrated in cell binding assays (Table 5). Type IV collagen also appeared to be an active component in the Matrigel, since its removal resulted in loss of bone cell organization (Vukicevic et al., 1990c). In addition, ascorbic acid, which promotes collagen synthesis, together with β-glycerophosphate, increased calcified bone nodule formation on Matrigel. These data demonstrate that osteoblastic cells interact with basement membrane components, and suggest that such a matrix may modulate the phenotype of bone cells. Moreover, basement membrane plays a significant role *in vitro* in formation of a network of cytoplasmic processes resembling the osteocyte network in bone.

Table 5. Attachment of Osteoblastic Cells to Laminin-Derived Peptides

Substrate	*Residue (chain)*	*Number of Wells*	*Cells per Field MC3T3-E1 Cells*
Laminin		12	32 ± 4
CDPGYIGSR	925-933 (B1)	16	35 ± 5
Ac-YIGER-*		10	0
FALRGDNP	1120-1127 (A)	12	23 ± 4
FALRDGNP*		8	0
CSRARKQAASIKVAVSADR	2091-2108 (A)	16	33 ± 3
CRKQAASIKVAVA	2094 ± 2105 (A)	8	29 ± 3
CIKVAVA	2099-2105 (A)	16	28 ± 4
KKSAVQARIVAS*		8	0

Notes: Peptides were synthesized to correspond to mouse laminin A and B1 chain cDNA sequences. The NH_2 terminal cysteine in the peptides is not present i the laminin chains but was included in the synthesis to facilitate coupling, and an amide group was present in all peptides at the carboxyl terminus. The peptides were applied to nitrocellulose at 1 mg/ml. Laminin (100 μg/ml) served as a positive control. Values given are means ± S.E.M. Asterisks indicate peptides at the carboxyl terminus. The peptides were applied to nitrocellulose at 1 mg/ml. Laminin (100 μg/ml) served as a positive control. Values given are means ± S.E.M. Asterisks indicate peptides with scrambled sequence used as control. Reprinted with permission: Vukicevic et al., Cell 63, 437, 1990.

Osteocytes do not directly contact basement membrane *in vivo*. The osteocyte is a developmental stage of the osteoprogenitor osteoblast lineage, and thus may retain a "memory" of the initial contact of osteoblast with laminin. Most osteons demonstrate a central blood vessel with the basement membrane in contact with the first circle of concentric osteoblasts. This initial contact may set into motion a ripple-like cascade of cell differentiation.

B. Growth Factors in Basement Membrane Regulate Osteoblastic Network Formation

Additional studies have demonstrated the presence of TGF-β, epidermal growth factor (EGF), insulin-like growth factor 1 (IGF-1), basic fibroblast growth factor (bFGF), and platelet-derived growth factor (PDGF) in the basement membrane, Matrigel (Vukicevic et al., 1992). These factors significantly contributed to the behavior of osteoblastic cells (MC3T3-E1) when cultured on Matrigel. Exogenous TGF-β blocked the migration and cellular network formation. Conversely, addition of TGF-β neutralizing antibodies to Matrigel stimulated the cellular network formation. bFGF, EGF, and PDGF all promoted cellular migration and organization on Matrigel. Addition of bFGF to MC3T3-E1 cells grown on Matrigel overcame the inhibitory effect of TGF-β. TGF-β cannot be removed from Matrigel by repeated 20% ammonium sulfate precipitations. Some TGF-β also remained bound to type IV collagen purified from reconstituted basement membrane, suggesting caution in the interpretation of experiments on cellular activity related to Matrigel, collagen type IV, and possibly other extracellular matrix components (Vukicevic et al., 1992). These data confirmed the above mentioned results of Paralkar et al. (1991) of TGF-β binding to type IV collagen of the basement membrane, suggesting that TGF-β may be a basement membrane component *in vivo*. All of these investigations provide a conceptual framework to explain the physiological role of blood vessels in osteogenesis, and moreover, provide evidence for the role of basement membrane macromolecules in the phenotypic modulation of osteoblastic cells.

V. CHALLENGES FOR THE FUTURE

What is the precise cellular and molecular basis of the interaction of endothelial elements and the osteoblastic lineage? Is there a causal role for basement membrane components and osteoprogenitor differentiation? What are the molecular domains in laminin which signal the osteoblastic lineage? What are the signal transducing mechanisms? These are exciting challenges for the future and provide opportunities for molecular insights into morphogenesis of bone. A better understanding of these molecular interactions will pave the path to solving baffling clinical problems, such as avascular necrosis of the femoral head, nonunion of fractures, and optimal biological fixation of the implant interface in the prostheses.

REFERENCES

Andres, J.I., Stanley, K., Cheifetz, S., & Massague, J. (1983). Membrane-anchored and soluble forms of betaglycan, a polymorphic proteoglycan that binds transforming growth factor-β. J. Cell Biol. 109, 3137–3145.

Bowers, G., Felton F., Middleton, C., Glynn, D., Sharp, S., Mellonig J., Corio, R., Emerson, J., Park, S., Suzuki, J., Ma, S., Romberg, E., & Reddi, A.H. (1991). Histologic comparison of regeneration in human intrabony defects when osteogenin is combined with demineralized freeze-dried bone allograft and with purified bovine collagen. J. Periodontol. 62, 690–702.

Burges, W.H. & Maciag, T. (1989). The heparin-binding (fibroblast) growth factor family of proteins. Annu. Rev. Biochem. 58, 575–606.

Carrington, J.L., Chen, P., Yanagishita, M., & Reddi, A.H. (1991). Osteogenin (bone morphogenetic protein-3) stimulates cartilage formation by chick limb bud cells *in vitro*. Dev. Biol. 146, 406–415.

Celeste, A.J., Iannazzi, J.A., Taylor, R.C., Hewock, R.M., Rosen, V., Wang, E.A., & Wozney, J.M. (1990). Identification of transforming growth factor beta family members present in bone-inductive protein purified from bovine bone. Proc. Natl. Acad. Sci. USA 87, 9843–9847.

Chen, P., Carrington, J.L., Hammonds, R.G. Jr., & Reddi, A.H. (1991). Stimulation of chondrogenesis in limb bud mesoderm cells by recombinant BMP-2B and modulation by TGF-β1 and TGF-β2. Exp. Cell Res. 195, 509–515.

Chen, T.L., Bates, R.L., Dudley, A., Hammonds, R.G. Jr., & Amento, E.P. (1991). Bone morphogenetic protein-2b stimulation of growth and osteogenic phenotypes in rat osteoblast-like cells: comparison with TGF-β1. J. Bone Miner. Res. 6, 1387–1393.

Damon, D.H., D'Amore, P.A., & Wagner, J.A. (1986). Sulfated glycosaminoglycans modify growth factor-induced neurite outgrowth in PC12 cells. J. Cell Physiol. 135, 293–300.

Foidart, J.M. & Reddi, A.H. (1980). Immunofluorescent localization of type IV collagen and laminin during endochondral bone differentiation and regulation by pituitary growth hormone. Dev. Biol. 75, 130–136.

Gupta, S.R., Ghany, M.A., Koland, J., & Racker, E. (1991). Heparin stimulates epidermal growth factor receptor-mediated phosphorylation of tyrosine and threonine residues. Proc. Natl. Acad. Sci, USA 88, 5954–5958.

Hammonds, R.G. Jr., Schwall, R., Dudley, A., Berkemeier, L., Lai, C., Lee, J., Cunningham, N.S., Reddi, A.H., Wood, I.W., & Mason, A.J. (1990). Bone-inducing activity of mature BMP–2B produced from a hybrid BMP-2A/2B precursor. Mol. Endocrinol. 4, 149–155.

Hiraki, Y., Inoue, H., Shigeno, C., Sanma, Y., Bentz, H., Rosen, D.M., Asada, A., & Suzuki, F. (1991). Bone morphogenetic proteins (BMP-2 and BMP-3) promote growth and expression of the differentiated phenotype of rabbit chondrocytes and osteoblastic MC3T3-El cells *in vitro*. J. Bone Miner. Res. 6, 1373–1385.

Jaworski, Z.F.G. (1976). Proceedings of the First Workshop on Bone Histomorphometry, pp. 1–395. Ottawa Press, Ottawa.

Khouri, R.K., Koudsi, B., & Reddi, A.H. (1991). Tissue transformation into bone *in vivo*. JAMA 266, 1953–1955.

Kimmel, D.B. & Jee, W.S.S. (1980). Bone cell kinetics during longitudinal bone growth in the rat. Calcif. Tissue Int. 32, 123–133.

Kodama, H., Amagai, Y., Dudo, H., Kasai, S., & Yamamoto, S. (1981). Establishment of a clonal osteogenic cell line from newborn mouse calvaria. Jpn. J. Oral Biol. 23, 899–901.

Li, Y.S., Milner, P.G., Chauhan, A.K., Watson, M.A., Hoffman, R.M., Kodner, C.M., Molbrandt, J., & Deuel, T.F. (1990). Cloning and expression of a developmentally regulated protein that induces mitogenic and neurite outgrowth activity. Science 250, 1690–1694.

Lillien, J.E., Sendtner, M., & Raff, M.C. (1990). Extracellular matrix-associated molecules collaborate with ciliary neurotrophic factor to induce type-2 astrocyte development. J. Cell Biology. 111, 635–644.

Luyten, F.P., Cunningham, N.S., Ma, S., Muthukumaran, R.G., Hammonds, R.G. Jr., Nevins, W.B., & Reddi, A.H. (1989). Purification and partial amino acid sequence of osteogenin, a protein initiating bone differentiation. J. Biol. Chem. 264, 13777–13380.

Luyten, F.P., Yu, Y.M., Yanagishita, M., Vukicevic, S., Hammonds, R.G. Jr., & Reddi, A.H. (1992). Natural bovine osteogenin and recombinant human bone morphogenetic protein-2B are equipotent in the maintenance of proteoglycan in bovine articular cartilage explant cultures. J. Biol. Chem. 267, 3691–3695.

Lyons, K.M., Pelton, R.W., & Hogan, B.L.M. (1989). Patterns of expression of murine Vgr-l and BMP-2a RNA suggest that transforming growth factor β-like genes coordinately regulate aspects of embryonic development. Genes Develop. 3, 1657–1668.

Lyons, K.M., Pelton, R.W., & Hogan, B.L.M. (1990). Organogenesis and pattern formation in the mouse: RNA distribution patterns suggest a role for bone morphogenetic protein-2A (BMP-2A). Development 109, 833–844.

Massague, J. (1990). The transformation growth factor-β family. Annu. Rev. Cell Biol. 6, 597–641.

Merenmies, J. & Rauvala, H. (1990). Molecular cloning of the 18-kDa growth-associated protein of developing brain. J. Biol. Chem 265, 16, 721–16, 724.

Ozkaynak, E., Rueger, D.C., Drier, E.A., Corbett, C., Ridge, R.J., Sampath, T.K., & Oppermann, H. (1990). Op-1 cDNA encodes an osteogenic protein in the TGF-β family. EMBO J. 9, 2085–2093.

Paralkar, V.M., Nandedkar, A.K.N., Pointer, R.H., Kleinman, H.K., & Reddi, A.H. (1990). Interaction of osteogenin, a heparin binding bone morphogenetic protein, with type IV collagen. J. Biol. Chem. 265, 17, 281–17, 284.

Paralkar, V.M., Vukicevic, S., & Reddi, A.H. (1991). Transforming growth factor beta 1 binds to collagen IV of basement membrane: Implications for development. Dev. Biol. 143, 303–308.

Paralkar, V.M., Hammonds, R.G.Jr., & Reddi, A.H. (1991). Identification and characterization of cellular binding proteins (receptors) for recombinant human bone morphogenetic protein-2B, an initiator or bone differentiation cascade. Proc. Natl. Acad. Sci. USA 88, 3397–3401.

Rathjen, P.D., Tuth, S., Willis, A., Health, J.K., & Smith, A.G. (1990). Differentiation inhibiting activity is produced in matrix-associated and diffusible forms that are generated by alternate promoter usage. Cell 62, 1105–1114.

Reddi, A.H. (1981). Cell biology and biochemistry of endochondral bone development. Collagen Rel. Res. 1, 209–226.

Reddi, A.H. & Huggins, C.B. (1972). Biochemical sequences in the transformation of normal fibroblasts in adolescent rats. Proc. Natl. Acad. Sci. USA 69, 1601–1605.

Reddi, A.H. & Kuettner, K.E. (1981). Vascular invasion of cartilage: correlation of morphology with lysozyme glycosaminoglycans, protease and protease inhibitory activity during endochondral bone development. Dev. Biol. 82, 217–223.

Rhodin, J.A.G. (1974). Histology. Oxford University Press, London, pp. 212–220.

Ripamonti, U. (1991). Bone induction in nonhuman primates. An experimental study on the baboon (Papio ursinus). Clin. Orthop. Rel. Res. 269, 284–293.

Ripamonti, U., Ma, S., Cunningham, N.S., Yeates, L., & Reddi, A.H. (1992). Initiation of bone regeneration in adult baboons by osteogenin, a bone morphogenetic protein. Matrix 12, 369–380.

Ripamonti, U. & Reddi, A.H. (1992). Growth and morphogenetic factors in bone induction: Role of osteogenin and related bone morphogenetic proteins in craniofacial and periodontal bone repair. Crit. Rev. Oral Biol. Med. 3, 1–14.

Roberts, R., Gallagher, J., Spooncer, E., Allen, T.D., Bloomfield, F., & Dexter, T.M. (1988). Heparan sulfate bound growth factors: a mechanism for stromal cell mediated hematopoiesis. Nature 332, 376–378.

Ruoslahti, E. & Yamaguchi, Y. (1991). Proteoglycans as modulators of growth factor activities. Cell 64, 867–869.

Sampath, T.K., Muthukumaran, N., & Reddi, A.H. (1987). Isolation of osteogenin, an extracellular matrix-associated bone-inductive protein, by heparin affinity chromatography. Proc. Natl. Acad. Sci. USA 84, 7109–7113.

Sampath, T.K. & Reddi, A.H. (1981). Dissociative extraction and reconstitution of extracellular matrix components involved in local bone differentiation. Proc. Natl. Acad. Sci. USA 78, 7599–7602.

Sampath, T.K. & Reddi, A.H. (1983). Homology of bone-inductive proteins from human, monkey, bovine and rat extracellular matrix. Proc. Natl. Acad. Sci. USA 80, 6591–6595.

Scott, B.L. & Pease, D.C. (1956). Electron microscopy of the epiphyseal apparatus. Anat. Rec. 126, 465–495.

Simmons, D.J. (1980). Fracture healing. In: Fundamental Clinical and Bone Physiology (Urist, M.R., Ed.). Lippincott, Philadelphia.

Tabin, C.J. (1991). Retinoids, homeoboxes, and growth factors: towards molecular models for limb development. Cell 66, 199–217.

Takuwa, Y., Ohse, C., Wang, E.A., Wozney, J.M., & Yamashita, K. (1991). Bone morphogenetic protein-2 stimulates alkaline phosphatase activity and collagen synthesis in cultured osteoblastic cells, MC3T3-El. Biochem. Biophys. Res. Commun. 174, 96–101.

Thompson, J.A., Anderson, K.D., Dipietro, J.D., Zwiebel, J.A., Zametta, M., Anderson, W.F., & Maciag, T. (1988). Site-directed neovessel formation *in vivo*. Science 241, 1349–1352.

Trueta, J. (1963). The role of the vessels in osteogenesis. J. Bone Joint. Surg. 45B, 402–411.

Urist, M.R. (1965). Bone: formation by autoinduction. Science 150, 893–899.

Vukicevic, S., Luyten, F., & Reddi, A.H. (1989). Stimulation of the expression of osteogenic and chondrogenic phenotypes *in vitro* by osteogenin. Proc. Natl. Acad. Sci. USA 86, 8793–8797.

Vukicevic, S., Paralkar, V.M., Cunningham, N.S., Gutkind, J.S., & Reddi, A.H. (1990a). Autoradiographic localization of osteogenin binding sites in cartilage and bone during embryonic development. Dev. Biol. 140, 209–214.

Vukicevic, S., Luyten, F., & Reddi, A.H. (1990b). Osteogenin inhibits proliferation and stimulates differentiation in mouse osteoblast-like cells (MC3T3-E1). Biochem. Biophys. Res. Commun. 166, 750–756.

Vukicevic, S., Luyten, F.P., Kleinman, H.K., & Reddi, A.H. (1990c). Differentiation of canalicular cell processes in bone cells by basement membrane matrix components: Regulation by discrete domains of laminin. Cell 63, 437–445.

Vukicevic, S., Kleinman H.K., Luyten, F.P., Roberts, A.B., Roche, N.S., & Reddi, A.H. (1992). Identification of multiple active growth factors in basement membrane Matrigel suggests caution in interpretation on cellular activity related to extracellular matrix components. Exp. Cell Res. 202, 1–8.

Wang, E.A., Rosen, V., Cordes, P., Hewick, R.M., Kriz, M.J., Luxenburg, D.P., Sibley, B.S., Wozney, J.M. (1990). Recombinant human bone morphogenetic protein induces bone formation. Proc. Natl. Acad. Sci. USA 87, 2220–2224.

Wozney, J.M., Rosen, V., Celeste, A.J., Mitsock, L.M., Whitters, R.W., Kriz, R.W., Hewick, R.M., Wang, E.A. (1988). Novel regulators of bone formation: molecular clones and activities. Science 242, 1528–1534.

Yamaguchi, A., Katagiri, T., Ikeda, T., Wozney, J.M., Rosen, V., Wang, E.A., Kahn, A.J., Suda, T., & Yoshiki, S. (1991). Recombinant human bone morphogenetic protein-2 stimulates osteoblastic maturation and inhibits myogenic differentiation *in vitro*. J. Cell Biol. 113, 681–687.

Yamaguchi, Y., Mann, D.M., & Ruoslahti, E. (1990). Negative regulation of transforming growth factor beta by the proteoglycan decorin. Nature 346, 281–284.

Yayon, A., Klagsbrun, M., Esko, D.E., Leder, P., & Ornitz, D.M. (1991). Cell surface, heparin-like molecules are required for binding of basic fibroblast growth factor to its high affinity receptor. Cell 64, 841–848.

INTEGRINS:

STRUCTURE, FUNCTION, AND BIOLOGICAL PROPERTIES

David A. Cheresh

Advances in Molecular and Cell Biology
Volume 6, pages 225–252.

ISBN: 1-55938-515-4

I. INTRODUCTION

Cellular adhesion mechanisms have been implicated in a wide variety of biological phenomena including thrombosis, wound healing, inflammation, embryogenesis, and cancer. Recently, significant progress has been made toward elucidating the molecular basis of the adhesive interactions involved in these biological processes. In particular, a family of cell adhesion receptors, termed integrins, has been defined that potentiate a wide variety of cell-cell and cell-matrix interactions (Hynes, 1987; Ruoslahti and Pierschbacher, 1987; Phillips et al., 1988; Ruoslahti and Giancotti, 1989; Albelda and Buck, 1990; Hemler, 1990; Springer, 1990; Cheresh, 1991). Integrins not only mediate cell adhesion events but are also involved in transmitting signals from the extracellular environment to the intracellular compartment (Schwartz et al., 1989, 1990; Altieri et al., 1990; Ingber et al., 1990; Miyauchi et al., 1991; Ng-Sikorski, 1991).

Integrins are heterodimeric molecules expressed on the cell surface that require divalent cations for functional activity (Pierschbacher and Ruoslahti, 1987; Phillips et al., 1988; Smith and Cheresh, 1988; Smith and Cheresh, 1991). The integrin family is composed of at least 18 members based on the expression of 12α and 7β subunits (Figure 1). Depending on the α/β subunit pairing, the intact heterodimer can recognize either one or multiple ligands present in the extracellular matrix or on the surface of opposing cells. In some cases, a given β subunit such as β1, can associate with multiple α subunits leading to cell adhesion to ligands such as fibronectin, collagen, and laminin among others (Wayner and Carter, 1987; Hemler, 1990). The β2 integrins are restricted to leukocytes and primarily mediate cell-cell interactions that are important in immune recognition and inflammation (Springer, 1990). The β3 integrins include two members characterized by αIIb expressed on platelets and αv expressed on a variety of cell types (Phillips et al., 1988; Cheresh et al., 1989a). The β3 integrins recognize fibrinogen, fibronectin, von Willebrand factor, and vitronectin among others (Cheresh and Spiro, 1987; Cheresh et al., 1989a; Charo et al., 1990; Smith et al., 1990a). However, the αv subunit has the rather unique ability to associate with a wide variety of β subunits (10–13) (Cheresh et. al., 1989b; Freed et al., 1989; Bodary and McLean, 1990; Krissanssen et al., 1990; Sheppard et al., 1990; Vogel et al., 1990). For example, it is now clear that αv, expressed on most every cell type, can associate with β1 (Bodary and McLean, 1990; Vogel et al., 1990), β3 (Cheresh and Spiro, 1987; Cheresh et al., 1989), and the newly defined β5 (Cheresh et al., 1989b; McLean et al., 1990, Ramaswamy and

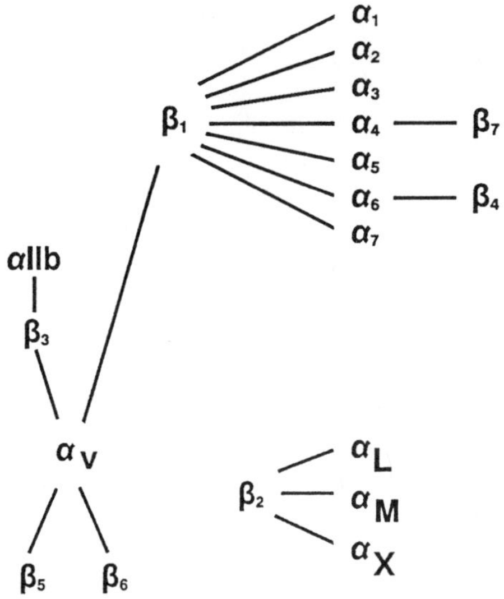

Figure 1. Integrin family of cell adhesion receptors. The integrin family of cell adhesion receptors is comprised of multiple heterodimers of α and β subunits associated in the following manner.

Hemler, 1990; Smith et al., 1990a) and β6 (Sheppard et al., 1990) subunits (see Figure 1). While a given cell may have a specific integrin repertoire, little is known about what regulates the coordinate expression of the individual integrin subunits within the cell. It is clear, however, that the overall adhesive phenotype of a given cell and its capacity to respond to its extracellular environment is primarily dictated by the combination of integrins expressed on its surface.

II. CELLULAR CONSEQUENCES OF INTEGRIN-LIGAND INTERACTION

A. Integrins Interact with the Actin-Cytoskeleton

Once cells make an integrin-mediated contact, the respective integrin typically clusters into focal contacts in the plane of the lipid bilayer leading to a change in cell shape and biological responsiveness. The formation of these adhesion structures depends on the ability of the integrins to associate with the actin-cytoskeleton within the cell (Chen et al., 1985; Damsky et al., 1985; Burridge et al., 1988; Dejanna et al., 1988; Singer et al., 1988; Carter et al., 1990; Wayner et al., 1991) through their interaction with cytoplasmic proteins such as vinculin, talin, and α-actinin (Geiger, 1979; Horwitz et al., 1986; Tapley et al., 1989; Nuckolls et al.,

1990; Otey et al., 1990, Wayner et al., 1991). This, in turn, results in an integrin-actin dependent change in cell shape or migratory behavior. Thus, once cells are attached, they may spread and/or migrate depending on the integrin/ligand interaction involved. It is important to note that cells change shape in response to the occupancy of a particular integrin. For example, on a fibronectin substrate, the cell fibronectin receptor α5β1 is found in focal contacts on the ventral side of the cell, colocalizing with vinculin and talin at the ends of actin filaments, whereas the vitronectin receptor αvβ3 is uniformly expressed over the entire cell membrane. Conversely, on vitronectin, only the vitronectin receptor associates with focal contacts and leads to cell spreading (Singer et al., 1988).

Recently it's been shown that the actual spacing of an immobilized adhesive ligand dictates whether cells will spread and form focal contacts and stress fibers. In fact, it was demonstrated that spacing of an RGD peptide at 440 nm was sufficient for integrin αvβ3-mediated fibroblast spreading, whereas 140-nm spacing of this ligand was required for focal contact and stress fiber formation (Massia and Hubbell, 1991). This study implies that well-defined focal contacts and stress fibers are not necessary for cell spreading. However, it is likely that spreading requires some form of integrin association with the actin cytoskeleton since cytochalasin, a drug which prevents actin polymerization, inhibits αvβ3-dependent cell spreading (Orlando and Cheresh, 1991).

The structural basis of integrin-focal contact formation is not well understood. However, based on mutant integrin expression studies, it appears that the cytoplasmic domain of the β subunit is critically involved in this process (Solowska et al., 1989; Hayashi et al., 1990; Reszka and Horwitz, 1991). Recently, a detailed mutational analysis of the β1 subunit was performed (Reszka and Horwitz, 1991). In this case, mutant avian β1 was transfected into NIH 3T3 cells, and the resulting cell surface heterodimer was monitored for expression in focal contacts. In general, there were three key regions identified based on the ability of missense mutations to reduce focal contact formation. Mutations in two of these three regions produced further reduction in focal contacts, suggesting that these regions may function cooperatively together in directing the localization of focal contacts. Two of these regions contained the NPXY sequence associated with movement of molecules into clathrin-coated pits. A tyrosine and serine known to be phosphorylated *in vivo* when substituted with nonphosphorylated, nonpolar residues failed to reduce focal contacts, indicating that phosphorylation of these sites may not be a prerequisite for focal contact localization of the β1 integrins.

B. Integrins Sometimes Fail to Interact with the Actin-Cytoskeleton

It has generally been assumed that integrins which potentiate cell adhesion eventually promote the assembly of the actin cytoskeleton, and thereby promote a change in cell morphology and migration. However, there have been two notable exceptions to this rule. The first example is that of the integrin α6β4, which was recently shown to be part of an adhesive structure known as the hemidesmosome complex (Stepp et al., 1990; Quaranta and Jones, 1991). This is a specialized adhesion structure present on the basal surface of epithelial cells that promotes a strong adhesive interaction with the basement membrane and underlying connective tissue (Schwarz et al., 1990). The hemidesmosome is thought to be anchored by intermediate filaments within the cell. Association with the intermediate filaments may be due to the expression of the β4 subunit whose cytoplasmic tail, approximately 1100 residues in length (Tamura et al., 1990), is unlike any other integrin β subunit cytoplasmic tail which typically range from 40 to 50 residues. While integrin α6β4 appears to be tissue-specific and localized to epithelial cells, its actual ligand-binding capability remains to be elucidated.

Another example of an integrin that apparently fails to associate with the actin-cytoskeleton is the vitronectin receptor integrin αvβ5. In this case, it was recently reported that although this receptor contributes to attachment to vitronectin, it fails to localize to focal contacts. This was of interest since on the same cell attached to vitronectin, integrin αvβ3 could be found in focal contacts associated with vinculin, talin, and the ends of actin filaments (Wayner et al., 1991). Based on these observations, it was concluded that αvβ3, but not αvβ5, was required for cell spreading on vitronectin. This suggests that integrins αvβ3 and αvβ5 potentiate distinct signals even though both receptors contribute to vitronectin-mediated cell adhesion. Furthermore, it is likely that the distinct distribution of these receptors is the consequence of structural differences in their β subunits since both receptors share a common α subunit. The primary amino acid sequence of β3 (Fitzgerald et al., 1987) and β5 (McLean et al., 1990; Ramaswamy and Hemler, 1990; Suzuki et al., 1990) show 56% identity. However, β5 contains a cytoplasmic tail that is distinct from that of β3. In fact, the cytoplasmic tail of β5 is only 10 to 20% homologous to β3 or to the other integrin β subunits (McLean et al., 1990; Ramaswamy and Hemler, 1990; Suzuki et al., 1990). Moreover, β5 contains a 10-amino acid residue extension to the carboxy-terminus which is not found on β3 or on any of the known integrin β subunits. Since the cytoplasmic tail of integrin β1 subunit has been implicated in focal contact formation (Solowska et al., 1989; Hayashi et al., 1990; Reszka and Horwitz, 1991), it is conceivable that the unique cytoplasmic portion of β5 prevents its association with the actin-cytoskeleton. In any event, the integrins αvβ3 and αvβ5 would likely differ in their ability to promote cellular migration on a vitronectin substrate.

C. Integrins Mediate Cellular Signals

Integrin-dependent cell adhesion regulates not only cell structure and morphology, but also proliferation, differentiation, and gene expression. However, little is known about the molecular basis underlying these events. It has generally been assumed that integrins exert their effect through the cytoskeleton, but recent work has suggested that signal transduction pathways normally associated with hormone receptors may also be involved (Schwartz et al., 1990). Multiple integrins have been shown to regulate the activation of the Na/H antiporter and hence,intracellular pH (Schwartz et al., 1991). They were also reported to promote either an increase or decrease in intracellular calcium, depending on the cellular system being examined (Altieri et al., 1990; Miyuachi et al., 1991). Integrins were found to regulate inositol lipid turnover and phosphorylation of intracellular proteins on tyrosine (Hirst et al., 1986; Tapley et al., 1989; Haimovich et al., 1991). In this regard, it is highly intriguing that several nonreceptor tyrosine kinases (e.g., $pp60^{src}$) and most of the intracellular phosphotyrosine is localized to integrin-containing focal contacts (Hirst et al., 1986).

Integrins are also recipients of intracellular signals. In myeloid and lymphoid cells, stimulation with either cytokines (Richter et al., 1990) or antibodies (Springer, 1990) induces increased avidity of integrins for their ligands, and these changes may be accompanied by alterations in phosphorylation and receptor clustering (Hirst et al., 1986; Tapley et al., 1989; Haimovich et al., 1991). Phosphorylation of integrins on tyrosine by tyrosine kinase oncogenes may account for the anchorage-independent growth properties of transformed cells and the generally reduced adhesion of transformed cells. The βs subunit, which may be related to or identical with β5, associates with αv on some cells and can be phosphorylated on serine; this event can be induced by phorbol esters (Freed et al., 1989). In general, however, both signaling by integrins and the effects of signals on integrins are poorly understood at either the molecular or cellular levels.

The structural basis of these signal transducing mechanism(s) may be related to events associated with integrin occupancy leading to a conformational change in the receptor. Parise et al. (1987) have demonstrated that the platelet receptor αIIbβ3, in the presence of an RGD peptide, undergoes an increased Stokes radius and susceptibility to protease digestion, suggesting a conformational change. Other reports indicate that this receptor changes shape upon ligand binding since novel antigenic epitopes can be specifically recognized on ligand-occupied receptor (Frelinger et al., 1988). Recent evidence suggests that the cytoplasmic tail of the β subunit may not only be involved in signal transduction, but may also potentiate ligand binding. Hibbs et al. (1991) demonstrated that the C-terminal cytoplasmic domain of the β2 subunit play a direct role in LFA-1 binding to its ligand ICAM. Specifically, when the β2 subunit was expressed in CHO cells with a deletion of the last several residues of its C-terminus, the resulting truncated LFA-1 heterodimer showed a significant reduction in its capacity to bind ICAM-1 (Hibbs et al., 1991). These results imply that the ligand-binding portion of the integrin, initially thought to only involve the N-terminal domains of α and β chains, also may depend on a functionally active cytoplasmic C-terminal domain. This concept may explain in part the molecular basis of the signaling mechanism that dictates cellular responsiveness to the extracellular environment.

The clustering of integrins has been proposed to promote an intracellular signal, as indicated for the fibronectin receptor α5β1 (Schwartz et al., 1990). In addition,

Werb and colleagues (1989), have demonstrated gene expression for collagenase and stromelysin depends on the capacity to cross-link and cluster integrin α5β1 receptor on the cell surface. This study demonstrated that antibody to the receptor and fragments of fibronectin could potentiate this biological effect. However, it is not clear whether the type of signal promoted by either antibody or ligand is the same. It is now apparent that signal mechanisms through many types of receptors depend on the clustering of receptors. Whether this is the cause or the effect of a signal remains to be determined.

III. THE MOLECULAR BASIS OF INTEGRIN-LIGAND INTERACTION

A. Integrins Recognize Specific Sites within Adhesive Proteins

A number of studies have indicated that integrin-ligand interactions depend on recognition of distinct sequences within the ligand. For example, the most widely studied sequence involves Arg-Gly-Asp which is present in a wide variety of extracellular adhesive ligands including fibronectin, vitronectin, fibrinogen, collagen, laminin, von Willebrand factor, thrombospondin, and osteopontin. Some integrins recognize one or more of these ligands in the context of the Arg-Gly-Asp sequence. For example, integrin αvβ3 has been shown to interact with each of these protein ligands (Cheresh and Spiro, 1987; Lawler et al., 1988; Charo et al., 1990; Kramer et al., 1990; Miyauchi et al., 1991), while other integrins are clearly more restricted in their ligand binding specificity. Integrin α5β1 (Pytela et al., 1985) exclusively binds fibronectin, while integrin αvβ5 recognizes vitronectin and αvβ1 binds fibronectin (Vogel et al., 1990) or vitronectin (Bodary and McLean, 1990); yet each of these receptors can be inhibited with the same Arg-Gly-Asp-containing peptide. Thus, it is likely that alternative regions within the ligand contribute to the specificity of these integrins. For example, α5β1 clearly depends on the recognition of the fibronectin RGD sequence in the context of a "synergy" sequence that is a distance of 14 and 28 Kd away (Obara et al., 1988; Yamada, 1991). Also, the α5β1 fibronectin receptor is best isolated in pure form on a 110-Kd proteolytic fragment of fibronectin that contains the Arg-Gly-Asp sequence and both synergy regions (Pytela et al., 1985). It is significant that this receptor is not retained on an affinity column consisting of a short Arg-Gly-Asp-containing peptide derived from fibronectin.

The conformational presentation of a given adhesive sequence within the context of the intact macromolecule is clearly important for integrin recognition leading to cell adhesion. For example, the α chain of fibrinogen contains two Arg-Gly-Asp sequences, one near the N-terminus and the second near the C-terminus (Doolittle et al., 1979). Experiments have demonstrated that αvβ3-bearing cells or the purified receptor interact with the C-terminal RGD sequence within the α chain of fibrinogen, not with the N-terminal RGD sequence on the same polypeptide. This was demonstrated using a monoclonal antibody to this sequence that blocked cell adhesion to intact fibrinogen, while a monoclonal antibody that reacted with the N-terminal Arg-Gly-Asp sequence failed to do so (Cheresh et al., 1989a). However, cells readily attached to peptides containing Arg-Gly-Asp that were derived from either the N- or C-terminal domains of the fibrinogen α chain, and these adhesion events were blocked by the corresponding antipeptide monoclonal antibody (Cheresh et al., 1989a). These results demonstrated that the RGD peptide sequence derived from the N-terminal domain of the α-chain of fibrinogen, although capable of supporting adhesion when present in a peptide of 17 residues, was unable to do so within the context of the intact fibrinogen macromolecule. A possible explanation of these results is based on the structural analysis of fibrinogen which has revealed that the Arg-Gly-Asp sequence near the N-terminus of the α-chain is contained within an alpha helical domain of the molecule, and thus, may not be appropriately exposed for integrin-mediated cell adhesion. In contrast, the C-terminal Arg-Gly-Asp sequence is not associated with an α-alpha helical domain of fibrinogen (Doolittle et al., 1979). These data imply that integrin-mediated ligand binding leading to cell adhesion not only depends on primary sequence recognition but also on secondary and tertiary structural requirements.

Sequences other than RGD can promote integrin-mediated binding events. For example, the platelet integrin αIIbβ3 binds to the fibrinogen gamma chain carboxy terminus represented by the sequence HHLGGAKQAGDV (Hawiger et al., 1982). In addition, a new adhesive sequence has been identified in the fibronectin molecule. In this case, it is located within the alternatively spliced type III connecting segment which contains a 25-amino acid adhesion domain, termed CS1. It has now been demonstrated that the minimal essential sequence for cell adhesion to this region of fibronectin involves the Leu-Asp-Val (Komoriya et al., 1991). In this case, integrin α4β1 expressed primarily by leukocytes (Wayner et al., 1989; Hemler, 1990) and melanoma cells (Mould et al., 1991) interacts with fibronectin within the context of Leu-Asp-Val. This adhesive site is also found in the counter-receptor, VCAM, expressed on activated endothelial cells (Osborn, 1990), and thus can be recognized by cells expressing the α4β1 integrin. Recently the sequence Asp-Gly-Glu-Ala in collagen has been identified as the recognition sequence for the integrin α2β1 (Staatz et al., 1991). It is of interest that each of these integrin recognition sequences contain the negatively charged animo acid, Asp. It is tempting to speculate that this apparent charge requirement plays a role in the cation-dependent recognition by integrins (see below, Figure 2).

B. Structural Basis for Integrin-Ligand Interaction: αvβ3 as a Model

The β3 integrins (αvβ3 and αIIbβ3) have the unique ability to specifically interact with small synthetic peptides containing the Arg-Gly-Asp sequence making them particularly suited for biochemical studies to define the structural basis of integrin-ligand interaction. Based on this property, we set out to define the sequence on integrin αvβ3 that represents the ligand (Arg-Gly-Asp) binding domain. This goal was facilitated by our ability to purify large quantities (approximately 1 mg per placenta) of αvβ3 from placental tissue by immunoaffinity chromatography (Smith and Cheresh, 1988, 1990). This receptor preparation was then allowed to bind to a peptide with the sequence GRGDSPK that was coupled to the photoactivatable cross-linker SASD prelabeled with ^{125}I (Smith and Cheresh, 1988). After the radiolabeled peptide-SASD conjugate was allowed to interact with pure αvβ3 in solution, it was then photolyzed to cross-link it to the Arg-Gly-Asp binding site on the receptor. This procedure resulted in the specific radiolabeling of both subunits of the receptor, with approximately 20% of the label on the α subunit and 80% on the β subunit. The specificity of this interaction was documented in two ways. First, cross-linking to both subunits could be completely inhibited with unlabeled Arg-Gly-Asp-containing peptide. Second, the cross-linking showed an absolute requirement for divalent cations in the concentration range known to be required by integrins on the cell surface to promote cell adhesion (Smith and Cheresh, 1988,1990). To identify the specific region(s) on the receptor bound to the Arg-Gly-Asp peptide-SASD conjugate, the individual radiolabeled α and β subunits were separated by SDS-PAGE, purified by electroelution from the acrylamide gel, and subjected to proteolysis. The resulting radiolabeled peptides were separated by SDS-PAGE and individually subjected to NH_2-terminal sequence analysis to identify the cross-linked domains. As shown in Figure 2, this allowed us to assign regions near the NH_2-terminus of both the α and β subunits as being proximal to an RGD binding site. It is of interest to point out that the α chain was radiolabeled at two sites from within residues 139–349 comprising the first two-three divalent cation-binding regions of the α-subunit. The β subunit was radiolabeled from residues 61–203. Based on primary sequence analysis, this region of β3 corresponds to one of the most highly conserved regions of all the known integrin β subunits. These results imply that the RGD-binding pocket of αvβ3 involved amino-terminal regions of both subunits, and that the divalent cation-binding domain of αv plays a direct role in the capacity of this receptor to bind ligand. These results likely extend to other members of the integrin family of cell adhesion receptors since studies performed with the platelet integrin αIIbβ3 showed a similar cross-linked region (D'Souza et al., 1988, 1990).

C. RGD Binding Leading to Molecular Stabilization

Although integrin recognition of sequences such as Arg-Gly-Asp are required for ligand occupancy, it is quite possible that additional molecular interactions actually potentiate ligand recognition specificity and signaling mechanisms. This is apparent since individual integrins can discriminate between distinct Arg-Gly-Asp-containing ligands. In addition, once cells spread on a substrate, they form highly strengthened interactions that are thought to depend on the polymerized actin

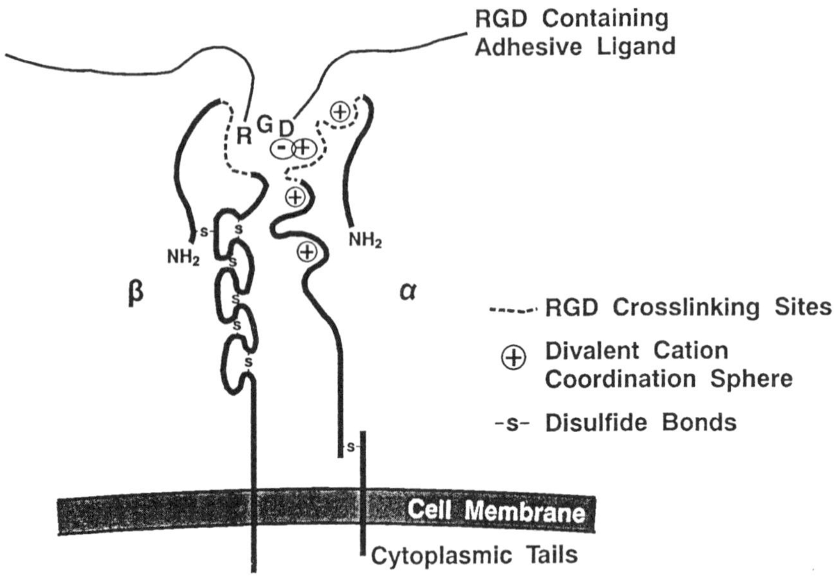

Figure 2. Model of integrin $\alpha v\beta 3$ associated with an RGD-containing ligand. The $\alpha v\beta 3$ integrin is depicted within the plane of a cell membrane. The receptor is associated with an RGD-containing ligand that binds to N-terminal regions of the α and β chains of this receptor.

cytoskeleton and the resulting focal contact formation (Lotz et al., 1989). However, little is known about the molecular event(s) that take place immediately after Arg-Gly-Asp recognition and before cell spreading. To address this question, we pretreated M21 human melanoma cells with cytochalasin B to prevent actin polymerization, and then allowed the treated cells to attach to immobilized ligand (vitronectin) or to an Arg-Gly-Asp containing peptide, both of which serve as ligands for integrin $\alpha v\beta 3$ on various cell types (Cheresh, 1987; Cheresh and Spiro, 1987). Following adhesion, cells were challenged with an excess of soluble ligand to determine the reversibility of these adhesion events. Under these conditions, cell attachment to the Arg-Gly-Asp-containing peptide could be readily dissociated with excess soluble ligand (Arg-Gly-Asp peptide), whereas cells attached to intact vitronectin could not be detached (Orlando and Cheresh, 1991). These results suggested that integrin-mediated cell attachment to an intact adhesive protein results in a highly stabilized interaction that is independent of the polymerized actin cytoskeleton (Figure 3). To substantiate these observations, integrin $\alpha v\beta 3$ was purified to homogeneity and its binding properties measured in a solid-phase receptor assay. Under these conditions, it was demonstrated that $\alpha v\beta 3$ bound vitronectin or fibronectin in a nondissociable manner while an RGD-containing peptide derived from vitronectin bound specifically, but was completely dissociable

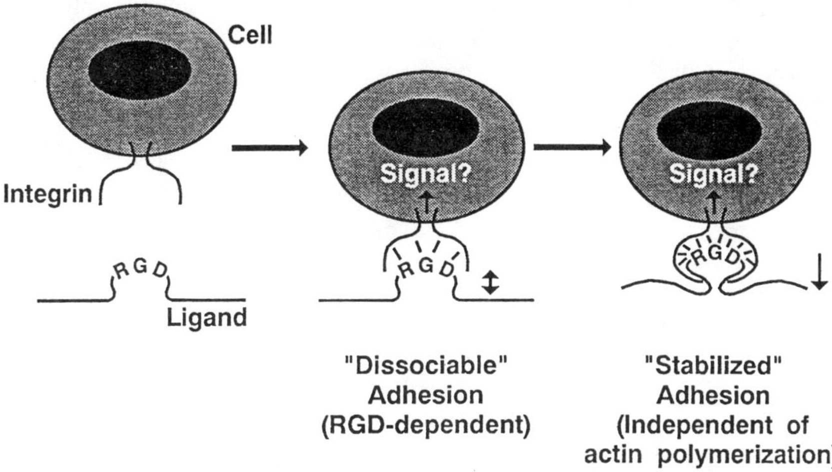

Figure 3. RGD binding leads to molecular stabilization between integrin and ligand. Model depicts cell surface-expressed integrin binding to a ligand resulting in a highly stabilized molecular interaction as described by Orlando and Cheresh (1991). The stable binding between the cell and the ligand likely depends on conformational changes involving the integrin and/or ligand. The integrin-mediated cellular signal(s) may result from the initial ligand binding event and/or this subsequent molecular stabilization event.

with a Kd of 9×10^{-7} M. In fact, it required at least 500-fold more soluble competitor to observe dissociation of the intact ligand from $\alpha v\beta 3$ (Orlando and Cheresh, 1991). Chemical modification of the receptor with mild glutaraldehyde treatment allowed vitronectin to bind in an RGD-dependent and dissociable manner, suggesting that receptor conformational changes or specific amino acid residues proximal to the ligand binding site(s) are involved in the stabilization event (Orlando and Cheresh, 1991). Thus, in the absence of cytoskeletal proteins or other cellular components, integrins have not only the capacity to recognize specific sequences, such as Arg-Gly-Asp, but also to participate in a highly stabilized protein-protein interaction that is independent of the Arg-Gly-Asp recognition event (Figure 3). At present, it is not clear whether the signal transduction mechanism(s) depends on this stabilized molecular interaction, or whether reversible Arg-Gly-Asp binding is sufficient to promote a molecular signal (Figure 3).

The integrin-ligand stabilization event may have significant biological effects on the cellular response to ligand binding. Thus, during thrombosis, activation of the integrin $\alpha IIb\beta 3$ results in the binding of fibrinogen to the platelet surface, an important preliminary step to clot retraction (Phillips et al. 1988) and an event that likely requires organization of the platelet cytoskeleton. It has been shown that the binding of fibrinogen to the platelet surface becomes nondissociable in a time-dependent manner (Marguerie et al., 1980). Thus, it is tempting to speculate that this nondissociable interaction may in some way potentiate the signal-transducing event(s) responsible for clot retraction. Recently, Du et al. (1991) demonstrated that Arg-Gly-Asp containing peptides could serve to activate the platelet integrin αIIbβ3 to a high-affinity fibrinogen binding receptor. At present, it is not clear whether this increased affinity is related to the findings that αvβ3 binds ligand in a nonreversible (stabilized) manner.

On nucleated cells, the highly stabilized interaction between integrin and ligand may increase our understanding of the mechanism(s) involved in cell migration on extracellular matrices. Although the actual role integrins play in cell migration is not completely defined, their interaction with matrix molecules was shown to provide traction and direction during cellular movement (McCarthy and Furcht, 1984). The stable association of the matrix proteins, vitronectin and fibronectin, with αvβ3 implies that alternative methods for the release of cells from substrata may be required other than simple dissociation. For example, release of the cell surface from matrix components may require pinching off membrane material, leaving behind focal contacts that contain the stable integrin-ligand complex. Alternatively, the receptor-ligand anchor may be cleared through the internalization/degradation pathway, since Mabs to the fibronectin receptor was shown to be endocytosed and rapidly became associated with coated pits (Raub and Kuentzel, 1989). In addition, it has been demonstrated that fibronectin-degrading proteases are localized to the cell membrane of transformed cells, suggesting that localized proteolysis may disrupt contacts between integrins and matrix proteins (Chen and Chen, 1987). To this end, Werb and coworkers have shown that occupancy and clustering of the fibronectin receptor induces the synthesis of two matrix-degrading enzymes; namely, collagenase and stromelysin (Werb et al., 1989). These results suggest that part of the integrin-mediated signaling response may involve the synthesis of degradative enzymes, which may be necessary to release the cell into a migratory state.

Integrin antagonists are currently under investigation as antithrombotic or antimetastatic reagents. The use of RGD peptides or other integrin inhibitors that block cellular adhesive mechanisms must be considered in the context of the highly stabilized interactions that occur between an integrin and its ligand. Thus, integrin antagonists may selectively perturb those interactions that are in a state of flux—as expected to occur with either migratory tumor cells or platelets during thrombosis—but which do not effect cells with previously stabilized integrin-ligand interactions.

D. Integrin Function Depends on Divalent Cations

It is well documented that cell-matrix interactions depend on the presence of divalent cations. In fact, qualitative and quantitative integrin binding events are highly dependent on particular cations. For example, manganese is known to increase the adhesiveness of certain integrins for their ligands (Gailet and Ruoslahti, 1988; Smith and Cheresh, 1988). The integrin α2β1 can be potentiated by magnesium and inhibited by calcium (Santoro, 1986; Staatz et al., 1989). A mutant version of the platelet integrin αIIbβ3 integrin that is unable to bind fibrinogen fails to undergo a cation-dependent conformational change (Loftus et al., 1990). These studies imply that divalent cations must interact in some way with the integrin and potentiate ligand binding events. It has been hypothesized that integrins bind divalent cations through an interaction with the α subunit, which typically contains at least three metal binding domains (see Figure 2). While these regions share homology with the helix-loop-helix calcium binding proteins (Phillips et al., 1988), they lack the α helical domains flanking each cation binding loop, and thus might be expected to bind cation with lower affinity. However, these regions are of crucial importance to integrin function, as stated above, since the cation-binding regions of both αv and αIIb were shown to directly bind synthetic peptide ligands (D'Souza et al., 1990; Smith and Cheresh, 1990). At present, the precise role that cations play in integrin function is not well defined. Recently, a study performed in our laboratory indicates that the cation coordination sphere is directly involved in the functional properties of integrin αvβ3 (Smith and Cheresh, 1991). In this case, we took advantage of the unique electron shell of cobalt which under normal conditions will support αvβ3-ligand binding events. However, Co(II), once oxidized to Co(III), no longer supported ligand binding, indicating that the coordination sphere of the cation is a critical determinant in integrin-ligand binding. This oxidative conversion to Co(III) resulted in a covalent linkage between the cation and receptor, permitting the affinity labeling of the receptor and resulting in an average of 3 to 4 moles of cobalt bound per mole of receptor. In addition, the apparent high affinity of manganese for αvβ3 allowed us to measure a direct interaction of $^{54}Mn^{+2}$ with the receptor under native conditions by equilibrium gel filtration. Interestingly, manganese abolished the cobalt oxidative affinity labeling of αvβ3, suggesting that these cations bind to the same or mutually exclusive sites. More importantly, these data provide a functional link between the coordination sphere of the cation and the ligand recognition properties of the integrin. These results are consistent with RGD cross-linking studies which directly implicate the divalent cation binding repeats in the integrin ligand binding pocket. The putative role of the divalent cation coordination sphere is shown in Figure 2. This model depicts that the negatively charged aspartic acid within the receptor-bound Arg-Gly-Asp sequence interacts electrostatically with the coordination sphere of the positively charged cation. Such a model would also take into account non-RGD integrin binding events. This is based on the observation that the negatively charged aspartic acid (D) is also found in various other adhesive sequences; that is, AGDV (fibrinogen, gamma chain), LDV (fibronectin, IIICS alternate splice site), and DGEA (collagen).

E. Integrin Activation and/or Regulation on the Cell Surface

There is considerable evidence that integrin function can vary within the plane of the membrane. In fact, integrin avidity for ligand can change in response to specific agonists. For example, upon activation, platelets become adhesive and aggregate in the presence of soluble fibrinogen, resulting in a platelet thrombus that becomes deposited at the site of injury. This involves the specific activation of integrin αIIbβ3 on the platelet surface (reviewed in Phillips et al., 1991). It is important that the activation process is a local event involving only those platelets in proximity to the injury. Recently, it has been shown that Arg-Gly-Asp-containing peptide ligands can serve to activate the integrin αIIbβ3 receptor (Du et al., 1991). In this case, it was shown that Arg-Gly-Asp binding leads to changes in this receptor that are associated with fibrinogen-binding function and platelet aggregation. This study demonstrated that the Arg-Gly-Asp sequence can be both a trigger of high-affinity ligand binding and an antagonist of αIIbβ3 function.

Another notable cellular adhesion event that can be activated involves β2 integrins on leukocytes (Springer, 1990). In this case, antigen recognition promotes cell-cell interactions that enable new adhesive interactions to take place reversibly in order to amplify the immune response. It appears that the β2 integrins which recognize the counter receptors ICAM-1 or ICAM-2 on other cells undergo specific conformational changes that result in an increase or decreased affinity (Springer, 1990). Antigen binding to the T-cell receptor can apparently trigger this event. Thus, leukocytes which express β2 integrins can become transiently adhesive, thereby enabling cell-cell binding events that are crucial to the immune response.

Recently it has been shown that a novel β1 containing integrin expressed by developing retinal neurons can be activated in the presence of a monoclonal antibody which promotes cell attachment to laminin and to collagen, but blocks it to vitronectin (Neugebauer and Reichardt, 1991). It was further demonstrated that developing retinal neurons, which had lost responsiveness to laminin, regain this ability in the presence of this antibody. Thus, the β1 class of integrins likely exists in multiple affinity states that can be modulated at the cell surface.

Integrin function can also be regulated by other molecules in the plane of the cell membrane. Specifically, antibodies to the oligosaccharide portion of cell-associated ganglioside were shown to perturb cell attachment events mediated by integrins αvβ3 (Cheresh et al., 1986; Cheresh et al., 1987), or α5β1 (Stallcup, 1988). Biochemical analysis revealed that gangliosides copurify with integrins on ligand affinity columns (Cheresh et al., 1987), suggesting a close interaction between the integrin and ganglioside on the cell surface. In fact, immunoelectron microscopic analysis revealed a close association of the GD2 ganglioside with integrin αvβ3 on M21 human melanoma cells (Cheresh and Klier, 1986). The association of gangliosides with integrins might be due to their mutual capacity to interact with divalent cations. In addition, integrin function can dramatically be modulated based on the lipid composition of the integrin-containing liposome (Conforti et al., 1990).

These data suggest that the regulation of integrin function can vary from cell to cell depending on the lateral association of other molecules in the plane of the membrane. In support of this contention is the finding that the α2β1 integrin functions as a laminin receptor in one cell type and a collagen receptor in another (Elices and Hemler, 1989; Kirchhofer et al., 1990). Thus, integrin function must be considered in the context of the cell membrane in which it is expressed.

IV. BIOLOGICAL CONSEQUENCES OF INTEGRIN-MEDIATED CELL ADHESION

A. Specific Integrins Potentiate Unique Cellular Functions

In some cases, integrins are preferentially expressed on certain cell types and impart unique biological functions to those cells. For example, as mentioned above, the hemidesmosome-associated integrin α6β4 appears to be preferentially associated with cells of epithelial origin (Quaranta and Jones, 1991). This integrin and its polar distribution may help provide anchorage and communication of the basal epithelial cell layer with its basement membrane.

Another example of an integrin restricted to a particular cell type involves the β2 family of integrins, members of which are uniquely expressed by leukocytes. This class of integrins, which includes three receptors defined by distinct α subunits (Springer, 1990), mediates a range of cell-cell and cell-substrate interactions. Thus, individuals with defects in the β2 gene have pronounced immune deficiency and are termed Leukocyte Adhesion Deficiency (LAD) (Anderson and Springer, 1987). Significant progress has been made regarding the structure and function of this family of integrins. This subject matter has been recently reviewed by Springer (1990).

The α4β1 integrin appears to be preferentially expressed on certain leukocytes and melanoma cells. This integrin serves as a fibronectin receptor capable of recognizing an alternatively spliced form (CS-1) of fibronectin (Wayner et al., 1989; Guan and Hynes, 1990) and the counter receptor, VCAM, expressed on activated endothelial cells (Osborn et al., 1989; Elices et al., 1990). Endothelial cells at the site of injury can become activated with cytokines, such as IL-1, resulting in their capacity to support leukocyte adhesion events (Osborn et al., 1989). Thus, α4β1 expression may explain, in part, the mechanism of leukocyte invasion of tissues during inflammation. However, it is clear that other adhesive mechanisms are involved between leukocytes and endothelial cells, including LFA-1 recognition of endothelial cell expressed ICAM (Dustin and Springer, 1988). Likewise, the α4β1 expression on melanoma cells may potentiate metastatic invasion of these tumor cells into the bloodstream by providing an adhesive substrate on the surface of endothelial cells.

Endothelial cells express integrin $\alpha v\beta 3$ as a major adhesive receptor (Cheresh, 1987). Endothelial cells (blood vessels) are capable of interacting with all known tissues and organs in order to provide a blood supply. Moreover, during angiogenesis associated with either wound healing, tumorigenicity, or inflammation, endothelial cells maintain the capacity to traverse virtually all basement membranes and extracellular matrix proteins. Endothelial cells are known to produce their own matrix, which includes the protein von Willebrand factor (Wagner et al., 1984). This protein typically accumulates in specialized compartments, known as Weibel–Palade bodies, which release their contents basolaterally (Wagner et. al., 1984). Interestingly, integrin $\alpha v\beta 3$ serves as a receptor for von Willebrand factor (Cheresh, 1987), indicating that endothelial cells not only remodel their matrix, but can specifically attach to it. In a recent study by Defilippi et al., 1991, it was shown that the integrin $\alpha 1\beta 1$ collagen/laminin receptor was preferentially expressed on microvascular endothelium, and thus could only be detected on small blood vessel and not large vessels such as aorta and umbilical and femoral veins. However, on cultured human umbilical vein endothelial cells, this integrin could be induced by treatment of the cells with tumor necrosis factor, α retinoic acid, or PMA. The induction of integrin $\alpha 1\beta 1$ on these cells increased their adhesiveness to collagen, laminin, and the laminin fragment P1, while having no effect on adhesion to either fibronectin or laminin fragment E8. Thus, integrin $\alpha 1\beta 1$ appears to be an inducible adhesion receptor, and under normal circumstances is preferentially expressed on a specific population of endothelial cells.

Epithelial cells can also secrete their own extracellular matrix. Recently, a novel protein was defined, termed epiligrin, which is specifically synthesized by epithelial cells and appears to promote tight adhesion via the integrin $\alpha 3\beta 1$ (Carter et al., 1991). However, this integrin is also capable of recognizing fibronectin, laminin, and collagen (Wayner et al., 1988). In any event, specialized cells not only have the ability to remodel their own extracellular matrix but often express specific integrin receptors allowing them to respond to the remodeled matrix.

B. Integrins: Role in Cancer

Numerous investigators have examined the adhesive properties of tumor cells and compared them to nontransformed cell types. In general, tumor cells are less adhesive and tend to be somewhat more migratory than their nontransformed counterparts. Tumor cell growth *in vitro* is independent of cell adhesion and is not inhibited by cell-cell contact. The capacity of the metastatic tumor cell to remodel and/or degrade the extracellular matrix and basement membrane may potentiate its migration from the primary tumor site. Therefore, the adhesive properties of tumor cells are likely to be an important factor governing their proliferative and metastatic properties *in vivo*.

Integrin-mediated cellular adhesion mechanisms have been proposed to play a role in the malignant growth of tumors (Humphries et al., 1986, 1988; Plantefaber and Hynes, 1989; Giancotti and Ruoslahti, 1990; Chan et al., 1991; Cheresh, 1991). In recent studies, overexpression of the fibronectin receptor α5β1 suppressed the transformed phenotype of chinese hamster ovary cells (Giancotti and Ruoslahti, 1990). In this case, associated with the α5β1 increase, was an increased fibronectin matrix deposition, loss of a anchorage-independent cell growth, and tumorigenicity in nude mice. Plantefaber and Hynes (1989) demonstrated that *ras*-transformed rodent cells showed a specific reduction in α5β1 compared to their nontransformed counterparts. Recently it was shown that human rhabdomyosarcoma cells acquired an increased metastatic capacity in nude mice when transfected with the collagen/laminin receptor integrin α2β1 (Chan et al., 1991). However, the expression of this integrin had no effect on the growth of the primary tumor in these animals.

Glioblastoma multiforme, the most malignant astroglial-derived tumor, grows as an adherent mass and locally invades brain. An examination of adult cerebral glioblastoma biopsy material revealed that αvβ3 is the primary integrin expressed by these cells (Gladson and Cheresh, 1991). Moreover, vitronectin was the only detectable tumor cell-associated adhesive ligand present in these tissues, and its expression correlated with late stage glial-derived tumors. In this regard, normal glial cells and early stage glioblastoma tumors failed to express vitronectin. In a cell attachment assay, cultured glioblastoma cells attached to the parenchyma of tumor cryostat sections at the sites of vitronectin expression, but failed to attach to normal brain. This adhesion event was inhibited by an RGD-containing peptide and antibodies against vitronectin and integrin αvβ3 (Gladson and Cheresh, 1991). These observations provide evidence for a cell adhesion mechanism that might potentiate glioblastoma cell invasion of normal brain.

C. Integrin αvβ3: A Functional Marker of Human Melanoma Cells

Melanoma is one of the most malignant tumors in man. Primary melanoma appears in the skin from where it can metastasize to almost any organ. Primary melanoma tumors invade either horizontally or vertically. Typically, horizontally invasive tumors can be surgically removed often without further consequence, whereas vertically invasive primary tumors frequently lead to widespread metastatic disease and death. In a recent study, Albelda and co-workers (Albelda et al., 1990) examined the expression of αvβ3 and other integrins on human melanoma biopsy material, including metastatic lesions and those derived from vertical and horizontal growth phase primary melanoma. In addition, normal skin melanocytes and various types of nevi were examined for expression of these integrins. The investigators reported in this study that 8/10 vertical growth phase primary lesions, and all metastatic lesions expressed αvβ3 as determined by immunohistological staining with Mabs, directed to the α and β subunits of this receptor. In every positive case, between 30 to 90% of the cells stained strongly with antibody directed against the β3 subunit. In contrast, the β1 integrin subunit could be detected on virtually all of the cells in these same tissues. In normal skin melanocytes, however, nevi and most importantly horizontal growth phase primary melanoma, β3 was not detectable even though αv was found to be present. It is likely that αv associates with β1, or with the newly defined β5 and/or β6 subunits. These findings suggest that the expression of the β3 subunit may be linked to the progression of human melanoma.

The capacity of β3 to associate with αv on melanoma cells provides a broad spectrum adhesive ability that may potentiate tumor cell malignancy *in vivo*. Experimental evidence supports this contention. Specifically, an antibody to the β3 subunit was reported to inhibit the growth of human melanoma cells in nude mice (Boukerche et al., 1989). Similarly, Arg-Gly-Asp-containing peptides, which are known to block the function of αvβ3, were shown to inhibit both melanoma tumor invasion *in vitro* (Gehlson et al., 1988) and the development of experimental metastasis in a murine melanoma model system (Humphries et al., 1986, 1988). These studies indicate that the expression of integrin αvβ3 on human malignant melanoma cells may play an important role in the proliferation and metastatic phenotype of this tumor. In support of this contention are recent results demonstrating that integrin αv gene expression is required for human melanoma tumorigenicity (Felding-Habermann et al., 1992). In this study, M21 cell variants (M21-L cells) selected for the absence of αv mRNA expression failed to express integrin αvβ3 (Cheresh and Spiro, 1987). M21-L cells not only lost the ability to attach to vitronectin but showed a dramatic reduction in tumorigenicity when transplanted subcutaneously into athymic nude mice compared to M21 parental cells, even though both cell types showed identical β1 integrin expression and similar growth properties *in vitro*. M21-L cells were then stably transfected with a cDNA encoding αv, which resulted in the functional expression of integrin αvβ3 on these cells and completely restored their tumorigenicity (Felding-Habermann et al., 1992). Thus, the integrin αvβ3 appears to be directly involved in the proliferation of human melanoma *in vivo*.

A unique property of integrin αvβ3 is its ability to promote cell adhesion to multiple Arg-Gly-Asp-containing ligands (Cheresh, 1987; Cheresh and Spiro, 1987). This may explain, in part, why melanoma cells invade such a wide array of diverse tissues including the dermis, lymph node, spleen, bone, brain, lung, and liver. In addition, the ability of melanoma cells to escape from the primary tumor site may be due to their ability to interact with fibrin/fibrinogen via αvβ3 (Cheresh and Spiro, 1987). Tumors are often surrounded by a fibrin coat which is produced by the host in response to the tumor (Dvorak, 1990). This may explain why many human tumor cell lines express plasminogen activator activity which results in the proteolytic digestion of fibrin/fibrinogen (Dano et al., 1985). This enzyme activates plasminogen, converting it to plasmin, which, in turn, degrades fibrin. Thus, human melanoma cells not only can attach to fibrinogen/fibrin but can degrade this matrix as well. Invasion of bone may be mediated by the ability of αvβ3 to recognize osteopontin or bone sialoprotein. Therefore, expression of αvβ3 by melanoma or other types of malignant cells could increase their metastatic potential.

D. Integrin Antagonists Perturb Biologically Relevant Adhesion Events

The use of integrin antagonists, such as RGD-containing peptides, may be useful as antimetastatic or antithrombotic agents. In fact, recently a family of naturally occurring cyclic RGD peptides isolated from viper venom have shown significant promise as integrin antagonists, showing 100- to 1000-fold more potency than linear RGD peptides (Huang et al., 1987; Chao et al., 1989; Garsky et al., 1989; Scarborough et al., 1991). These peptides, termed dis-integrins, are naturally cyclic due to multiple disulfide bonds, and typically contain from 50 to 70 amino acids. At least, in some instances, these peptides bind with high specificity to the platelet receptor αIIbβ3 (Scarbourough et al., 1991). Recent structural evidence has accumulated for one member of this family termed, Kistrin (Adler et al., 1991). Based on detailed two-dimensional nuclear magnetic resonance spectroscopic measurements, it is apparent that the RGD site is located at the apex of a long loop across the surface of the protein. These investigators concluded that the relative high affinity of this peptide and other viper venom peptides for integrins is likely due to favorable conformational restraints of the RGD sequence, and not due to the presence of major secondary binding determinants. It is of interest that the viper venom peptide Eichistatin binds to integrin αvβ3 on osteoclasts and prevents their adhesion to bone (Sato et al., 1990), resulting in the blocking of bone resorption. Linear RGD-containing peptides have been used to block experimental metastasis of murine melanoma cells *in vivo* (Humphries et al., 1986;1988). In this case, it was demonstrated that lung colonization was drastically reduced in the presence of the peptide. Taken together, the results of these studies indicate that integrin antagonists may be useful for the treatment of diseases, such as vascular occlusion, cancer, and osteoporosis. However, additional structural evidence is necessary in order to design highly specific active compounds that serve to inhibit distinct cellular adhesion events. The cyclic Arg-Gly-Asp peptides isolated from viper venom should be useful tools to further our knowledge on the structural and molecular basis of integrin function. This information could lead to development of synthetic organic compounds that have high affinity for particular integrins. Such compounds may ultimately prove to be potent inhibitors of inflammation, metastasis, and thrombosis, especially if they are found to be nontoxic and possess favorable pharmacological properties.

E. Integrins are Involved in Differentiation and Development

Cellular adhesive mechanisms have been implicated in embryonic development and differentiation. Events during amphibian gastrulation and chick neural cell migration can be perturbed by integrin antagonists such as Arg-Gly-Asp or anti-integrin antibodies (Boucaut et al., 1984; Bronner-Fraser, 1986). During this stage of development, fibronectin levels increase (Lee et al., 1984) as do mRNAs corresponding to multiple integrin β subunits (DeSimone and Hynes,

1988). Integrins have been identified as functionally relevant in *Drosophila* development (Bogaert et al., 1987; MacKrell et al., 1988; Leptin et al., 1989; Volk et al., 1990; Zusman et al., 1990). A gene required for embryonic morphogenesis (myospheroid) encodes a gene homologous to the integrin β subunit (MacKrell et al., 1988). Embryos possessing mutants in this gene die late in embryogenesis as a result of various morphogenetic defects. In fact, *Drosophila* expresses at least two integrins characterized by distinct α subunits. Volk et al. (1990) proposed that integrins may help to coordinate the differentiation of the *Drosophila* sarcomeric cytoarchitecture of muscle fibers within their microenvironment, and are thus essential for proper integration of muscle cells into tissues. Zusman et al. (1990) generated flies that were mosaic for wild-type and mutant alleles of *lethal(1)myospheroid* using gynandromorphs and radiation-induced somatic crossing over. The developmental defects observed demonstrated broad requirements for PS integrins in ventrally derived structures. Smaller clones induced during larval development resulted in blister and vein defects in the wings and aberrant development of photoreceptor cells. Thus, PS integrins are required for the close association between the dorsal and ventral wing epithelia, and for the proper arrangement of photoreceptor cells. Although many other adhesive functions appeared normal, much work is needed to determine the distribution of integrins and matrix proteins during development. There is little doubt, however, that these adhesive interactions will provide crucial signals during embryogenesis.

Recently it was shown that murine embryonic stem cells (ES cells) express an alternatively spliced laminin receptor α6 subunit containing a distinct cytoplasmic tail (Cooper et al., 1991). When ES cells were cultured under conditions that permitted their differentiation, they lost the α6 alternative cytoplasmic tail and acquired the cytoplasmic tail commonly found in a wide variety of differentiated cell types. These data imply that the signal transduction mechanism during embryogenesis may be altered by alternative splicing of the integrin α6 subunit cytoplasmic tail. However, at present it is not clear what functional properties this alternative splicing imparts on this receptor. Since this integrin normally recognizes laminin, the switch between these forms of α6 might be of relevance to the developmental role of laminin. This form of alternative splicing has also been identified for the α3 subunit (Tamura et al., 1991) which serves as a receptor for multiple adhesive ligands. Thus, alternative exon usage might be cell type specific and related to development. It may also be a general property of integrins and considerably add to the structural and functional diversity of this family of adhesion receptors.

Evidence suggests that integrins are also involved in the regulation of differentiation in the adult organism. Nicholson and Watt (1991) demonstrated that expression of fibronectin and the fibronectin receptor α5β1 were decreased during terminal differentiation of human keratinocytes. During terminal differentiation, keratinocytes become less adhesive to extracellular matrix proteins (Adams and Watt, 1990), and this likely potentiates cellular movement away from the basal layer as differentiation takes place. Thus, a feedback loop has been proposed where reduced contact with the matrix causes cells to initiate differentiation and simultaneous migration (Watt, 1988). Fibronectin is a relatively minor component of the normal epidermal basement membrane, yet it is abundantly expressed during wound healing and promotes keratinocyte migration (Clark, 1990). The fact that fibronectin prevents the differentiation of keratinocytes suggests that integrins which potentiate adhesion to this protein may play a role in this process.

V. INTEGRINS: FUTURE STUDIES

Many integrins have been identified (see Figure 1) and their primary structure elucidated. It seems likely that additional subunits will be identified in the near future. Perhaps further examples of integrin isoforms as those described for $\alpha3$ and $\alpha6$ will be uncovered (Cooper et al., 1991; Tamura et al., 1991).

There are now a battery of highly specific antibodies and cDNAs available as molecular probes to study the structure, function, and biological activity of integrins. Cell models must be developed that specifically lack individual integrin subunits. Such models will facilitate studies involving transfection of mutant, truncated, or chimeric integrin subunits into such cells, and thereby help define the structural basis of integrin-ligand interactive domains as well as provide insight into the molecular mechanism of integrin subunit assembly. At present, it is still unclear how a given cell regulates its integrin repertoire which defines its adhesive phenotype. While it is clearly established that integrins mediate cell signals, future studies will address the molecular basis of these signals and the consequences to the cell in terms of second messengers, gene activation, and cellular proliferation. By understanding these events at the molecular level, it may be possible to comprehend why tumor cells are anchorage-independent and normal cells require anchorage for growth and differentiation. Moreover, this may lead to an explanation as to why one cell type migrates on a given matrix while another fails to do so. Studies mentioned above should lead to more complex experiments that will deal with the role of integrins within the intact organism. To accomplish these goals, transgenic expression or integrin subunit "gene knockout" experiments will be helpful in defining the precise role of integrins during development.

Ultimately it will be crucial to obtain three-dimensional X-ray crystallographic analysis of integrins to understand the true structural basis of integrin-ligand interaction. These studies should also allow for a molecular understanding of the activation event(s) and the conformational changes leading to signal transduction. However, one must always consider the structure of integrins within the context of the cell membrane since it is this environment in which integrins are designed to function and provide cell biological activity. Based on these studies, it may ultimately be feasible to design specific integrin antagonists that regulate or inhibit cellular adhesive events associated with thrombosis, inflammation, microbial invasion, and cancer.

REFERENCES

Adams, J.C. & Watt, F.M. (1990). Changes in keratinocyte adhesion during terminal differentiation: Reduction in fibronectin binding precedes α5β1 loss from the cell surface. Cell 63, 425–435.

Adler, Lazarus, R.A., Dennis, M.S., & Wagner, G. (1991). Solution structure of kistrin, a potent platelet aggregation inhibitor and GPIIb-IIIa antagonist. Science 253, 445–448.

Albelda, S.M. & Buck, C.L. (1990). Integrins and other cell adhesion molecules. FASEB J. 4, 2868–2880.

Albeda, S.M., Mette, S.A., Elder, D.E., Stewart, R., Damjanovich, L., Herlyn, M., & Buck, C.A. (1990). Integrin distribution in malignant melanoma: Association of the β3 subunit with tumor progression. Cancer Res. 50, 6757–6764.

Altieri, D.C;, Wiltse, W.L., & Edgington, T.S. (1990). Signal transduction initiated by extracellular nucleotides regulates the high affinity ligand recognition of the adhesive receptor CD11b/CD18. J. Immunol. 154, 662–670.

Anderson, D.C. & Springer, T.A. (1987). Leukocyte adhesion deficiency: An inherited defect in the Mac-1, LFA-1, and p150, 95 glycoproteins. Annu. Rev. Med. 38, 175–194.

Bodary, S.C. & McLean, J.W. (1990). The integrin β1 subunit associates with the vitronectin receptor αv to form a novel vitronectin receptor in a human embryonic kidney cell line. J. Biol. Chem. 265, 5938–5941.

Bogaert, T., Brown, N., & Wilcox, M. (1987). The drosophila PS2 antigen is an invertebrate integrin that, like the fibronectin receptor, becomes localized to muscle attachments. Cell 51, 929–940.

Boucaut, J., Darribere, T., Poole, T.J., Aoyama, H., Yamada, K., & Thiery, J.P. (1984). Biologically active synthetic peptides as probes of embryonic development: A competitive peptide inhibitor of fibronectin function inhibits gastrulation in amphibian embryos and neural crest cell migration in avian embryos. J. Cell Biol. 99, 1822–1830.

Boulcerche, H., Berthier-Vergnes, O., Bailly, M., Dore, J.F., Leung, L.K., & McGregor, J.L. (1989). A monoclonal antibody (LYP18) directed against the blood platelet glycoprotein IIb/IIIa complex inhibits human melanoma growth *in vivo*. Blood 74, 909–912.

Bronner-Fraser, M. (1986). An antibody to a receptor for fibronectin and laminin perturbs cranial neural crest development *in vivo*. Dev. Biol. 117, 528–536.

Burridge, K., Fath, K., Kelly, T., Nuckolls, G., & Turner, C. (1988). Focal adhesions: Transmembrane junctions between the extracellular matrix and the cytoskeleton. Annu. Rev. Cell Biol. 4, 487–525.

Carlos, T.M., Schwartz, B.R., Kovach, N.L., Yee, E., Rosa, M., Osborn, L., Chi-Rosso, G., Newman, B., Lobb, R., & Harlan, J.M. (1990). Vascular cell adhesion molecule-1 (VCAM-1) mediates lymphocyte adherence to cytokine-activated cultured human endothelial cells. Blood 76, 965–971.

Carter, W.G., Ryan, M.C., & Gahr, P.J. (1991). Epiligrin, a new cell adhesion ligand for integrin α3β1 in epithelial basement membranes. Cell 65, 599–610.

Carter, W.G., Wayner, E.A., Bouchard, T.S., & Kaur, P. (1990). The role of integrins α2β1 and α3β1 in cell-cell and cell-substrate adhesion of human epidermal cells. J. Cell Biol. 110, 1387–1404.

Chan, B.M., Matsuura, N., Takada, Y., Zetler, B.R., & Hemler, M.E. (1991). *In vitro* and *in vivo* consequences of VLA-2 expression on rhabdomyosarcoma cells. Science 251, 1600–1602.

Chao, B.H., Jakubowski, J.A., Savage, B., Chow, E.P., Manec, U.M., Harker, L.A., & Maraganore, J.M. (1989). *Agkistrodon piscivorus piscivorus* platelet aggregation inhibitor: A potent inhibitor of platelet activation. Proc. Natl. Acad. Sci. USA 86, 8050–8054.

Charo, I.F., Nannizzi, L., Smith, J.W., & Cheresh, D.A. (1990). The vitronectin receptor αvβ3 binds fibronectin and acts in concert with α5β1 in promoting cellular attachment and spreading on fibronectin. J. Cell Biol. 111, 2795–2800.

Chen, J.M. & Chen, W.T. (1987). Fibronectin-degrading proteases from the membranes of transformed cells. Cell 48, 193–203.

Chen, W.T., Hasegawa, T., Hasegawa, C., Weinstock, C., & Yamada, K.M. (1985). Development of cell surface lineage complexes in cultivated fibroblasts. J. Cell Biol. 100, 1103–1114.

Cheresh, D.A. (1987). Human endothelial cells synthesize and express an arg-gly-asp-directed receptor involved in attachment to fibrinogen and von Willebrand factor. Proc. Natl. Acad. Sci. USA 84, 6471–6475.

Cheresh, D.A. (1991). Structure, function and biological properties of integrin $\alpha_v\beta_3$ on human melanoma cells. Cancer and Metastasis Reviews 10, 3–10.

Cheresh, D.A., Berliner, S.A., Vicente, V., & Ruggeri, Z.M. (1989). Recognition of distinct adhesive sites on fibrinogen by related integrins on platelets and endothelial cells. Cell 58, 945–953.

Cheresh, D.A. & Klier, F.G. (1986). Disialoganglioside GD2 distributes preferentially into substrate-associated microprocesses on human melanoma cells during their attachment to fibronectin. J. Cell Biol. 102, 1887–1897.

Cheresh, D.A., Pierschbacher, M.D., Herzig, M.A., & Mujoo, K. (1986). Disialogangliosides GD2 and GD3 are involved in the attachment of human melanoma and neuroblastoma cells to extracellular matrix proteins. J. Cell Biol. 102, 688–696.

Cheresh, D.A., Pytela, R., Klier, F.G., Pierschbacher, M.D., Ruoslahti, E., & Reisfeld, R.A. (1987). An arg-gly-asp directed receptor on the surface of human melanoma cells exists in a functional complex with the disialoganglioside GD2. J. Cell Biol. 105, 1163–1173.

Cheresh, D.A., Smith, J.W., Cooper, H.M., & Quaranta, V. (1989). A novel vitronectin receptor integrin ($\alpha v\beta x$) is responsible for distinct adhesive properties of carcinoma cells. Cell 57, 59–69.

Cheresh, D.A. & Spiro, R.C. (1987). Biosynthetic and functional properties of an arg-gly-asp-directed receptor involved in human melanoma cell attachment to vitronectin, fibrinogen and von Willebrand factor. J. Biol. Chem. 262, 17703–17711.

Clark, R.A.F. (1990). Fibronectin matrix deposition and fibronectin expression in healing and normal skin. J. Invest. Derm. 94, 1285–1349.

Conforti, G., Zanetti, A., Pasquali-Ronchetti, I., Quaglino, D., Neyroz, P., & Dejanna, E. (1990). Modulation of vitronectin receptor binding by membrane lipid composition. J. Biol. Chem. 265, 4011–4019.

Cooper, H.M., Tamura, R.N., & Quaranta, V. (1991). The major laminin receptor of mouse embryonic stem cells is a novel insoform of $\alpha 6\beta 1$ integrin.

D'Souza, S.E., Ginsberg, M.H., Burlce, T.A., & Plow, E.F. (1990). The ligand binding site of the platelet integrin receptor GPIIb-IIIa is proximal to the second calcium binding domain of its a subunit. J. Biol. Chem. 265, 3440–3446.

D'Souza, S.E., Ginsberg, M.H., Lam, S., & Plow, E. (1988). Chemical crosslinking of arginyl-glycyl-aspartic acid peptides on adhesion receptors on platelets. Science 242, 91–93.

D'Souza, S.E., Ginsberg, M.H., & Plow, E.F. (1991). Arginyl-glycyl-aspartic acid (RGD): A cell adhesion motif. TIBS 16, 246–250.

Damsky, C.M., Knudson, K.A., Bradley, D., Buck, C.A., & Horwitz, A. (1985). Distribution of the CSAT cell-matrix antigen on myogenic and fibroblastic cells in culture. J. Cell Biol. 100, 1528–1539.

Dano, K., Andreason, P.A., Grohndahl-Hansen, J., Kristensen, P., Nielsen, L.S., & Skriver, L. (1985). Plasminogin activators, tissue degradation and cancer. Adv. Cancer Res. 44, 139–266.

Defilippi, P., van Hinsbergh, V., Bertolotto, A., Rossino, P., Silengo, L., & Tarone, G. (1991). Differential distribution and modulation of expression of alpha1/beta1 integrin on human endothelial cells. J. Cell Biol. 114, 855–863.

Dejana, E., Colella, S., Conforti, G., Abbadini, M., Gaboli, M., & Marchisio, P.C. (1988). Fibronectin and vitronectin regulate the organization of their respective arg-gly-asp adhesion receptors in cultured human endothelial cells. J. Cell Biol. 107, 1215–1223.

DeSimone, D.W. & Hynes, R.O. (1988). *Xenopus laevis* integrins. J. Biol. Chem. 263, 5333–5340.

Doolittle, R.F., Watt, K.W.K., Cottrell, B.A., Strong, D.D., & Riley, M. (1979). The amino acid sequence of the a chain of human fibrinogen. Nature 280, 464–468.

Du, X., Plow, E.F., Frelinger, III, A.L., O'Toole, T.E., Loftus, J.C., & Ginsberg, M.H. (1991). Ligands "activate" integrin $\alpha_{IIb}\beta_3$ (Platelet GPIIb-IIIa). Cell 65, 409–416.

Dustin, M.L. & Springer, T.A. (1988). Lymphocyte function-associated antigen-1 (LFA-1) interaction with intercellular adhesion molecule-1 (ICAM-1) is one of at least three mechanisms for lymphocyte adhesion to cultured endothelial cells. J. Cell Biol. 107, 321–331.

Dvorak, H.F. (1990). Lealcy tumor vessels: Consequences for tumor stroma generation and for solid tumor therapy. Prog. Clin. Biol. Res. 354, 317–330.

Elices, M.J. & Hemler, M.E. (1989). The human integrin VLA-2 is a collagen receptor on some cells and a collagen/laminin receptor on others. Proc. Natl. Acad. Sci. USA 86, 9906–9910.

Elices, M.J., Osborn, L., Takada, Y., Crouse, C., Luhowskyi, S., Hemler, M.E., & Lobb, R.R. (1990). VCAM-1 on activated endothelium interacts with the leukocyte integrin VLA-4 at a site distinct from the VLA-4/fibronectin binding site. Cell 60, 577–584.

Erle, D.J., Rüegg, C., Sheppard, D., & Pytela, R. (1991). Complete amino acid sequence of an integrin β subunit (β_7) identified in leukocytes. J. Biol. Chem. 266, 11009–11016.

Felding-Habermann, B., Mueller, B.M., Romadahl, C., & Cheresh, D.A. (1992). Requirement of integrin αv gene expression for human melanoma-tumorigenicity. 89, 2018–2022.

Fitzgerald, L.A., Steiner, B., Rall, Jr., S.C., Lo, S., & Phillips, D.R. (1987). Protein sequence of endothelial glycoprotein IIIa derived from a cDNA clone. J. Biol. Chem. 262, 3936–3939.

Freed, E., Gailit, J., van da Geer, P., Ruoslahti, E., & Hunter, T. (1989). A novel integrin β subunit is associated with the vitronectin receptor α subunit (αv) in a human osteocarcinoma cell line as is a substrate for protein kinase C. J. Eur. Mol. Biol. Organ. 8, 2955–2965.

Frelinger, A.L., Lam, S.C.-T., Plow, E.F., Smith, M.A., Loftus, J.C., & Ginsberg, M.H. (1988). Occupancy of an adhesive glycoprotein receptor modulates expression of an antigenic site involved in cell adhesion. J. Biol. 7263, 12397–12402.

Gailit, J. & Ruoslahti, E. (1988). Regulation of the fibronectin receptor affinity by divalent cations. J. Biol. Chem. 263, 12927–12932.

Garsky, V.M., Lumma, P.K., Freidinger, R.M., Pitzenberger, S.M., Randall, W.C., Veber, D.F., Gould, R.J., & Friedman, P.A. (1989). Chemical synthesis of echistatin, a potent inhibitor of platelet aggregation from *Echis carinatus*: Synthesis and biological activity of selected analogs. Proc. Natl. Acad. Sci. USA 86, 4022–4026.

Gehlson, K.R., Argraves, W.S., Pierschbacher, M.D., & Ruoslahti, E. (1988). Inhibition of *in vitro* tumor invasion by arg-gly-asp-containing peptides. J. Cell Biol. 106, 25–930.

Geiger, B. (1979). A 130-K protein from chicken gizzard: Its localization at the termini of microfilament bundles in cultured chicken cells. Cell 18, 193–205.

Giancotti, F.G. & Ruoslahti, E. (1990). Elevated levels of the α5β1 receptor suppresses the transformed phenotype of Chinese hamster ovary cells. Cell 60, 849–859.

Gladson, C.L. & Cheresh, D.A. (1991). Glioblastoma expression of vitronectin and the αvβ3 integrin: Adhesion mechanism for transformed glial cells. J. Clin. Invest. (In press).

Guan, J.-L. & Hynes, R.O. (1990). Lymphoid cells recognize an alteratively spliced segment of fibronectin via the integrin receptor α4β1. Cell 60, 53–61.

Haimovich, B., Ancskievich, B.J., & Boettiger, D. (1991). Cellular partitioning of βl integrin and their phosphorylated forms is altered after transformation by Rous sarcoma virus or treatment with cytochalasin D. Cell Regulation 2, 271–283.

Hawiger, J., Timmons, S., Kloczewiek, M., Strong, D.D., & Doolittle, R.P. (1982). Gamma and alpha chains of human fibrinogen possess sites reactive with human platelet receptors. Proc. Natl. Acad. Sci. USA 79, 2068–2071.

Hayashi, Y., Haimovich, B., Reszka, A., Boefflger, D., & Horwitz, A. (1990). Expression and function of checken integrin β1 subunit and its cytoplasmic domain in mouse NIH 3T3 cells. J. Cell Biol. 110, 175–184.

Hemler, M.E. (1990). VLA proteins in the integrin family: Structures, functions and their role on leukocytes. Ann. Rev. Immunol. 8, 365–400.

Hibbs, M.L., Xu, H., Stacker, S.A., & Springer, T.A. (1991). Regulation of adhesion of ICAM-1 by the cytoplasmic domain of LFA-1. Science 251, 1611–1613.

Hirst, R., Horwitz, A.F., Buck, C., & Rohrschneider, L. (1986). Phosphorylation of the fibronectin receptor complex in cells transformed by oncogenes that encode tyrosine kinases. Proc. Natl. Acad. Sci. USA 83, 6470–6474.

Horwitz, A., Duggan, K., Buck, C., Beckerle, M.C., & Burridge, K. (1986). Interaction of plasma membrane fibronectin receptor with talin, a transmembrane linkage. Nature (Lond.) 320, 531–533.

Huang, T.-F., Holt, J.C., Kirby, E.P., & Niewiarowski, S. (1989). Trigramin: Primary structure and its inhibition of von Willebrand factor binding to glycoprotein IIb/IIIa complex on human platelets. Biochemistry 28, 661–666.

Huang, T.-F., Holt, J.C., Lukasiewicz, H., & Niewiarowski, S. (1987). Trigramin: A low molecular weight peptide inhibiting fibrinogen interaction with platelet receptors expressed on glycoprotein IIb-IIIa complex. J. Biol. Chem. 262, 16157–16163.

Humphries, M.J., Olden, K., & Yamada, K.M. (1986). A synthetic peptide from fibronectin inhibits experimental metastasis of murine melanoma cells. Science 233, 467–470.

Humphries, M.J., Yamada, K.M., & Olden, K. (1988). Investigation of the biological effects of the anti-cell adhesion synthetic peptides that inhibit experimental metastasis of B16–F10 murine melanoma cells. J. Clin. Invest. 81, 782–790.

Hynes, R.O. (1987). Integrins: A family of cell surface receptors. Cell 48, 549–554.

Ingber, D.E., Prusty, D., Frangioni, J.J., Cragoe, E.J., Lechene, C.P., & Schwartz, M.A. (1990). Control of intracellular pH and growth by fibronectin in capillary endothelial cells. J. Cell Biol. 110, 1803–1811.

Kirchhofer, D., Grzesiak, J., & Pierschbacher, M.D. (1991). Calcium as a potential physiological regulator of integrin-mediated cell adhesion. J. Biol. Chem. 266, 4471–4477.

Kirchhofer, D., Languino, L.R., Ruoslahti, E., & Pierschbacher, M.D. (1990). $\alpha_2\beta_1$ integrins from different cell types show different binding specificities. J. Biol. Chem. 265, 615–618.

Komoriya, A., Green, L.J., Mervic, M., Yamada, S.S., Yamada, K.M., & Humphries, M.J. (1991). The minimal essential sequence for a major cell type-specific adhesion site (CS1) within the alternatively spliced type III connecting segment domain of fibronectin is leucine-aspartic acid-valine. J. Biol. Chem. 266, 15075–15079.

Kramer, R.H., Cheng, Y.-F., & Clyman, R. (1990). Human microvascular endothelial cells use β1 and β3 integrin receptor complexes to attach to laminin. J. Cell Biol. 111, 1233–1243.

Krissansen, G.W., Elliot, M.J., Lucas, C.M., Stomski, F.C., Berndt, M.C., Cheresh, D.A., Lopez, A.F., & Burns, G.F. (1990). Identification of a novel integrin β subunit expressed on cultured monocytes (macrophages): Evidence that one α subunit can associate with multiple β subunits. J. Biol. Chem. 265, 823–830.

Laffón, A., García-Vicuña, R., Humbría, A., Postigo, A.A., Corbí, A.L., de Landázuri, M.O., & Sánchez-Madrid, F. (1991). Upregulated expression and function of VLA-4 fibronectin receptors on human activated T cells in rheumatoid arthritis. J. Clin. Invest. 88, 546–552.

Lawler, J., Weistein, R., & Hynes, R.O. (1988). Cell attachment to thrombospondin: The role of RGD and integrin receptors. J. Cell Biol. 107, 2351–2361.

Lee, G., Hynes, R.O., & Kirschner, M. (1984). Temporal and spatial regulation of fibronectin in early xenopus development. Cell 36, 729–740.

Leptin, M., Bogaert, T., Lehmann, R., & Wilcox, M. (1989). The function of PS integrins during drosophila embryogenesis. Cell 56, 401–408.

Lotz, M.M., Burdsal, C.A., Erickson, H.P. & McClay, D.R. (1989). Cell adhesion to fibronectin and tenascin: Quantitative measurements of initial binding and subsequent strengthening response. J. Cell Biol. 109, 1795–1805.

MacKrell, A.J., Blumberg, B., Haynes, S.R., & Fessler, J.H. (1988). The lethal myospheroid gene of drosophila encodes a membrane protein homologous to vertebrate integrin β subunits. Proc. Natl. Acad. Sci. USA 85, 2633–2637.

Massia, S.P. & Hubbell, J.A. (1991). An RGD spacing of 440 nm is sufficient for integrin $\alpha v\beta 3$-mediated fibroblast spreading and 140 nm for focal contact and stress fiber formation J. Cell Biol. 114, 1089–1100.

McCarthy, J.B. & Furcht, L.T. (1984). Laminin and fibronectin promote the haptotactic migration of β16 mouse melanoma cells *in vitro*. J. Cell Biol. 98, 1474–1480.

McLean, J.W., Vestal, D.J., Cheresh, D.A., & Bodary, S.C. (1990). cDNA sequence of the human integrin β5 subunit. J. Biol. Chem. 265, 17126–17131.

Miyauchi, A., Alvarez, J., Greenfield, E.M., Teti, A., Grano, M., Colucci, S., Zambonin-Zallone, A., Ross, F.P., Teitelbaum, S.L., Cheresh, D., & Hruska, K.A. (1991). Recognition of osteopontin and related peptides by an $\alpha_v\beta_3$ integrin stimulates immediate cell signals in osteoclasts. J. Biol. Chem. (In press).

Mould, A.P., Komoriya, A., Yamada, K.M., & Humphries, M.J. (1991). The CS5 peptide is a second site in the mcs region of fibronectin recognized by the integrin $\alpha_4\beta_1$: Inhibition of $\alpha_4\beta_1$ function by RGD peptide homologues. J. Biol. Chem. 266, 3579–3585.

Neugebauer, K.M. & Reichardt, L.F. (1991). Cell-surface regulation of β_1-integrin activity on developing retinal neurons. Nature 350, 68–71.

Ng-Sikorski, J., Andersson, R., Patarroyo, M., & Andersson, T. (1991). Calcium signaling capacity of the CD11b/CD18 integrin on human neutrophils. Exp. Cell Res. 195, 504–508.

Nicholson, L.J. & Watt, F.M. (1991). Decreased expression of fibronectin and the $\alpha_5\beta_1$ integrin during terminal differentiation of human keratinocytes. J. Cell Sci. 98, 225–232.

Nuckolls, G.H., Turner, C.E., & Turner, K. (1990). Functional studies of the domains of talin. J. Cell Biol. 110, 1635–1644.

Obara, M., Kang, M.S., & Yamada, K.M. (1988). Site directed mutagenesis of the cell-biding domain of human fibronectin: Separable synergistic sites mediate adhesive function. Cell 53, 649–657.

Orlando, R.A. & Cheresh, D.A. (1991). Arg-gly-asp binding leading to molecular stabilization between integrin $\alpha_v\beta_3$ and its ligand. J. Biol. Chem. 266, 19543–19550.

Osborn, L., Hession, C., Tizard, R., Vassallo, C., Luhowskyi, S., Chi-Rosso, G., Lobb, R. (1989). Direct expression cloning of vascular cell adhesion molecule 1, a cytokine-induced endothelial protein that binds to lymphocytes. Cell 59, 1203.

Otey, C.A., Pavalko, F.M., & Burridge, K. (1990). An interaction between α-actinin and the β1 integrin subunit *in vitro*. J. Cell Biol. 11, 721–729.

Parise, L.V., Helgerson, S.L., Steiner, B., Nannizzi, L., & Phillips, D.R. (1987). Synthetic peptides from fibrinogen and fibronectin change the conformation of purified platelet glycoproteins IIb-IIa. J. Biol. Chem. 262, 12597–12601.

Phillips, D.R., Charo, I.F., Parise, L.V., & Fitzgerald, L.A. (1988). The platelet membrane glycoprotein IIb-IIIa complex. Blood 71, 831–842.

Phillips, D.R., Charo, I.F., & Scarborough, R.M. (1991). GPIIb-IIIa: The responsive integrin. Cell 65, 359–362.

Plantefaber, L.C. & Hynes, R.O. (1989). Changes in integrin receptors on oncogenically transformed cells. Cell 56, 281–290.

Pytela, R., Pierschbacher, M.D., & Ruoslahti, E. (1985). Identification and isolation of a 140 kd cell surface glycoprotein with properties expected of a fibronectin receptor. Cell 40, 191–198.

Quaranta, V. & Jones, J.C.R. (1991). The internal affairs of an integrin. Trends Cell Biol. 1, 2–4.

Ramaswamy, H. & Hemler, M.E. (1990). Cloning, primary structure and properties of a novel human integrin β subunit. J. Eur. Mol. Biol. Organ. 9, 1561–1568.

Raub, T.J. & Kuentzel, S.L. (1989). Kinetic and morphological evidence for endocytosis of mammalian cell integrin receptors by using anti-fibronectin receptor β subunit monoclonal antibody. Exp. Cell Res. 184, 407–426.

Reszka, A.A. & Horwitz, A.F. (1991). Identification of avian integrin β1 subunit adhesion plaque localization sequences using site directed mutagenesis. J. Cell Biol. (In press).

Richter, J., Ng, S.J., Olsson, I., & Andersson, T. (1990). Tumor necrosis factor-induced degranulation in adherent human neutrophils is dependent on CDIIb/CD18-integrin-triggered oscillations of cytosolic free Ca^{2+}. Proc. Natl. Acad. Sci. USA 87, 9472–9476.

Ross, F.P., Alvarez, J.I., Chappel, J., Sander, D., Butler, W.T., Parach-Carson, M.C., Mintz, K.A., Robey, P.G., Teitelbaum, S.L., & Cheresh, D.A. (1991). Interactions between the bone matrix proteins osteopontin and bone sialoprotein and the osteoclast integrin $\alpha_v\beta_3$ potentiate bone resorption. Proc. Natl. Acad. Sci. USA (Submitted).

Ruoslahti, E. & Giancotti, F.G. (1989). Integrins and tumor cell dissemination. Cancer Cells 1, 119–126.

Ruoslahti, E. & Pierschbacher, M.D. (1987). New perspectives in cell adhesion: RGD and integrins. Science 238, 491–497.

Santoro, S.A. (1986). Identification of a 160,000 dalton platelet membrane protein that mediates the initial divalent cation-dependent adhesion of platelets to collagen. Cell 48, 549–554.

Sato, M., Sardana, M.K., Grasser, W.A., Garsky, V.M., Murray, J.M., & Gould, R.J. (1990). Echistatin is a potent inhibitor of bone resorption in culture. J. Cell Biol. 111, 1713–1723.

Scarborough, R.M., Rose, J.W., Hsu, M.A., Phillips, D.R., Fried, V.A., Campbell, A.M., Nannizzi, L., & Charo, I.F. (1991). Barbourin: A GPIIb-IIIa-specific integrin antagonist from the venom of *Sistrurus M. Barbouri*. J. Biol. Chem. 266, 9359–9362.

Schwartz, M.A., Both, G., & Lechene, C. (1989). Effect of cell spreading on cytoplasmic pH in normal and transformed fibroblasts. Proc. Natl. Acad. Sci. USA 86, 4525–4529.

Schwartz, M.A., Cragoe, E.J., & Lechene, C. (1990). pH regulation in spread and round cells J. Biol. Chem. 265, 1327–1332.

Schwartz, M.A., Ingber, D.E., Lawrence, M., Springer, T.A., & Lechene, C. (1991). Multiple integrins share the ability to induce elevation of intracellular pH. Exp. Cell Res. 195, 533–535.

Schwartz, M.A. & Lechene, C. (1991). Adhesion of 3T10 1/2 cells is required for protein kinase C-dependent activation of the Na-H antiporter by PDGF. (Submitted).

Schwarz, M.A., Owaribe, K., Kartenbeck, J., & Franke, W.W. (1990). Desmosomes and hemidesmosomes: Constitutive molecular components. Annu. Rev. Cell Biol. 6, 461–491.

Sheppard, D., Rozzo, C., Starr, L., Quaranta, V., Erle, D.J., & Pytela, R. (1990). Complete amino acid sequence of a novel integrin β subunit (β6) identified in epithelial cells using the polymerase chain reaction. J. Biol. Chem. 265, 11502–11507.

Singer, I.I., Scott, S., Kawka, D.W., Kazazis, D.M., Gailit, J., & Ruoslahti, E. (1988). Cell surface distribution of fibronectin and vitronectin receptors depends on substrate composition and extracellular matrix accumulation. J. Cell Biol. 106, 2171–2182.

Smith, J.W. & Cheresh, D.A. (1988). The RGD binding domain of the vitronectin receptor: Photoaffinity cross-linking implicates residues 61-203 of the β subunit. J. Biol. Chem. 263, 18726–18731.

Smith, J.W. & Cheresh, D.A. (1990). Integrin (αvβ3)-ligand interaction: Identification of a divalent cation-dependent, heterodimeric RGD binding site on the vitronectin receptor. J. Biol. Chem. 256, 2168–2172.

Smith, J.W. & Cheresh, D.A. (1991). Labeling of integrin $\alpha_v\beta_3$ with ^{58}Co(III). J. Biol. Chem. 266,11429–11432.

Smith, J.W., Ruggeri, Z.M., Kunicki, T.J., & Cheresh, D.A. (1990). Interaction of integrins $\alpha_v\beta_3$ and glycoprotein IIb-IIIa with fibrinogen. J. Biol. Chem. 265, 12267–12271.

Smith, J.W., Vestal, D.J., Irwin, S.V., Burke, T.A., & Cheresh, D.A. (1990). Purification and functional characterization of integrin αvβ5: An adhesion receptor for vitronectin. J. Biol. Chem. 265, 11088–11013.

Solowska, J., Guan, J.-L., Marcantonio, E.E., Trevithick, J.E., Buck, C.A., & Hynes, R.O. (1989). Expression of normal and mutant avian integrin subunits in rodent cells. J. Cell Biol. 109, 853–861.

Springer, T.A. (1990). Adhesion receptors of the immune system. Nature 346, 425–434.

Staatz, W.D., Fok, K.F., Zutter, M.M., Adams, S.T., Rodriguez, B.A., & Santoro, S.A. (1991). Identification of a tetrapeptide recognition sequence for the $\alpha_2\beta_1$ integrin in collagen. J. Biol. Chem. 266, 7363–7367.

Staatz, W.D., Rajpara, S.M., Wayner, E.A., Carter, W.G., & Santoro, S.A. (1989). The membrane glycoprotein Ia-IIa (VLA-2) complex mediates the Mg^{++}-dependent adhesion of platelets to collagen. J. Cell Biol. 108, 1917–1924.

Stallcup, W.B. (1988). Involvement of gangliosides and glycoprotein fibronectin receptors in cellular adhesion to fibronectin. Exp. Cell Res. 177, 90–102.

Stepp, M.A., Spurr-Michaud, S., Tisdale, A., Elwell, J., & Gipson, I.K. (1990). Proc. Natl. Acad. Sci. USA 87, 8970–8974.

Suzuki, S., Huang, Z.-S., & Tanihara, H. (1990). Cloning of an integrin β subunit exhibiting high homology with integrin β subunit. Proc. Natl. Acad. Sci. USA. 87, 5354–5358.

Suzuki, S. & Naitoh, Y. (1990). Amino acid sequence of a novel integrin β4 subunit an primary expression of the mRNA in epithelial cells. J. Eur. Mol. Biol. Organ. 9, 757–763.

Tamura, R.N., Cooper, H.M., Collo, G., & Quaranta, V. (1991). Cell-type specific integrin variants with alternative α chain cytoplasmic domain. Proc. Natl. Acad. Sci. USA (In press).

Tamura, R.N., Rozzo, C., Starr, L., Chambers, J., Reichardt, L.P., Cooper, H.M., & Quaranta, V. (1990). Epithelial integrin α6β4: Complete primary structure of α6 and variant forms of β4. J. Cell Biol. 111, 1593–1604.

Tapley, P., Horwitz, A.F., Buck, C.A., Burridge, K., Duggan, K., Hirst, R., & Rohrschneider, L. (1989). Analysis of the avian fibronectin receptor (integrin) as direct substrate for pp60 *v-src*. Oncogene 4, 325–333.

Vogel, B.E., Tarone, G., Giancotti, F.G., Gailit, J., & Ruoslahti, E. (1990). A novel fibronectin receptor with an unexpected subunit composition (αvβ1). J. Biol. Chem. 265, 5934–5937.

Volk, T., Fessler, L.I., & Fessler, J.H. (1990). A role for integrin in the formation of sarcomeric cytoarchitecture. Cell 63, 525–536.

Wagner, D.D., Urban-Pickering, M., & Marder, V.J. (1984). Von Willebrand protein binds to extracellular matrices independently of collagen. Proc. Natl. Acad. Sci. USA 81, 71–475.

Watt, F.M. (1988). The epidermal keratinocyte. BioEssays 8, 163–167.

Wayner, E.A. & Carter, W.G. (1987). Identification of multiple cell adhesion receptors for type VI collagen and fibronectin in human fibrosarcoma cells possessing unique α and common β subunits. J. Cell Biol. 105, 1873–1884.

Wayner, E.A., Carter, W.G., Piotrowicz, R.S., & Kunicki, T.J. (1988). The function of multiple extracellular matrix receptors in mediating cell adhesion to extracellular matrix: Preparation of monoclonal antibodies to the fibronectin receptor that specifically inhibit cell adhesion to fibronectin and react with platelet glycoproteins Ic-IIa. J. Cell Biol. 107, 1881–1891.

Wayner, E.A., Garcia-Pardo, A., Humphries, M.J., MacDonald, J.A., & Carter, W.G. (1989). Identification and characterization of the T lymphocyte adhesion receptor for an alternative cell attachment domain (CS-1) in plasma fibronectin. J. Cell Biol. 109, 1321–1330.

Wayner, E.A., Orlando, R.A., & Cheresh, D.A. (1991). Integrins α_v,β_3 and $\alpha_v\beta_5$ contribute to cell attachment to vitronectin but differentially distribute on the cell surface. J. Cell Biol. 113, 919–929.

Werb, Z., Tremble, P.M., Behrendtsen, O., Crowley, E., & Damsky, C.H. (1989). Signal transduction through the fibronectin receptor induces collagenase and stromelysin gene expression. J. Cell Biol. 109, 877–889.

Yamada, K.M. (1991). Adhesive recognition sequences. J. Biol. Chem. 266, 12809–12812.

Zusman, S., Patel-King, R.S., Ffrench-Constant, C., & Hynes, R.O. (1990). Requirements for integrins during *Drosophila* development. Development 108, 391–402.

THE ROLE OF THE EXTRACELLULAR MATRIX IN TUMOR GROWTH

Rafael Fridman

I. INTRODUCTION

The development of a malignant tumor depends upon the action of multiple factors leading to genotypic alterations and to the expression of a malignant phenotype (Weinberg, 1989; Bishop, 1991). Although critical early in the history of the neoplasm, genotypic alterations are not the only determinant in tumor progression (Nicolson, 1987). Similarly to the processes occurring during embryonic develop-

Advances in Molecular and Cell Biology
Volume 6, pages 253–269.

ISBN: 1-55938-515-4

ment and cell differentiation (Grobstein, 1967), the emergence of a malignant population of cells is also a consequence of inductive and restrictive signals which originate in the host environment (Nicolson, 1987). Among the external factors influencing neoplastic development is the extracellular matrix (ECM) produced by the stroma and the tumor cells. It is well known that the ECM is not merely a physical barrier against the invading tumors, but rather actively participates in tumor progression by regulating the expression of genes involved in malignancy. With the discovery of oncogenes and tumor suppressor genes, alterations in the interaction of tumor cells with the ECM have been partly attributed to the activation and/or inactivation of a particular group of genes able to regulate and control tumorigenesis and metastatic properties (Mareel and Van Roy, 1986; Liotta et al., 1991; Nicolson, 1991), and to allow tumor cells to circumvent microenvironmental control (Nicolson, 1991). Yet a growing line of evidences suggest that tumor cells are not completely autonomous and maintain a certain responsiveness to external stimuli which affect the expression of the malignant phenotype (De Cosse et al., 1973; Hayashi and Cunha, 1991). The molecular composition, and the organization of the ECM, as well as the biologically active molecules entrapped in the matrix, play an important role in tumor growth and in the expression of the malignant phenotype. The action of certain growth factors (Vlodavsky et al., 1991; Yayon et al., 1991) and oncogene products (Kiefer et al., 1991), both critical for tumorigenesis and angiogenesis, can also be regulated by ECM components. The ECM is a dynamic structure constantly modified by the surrounding cells through the production of enzymes, protease inhibitors, growth factors, and ECM components. The qualitative and quantitative changes accumulated on the ECM during tumor progression may generate aberrant influences on the already unstable population of transformed cells, further contributing to the process of tumor progression and to the conversion of a benign tumor into a malignant invasive carcinoma. Thus, it is possible that changes in the tumor ECM may actually precede the expression of a malignant phenotype and eventually may promote it (Beresford, 1988). On the other hand, the ECM of the developing tumor, which has not yet accumulated significant changes in structure and composition, may continue to emit "normal" signals to the premalignant cells. This may suppress or delay the appearance of the full malignant phenotype. This chapter discusses recent advances on the biochemical interactions of bioactive molecules with the ECM, and the relevance of such interactions on the stimulation of tumor cell proliferation by the ECM.

II. TUMOR CELL-EXTRACELLULAR MATRIX INTERACTIONS

Considerable knowledge now exists on the composition and structure of the ECM and the biochemical properties of the ECM components (Martin and Timpl, 1987; Yamada, 1989; Beck et al., 1990; Labat-Robert et al., 1990; Sage and Bornstein,

1991). Besides its role as a physical support for cells and tissues, the ECM actively regulates many cellular functions, and most importantly, it plays a pivotal role in the expression of the differentiated phenotype (Kleinman et al., 1987; Streuli and Bissell, 1991) and in malignant processes (Liotta, 1989; Starkey, 1990; Sweeny et al., 1991). Almost all types of cells interact with the ECM and are influenced by it. There is also growing evidence for a role of the ECM on immune responses through its interactions with lymphocytes and other immunocompetent cells (Levesque et al., 1991; Shimizu and Shaw, 1991; Sousa et al., 1991). Various components of the ECM, including laminin, fibronectin, and collagen type IV, have been shown to promote cell adhesion, motility, growth, and cell differentiation. Other components including SPARC/osteonectin, tenascin, and thrombospondin can act also as antiadhesive proteins for many cell types (Sage and Bornstein, 1991), thus modulating the responses of cells to the ECM.

The great diversity of cellular functions that can be influenced by the ECM would be consistent with a dynamic structure able to adapt properly to the particular event in progress. The temporal and spatial deposition of specific molecules into the matrix might be a mechanism by which cells can control the differential expression of genes during various stages in development, and possibly during neoplastic growth. The antiadhesive properties of tenascin, for example, may facilitate cell dispersion and invasion by interfering with the adhesion of cells to fibronectin (Chiquet-Ehrismann et al., 1988; Herlyn et al., 1991). Since tenascin expression in fibroblasts cells can be stimulated by transforming growth factor-β (TGF-β) (Pearson et al., 1988), the production of TGF-β by the tumor cells may stimulate the stroma to produce tenascin (Herlyn et al., 1991). Indeed, the tumor stroma has been shown to contain high levels of tenascin which is rarely present in normal tissues (Chiquet-Ehrismann et al., 1986; Inaguma et al., 1988).

Since human malignant gliomas are particularly rich in tenascin, a promising therapeutic model for this type of tumor was developed using radiolabeled monoclonal antibodies against tenascin (ShengLee et al., 1988). SPARC, another ECM protein, has also been shown to affect cell adhesion. Studies by Sage et al. (1989) have shown that addition of exogenous SPARC inhibited the spreading, but not the adhesion of fibroblasts, smooth muscle, and endothelial cells to various type of collagens. The formation of a complex between SPARC and collagen type III caused an inhibition of endothelial cell spreading (Sage et al., 1989). SPARC also has been shown to inhibit cell proliferation by retarding the progression of cells through the cell cycle (Sage and Bornstein, 1991). Thus, the adhesion, migration, and proliferation of cells in both normal and malignant situations can also be regulated by the production of ECM components with opposite biological functions.

Changes in cell shape caused by the ECM have profound effects on gene expression. The studies of Streuli and Bissell (1991) with mammary epithelial cells cultured on the reconstituted basement membrane, Matrigel, have shown that on Matrigel these cells adopt an alveolar-like configuration and secrete milk proteins into the lumen. Bissel et al. (1982) proposed that cell surface receptors for ECM components may regulate gene expression via molecular interactions of the receptors with the cytoskeleton, which in turn transmit the signal to the nucleus. The identification of cytoskeletal proteins, such as profilin (Goldschmit-Clermont and Janmey, 1991) and gelsolin (Janmey and Stossel, 1987), which are able to interact with both actin and membrane phosphoinositides, may help to understand the connection between changes in cell shape and signal transduction. An elegant model to investigate the molecular mechanisms involved in the regulation of transcriptional regulatory factors by the ECM has been recently reported by DiPersio et al. (1991). Normal hepatocytes transfected with plasmids containing the transcriptional enhancer of the albumin gene demonstrated enhancer activation when cultured on an ECM substrate. Interestingly, the tensile properties of the substratum, rather than its molecular composition in conjunction with changes in cell shape, appeared critical for enhancer activation (DiPersio et al., 1991). A relationship between ECM receptors (integrins) and transcriptional activation has been suggested from studies with the transactivation factor *Tat*. This protein, encoded by the human immunodeficiency virus, type I (HIV-1), contains the highly conserved tripeptide sequence Arg-Gly-Asp (RGD) present in various adhesive ECM proteins and important for integrin-mediated cell adhesion. *Tat* protein released by HIV-1 infected cells can be internalized by other cells and stimulate growth (Ensoli et al., 1990). Brake et al. (1990) have shown that purified *Tat* protein stimulates adhesion of various cell types. However, the role of the RGD sequence in the uptake of *Tat* by uninfected cells is still controversial. Indeed, Mann and Frankel (1991) have shown that truncated *Tat* proteins lacking the RGD sequence can still be internalized and exert transactivating activity. Interestingly, uptake of *Tat* protein was inhibited by polyanionic substances such as heparin and dextran sulfate (Mann and Frankel, 1991). It will be interesting to determine whether heparan sulfate proteoglycan and/or other ECM components play a role on the internalization and transactivating actions of *Tat* protein. Although the mechanism of *Tat* uptake, as well as its physiological relevance, are still unclear; these findings further support the notion that the interaction of biologically active molecules with ECM components may be important in the regulation of gene expression.

Another important function of the ECM is to participate in the regulation of cell growth. It has long been recognized that cells in contact with the ECM demonstrate an enhanced rate of proliferation. In stratified epithelium, the cells that are in contact with the basement membrane are those able to proliferate (Weinstein and Van Scott, 1965). *In vitro* studies have shown an enhanced rate of cell proliferation when the cells were seeded on ECM-coated dishes (Vlodavsky et al., 1980; Levesque et al., 1991). The effects of the ECM on cell proliferation are complex, involving changes in cell shape (Folkman and Moscona, 1978), direct action of ECM components with mitogenic properties (Terranova et al., 1986; Panayotou et al., 1989), and the presence of specific growth factors associated with normal constituents of the matrix (Ruoslahti and Yamaguchi, 1991). Heparan sulfate proteoglycan has been shown to bind fibroblast growth factor (FGF) and to act as a storage depot for this growth factor in the ECM (Vlodavsky et al., 1986; Saksela et al., 1988; Vlodavsky et al., 1991). Moreover, production of enzymes, such as heparanase, which degrades the sugar moieties of heparan sulfate proteoglycan, has been shown to cause the release of FGF from the ECM, making it available for the tumor cells (Bashkin et al., 1989; Vlodavsky et al., 1991). The released FGF may also stimulate the proliferation of endothelial cells increasing the vascularization of the tumor. Since changes in heparan sulfate have been shown to occur after neoplastic transformation (Iozzo, 1988), any alteration in the structure and amounts of heparan sulfate deposited in the ECM or on the surface of the tumor cells may consequently affect the stability and availability of FGF. Indeed, binding of basic FGF to its high-affinity receptor can occur only when the growth factor is bound to cell surface heparan sulfate, or to soluble heparin (Yayon et al., 1991). Kiefer et al. (1991) recently demonstrated that 3T3 cells transformed with the Int-2/Fgf-3 cDNA deposit Int-2 related products, a member of the FGF family, into the ECM in association with heparan sulfate. In the presence of heparin, the transformed cells showed reduced proliferation, decreased agar colony formation, and a more normal appearance. Besides heparan sulfate proteoglycan, other components of the matrix are also known for their ability to interact with growth factors and to modulate mitogenic activity. SPARC has been shown to form a complex with platelet-derived growth factor that inhibits receptor binding (Sage and Bornstein, 1991). Similarly, thrombospondin, an extracellular protein secreted by activated platelets and produced by many mesenchymal and epithelial cells *in vitro* has been shown to bind TGF-β, although the physiological role of such interaction remains unknown (Murphy-Ulrich et al., 1990). TGF-β also binds to collagen type IV (Paralkar et al., 1991).

The successful establishment of a cancer growth also depends on the ability of the malignant cells to degrade the ECM. Tumor cells produce a variety of proteolytic enzymes including plasminogen activators, metalloproteinases, and trypsin-like enzymes that all participate in ECM degradation. The level of expression of the metalloproteinases involved in the degradation of collagen type IV, for example, has been correlated with the expression of a malignant phenotype and with enhanced metastatic potential (Liotta et al., 1991). Many matrix-degrading enzymes are secreted in an inactive form that requires removal of the proenzyme region to acquire enzymatic activity. Furthermore, the latent form of the type IV collagenases forms a complex with their specific inhibitor (Matrisian, 1990; Woessner, 1991). The regulation of the enzymatic activity of the metalloproteinases, plasmin, and urokinase involves the action of both activators and specific inhibitors which are secreted into the extracellular space. The production of proteases by the tumor cells can be both constitutive or inducible depending on the state of progression of the tumor cells and on the availability of potential inductors. On the other hand, the action of activators or inhibitors may be controlled by other cell types and by a localized deposition of these molecules in those sites requiring regulation of protease activity.

Table 1. Tumor Cell Activities Regulated by the ECM

1.	Spreading, adhesion and detachment
2.	Cell growth
3.	Cell motility
4.	Induction of proteolytic enzymes
5.	Action of growth factors, proteases and protease inhibitors through interactions with ECM components
6.	Gene expression
7.	Cell-cell communication

Several proteases and protease inhibitors have been localized in the ECM, including plasminogen activators (Moscateli and Rifkin, 1988) and plasminogen activator inhibitor type 1 (PAI-1) (Mirumo and Loskutoff, 1989). Vitronectin, a plasma and ECM protein, has been shown to bind PAI-1. The binding of PAI-1 to vitronectin appears to increase the half-life of the inhibitor and to protect it from oxidation (Seiffert et al., 1990). Two metalloproteinases, the 72-kDa and the 92-kDa gelatinases, capable of degrading basement membrane collagen, posses a gelatin-binding domain homologous to that present in fibronectin (Collier et al., 1988; Wilhelm et al., 1989). Since both enzymes are secreted, this domain may be involved in the binding of the enzymes to the ECM. Matrigel, a reconstituted basement membrane, contains a 92-kDa gelatinase; however, it is yet unknown whether it is associated with a particular ECM component (unpublished observation). Table 1 summarizes the cellular activities of tumor cells that can be influenced and/or regulated by the ECM.

III. TUMORIGENESIS AND THE EXTRACELLULAR MATRIX

The ability of the ECM to influence the phenotype of cells has been elegantly shown in the works of Greenburg and Hay (1982, 1986, 1988) in which differentiated epithelial cells in contact with a three-dimensional matrix were able to transform into mesenchymal cells. These studies demonstrated that even well-differentiated cells can drastically change their phenotype in response to the ECM. Hay (1990) postulated that when cells emigrate from the epithelium, the ECM activates master regulatory gene/s that induce phenotypic transformation. This activity of the ECM appears to be mediated by integrins and syndecan, an heparan sulfate proteoglycan present on the surface of most epithelial cells (Hay, 1990). In kidney development, the conversion of mesenchyme to epithelium has been suggested to be controlled by a differential expression of laminin chains (Ekblom, 1989).

A similar situation of phenotypic transformation may occur in neoplasia in which premalignant tumor cells in contact with a different environment may undergo profound phenotypical and behavioral changes (Beresford, 1988). A large number of studies have demonstrated the influence of the environment (stroma and/or ECM) on tumorigenesis. It has been shown that teratocarcinoma cells implanted into the mouse blastocyst can develop into normal tissues (Mintz et al., 1975; Pierce et al., 1979). Loss of malignant properties was observed in basal cell carcinomas grown in association with normal stroma (Cooper and Pinkus, 1977) and human colon carcinoma cells cocultured with fetal rat mesenchyme differentiated into gland-like structures (Fukamachi et al., 1986). Hayashi and Cunha (1991) reported a loss of tumorigenicity and induction of secretory cytodifferentiation in Dunning prostatic adenocarcinoma cells grown in association with the mesenchyme of rat seminal vesicles. In other systems, the stroma has been shown to cause an enhancement of the malignant properties. For example, enhanced tumorigenicity of human cancer cells including prostate (Gleave et al., 1990; Camps et al., 1991), breast, and bladder tumor cells (Camps et al., 1991) was observed after coinoculation of the cells with fibroblasts. These effects were probably mediated by soluble factors produced by the fibroblasts; however, irradiated fibroblasts were also able to stimulate tumor growth suggesting a role for the ECM in these effects.

The study of the biological effects of the ECM in general, and the basement membrane in particular, have been facilitated by the use the Engelbreth–Holm–Swarm (EHS) murine tumor known to produce a multilaminar basement membrane-like matrix (Orkin et al., 1977). Biochemical analysis of the EHS basement membrane demonstrated the presence of laminin, collagen type IV, heparan sulfate proteoglycan, and entactin, all ubiquitous components of naturally occurring basement membranes (Kleinman et al., 1986). The EHS tumor became an excellent source for the isolation of specific basement membrane components and the production of Matrigel, a liquid extract of basement membrane components that forms a gel at 37 °C (Kleinman et al., 1986). Laminin was first isolated in intact form from neutral extracts of the EHS tumor and found to be composed of three distinct polypeptide chains designated A, B1, and B2 organized in a cross-shaped structure (Martin and Timpl, 1987; Beck et al., 1990). Today, there is considerable evidence that this particular organization of laminin is not universal, and other laminin isoforms have been described (Beck et al., 1990). The variability in the structure of laminin suggests that the basement membrane is not a single entity, but a dynamic network of proteins adapted to the functions and organization of each specific tissue. In the case of the EHS basement membrane, the matrix deposited by this particular type of tumor may reflect in its nature and composition some aspect of the malignant phenotype, including imbalance deposition of ECM components, presence of growth factors, enzymes, and other tumor-specific molecules. Despite its tumor origin, however, Matrigel has been shown to promote normal differentiation in many systems (Kleinman et al., 1986, 1987, 1991).

A strong line of evidence supporting the importance of the ECM in tumor growth came from experiments performed by several groups including ours on the effect of the ECM on the proliferation of tumor cells *in vivo*.

In these experiments, tumor cells were mixed with liquid Matrigel and then injected subcutaneously into athymic mice (Fridman et al., 1990). Using a series of Small Cell Lung Carcinoma (SCLC) cell lines, we showed that coinjection with Matrigel dramatically accelerated the growth of both classic and variant SCLC cell lines, two subtypes of SCLC cells. In the case of a variant SCLC cell line, tumor growth in the presence of Matrigel could be observed with as little as 25,000 cells per mouse, a cell inocula too small to produce tumors even 4 months after injection. Some of the classic SCLC cell lines formed tumors only in the presence of Matrigel, while others grew very slowly and they never reached the size of tumors obtained with Matrigel. The specific constitution of Matrigel appeared to be essential for cell proliferation since coinjection of SCLC cells with collagen type I did not result in tumor growth in spite of the ability of the collagen to polymerize into a gel. Small amounts of Matrigel were also sufficient to stimulate SCLC tumor growth since same size tumors were obtained with a 10-fold dilution of the original Matrigel solution (5 mg/mouse). A similar finding was observed with B16F10 melanoma cells in which case 100 to 500 μg of Matrigel/mouse were sufficient to support enhanced tumor growth (Fridman et al., 1991). These observations suggested that the component(s) of the Matrigel responsible for its effect on tumor growth was probably in a high concentration.

Indirect evidence suggesting a role for laminin in tumor growth and metastasis came from experiments using a synthetic peptide from the B1 chain of laminin, Tyr-Ile-Gly-Ser-Arg (YIGSR). The YIGSR sequence is a major site on laminin involved in receptor binding and in cell attachment (Graf et al., 1987). The YIGSR peptide has been shown to inhibit experimental metastasis formation (Iwamoto et al., 1987; Kawasaki et al., 1991), tumor growth, and angiogenesis (Sakamoto et al., 1991).

We have shown that presence of the YIGSR peptide in the mixture of SCLC cells and Matrigel significantly inhibited tumor growth (Figure 1). The effect was specific since another synthetic peptide with only one amino acid changed, had no effect. The presence of YIGSR inhibited the growth of the SCLC cell lines capable to attach and spread on laminin-coated dishes, but had no effect on one SCLC cell line unable to attach to laminin yet responsive to Matrigel when coinjected into athymic mice (Fridman et al., 1990). Thus, both laminin and other Matrigel constituents may play a role in the stimulation of tumor growth. However, coinjection of cells with purified laminin did not result in accelerated growth probably due to the inability of laminin to form a gel and/or its instability *in vivo*. The inhibitory effect of the YIGSR peptide on tumor growth could be a consequence of an inhibition of tumor-induced angiogenesis by the peptide as previously shown by Sakamoto et al. (1991).

In our studies (Fridman et al., 1990), it was interesting to observe that a single dose of the YIGSR peptide, administered at the time of injection in the mixture of cells and Matrigel, was sufficient to cause inhibition of growth. Probably, the gel formed by Matrigel provides a protective and stabilizing environment for both the

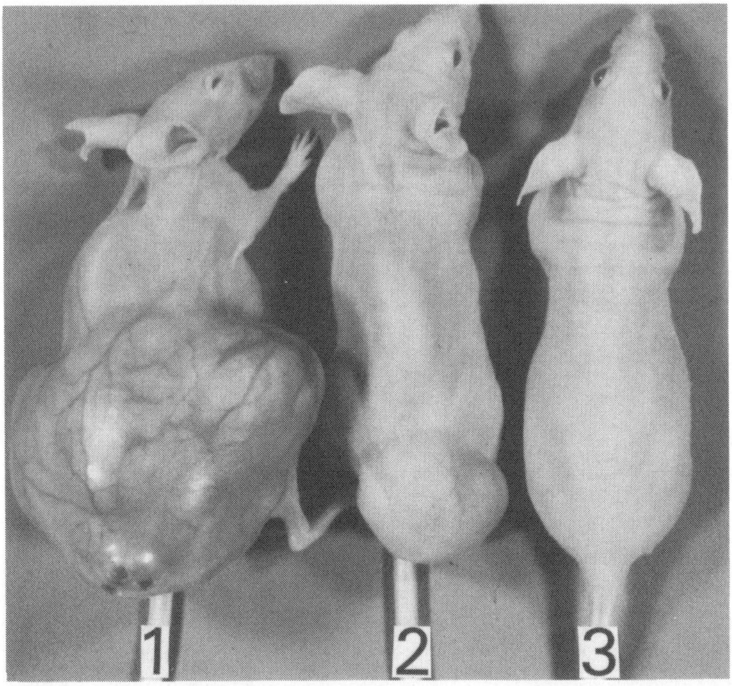

Figure 1. Effect of a synthetic YIGSR peptide on the proliferation of SCLC cells coinjected with Matrigel in athymic mice. **1.** SCLC cells + Matrigel; **2.** SCLC cells + Matrigel + YIGSR; **3.** SCLC cells alone. The photograph was taken 45 days after injection.

cells and the synthetic peptide. A recent study demonstrated that nanogram amounts of basic FGF coinjected with Matrigel and heparin result in a strong angiogenic response that does not occur in the absence of Matrigel, or in the absence of FGF (Passaniti et al., 1992). While Matrigel itself does not appears to be angiogenic, it can support vascularization if the appropriate factors are present. The growth of tumors in the presence of Matrigel may be due to the stabilization of angiogenic factors produced by the tumor cells (Vlodavsky et al., 1991). Table 2 lists some of the possible mechanisms by which Matrigel elicits its stimulation of cell proliferation *in vivo*.

The ability of Matrigel to support tumor growth is not limited to a specific type of tumor cells. A stimulatory effect of Matrigel has been reported on a variety of tumor cell lines, including B16F10 melanoma cells, human submandibular carcinoma A253 cells, human epidermoid carcinoma KB cells (Fridman et al., 1991), B16F6 melanoma cells, CBA/T6T6 mouse fibrosarcoma cells, rat Walker 256 adenocarcinoma cells (Vukicevic et al., 1991), and human HT1080 fibrosarcoma cells (Sweeny et al., 1991). Matrigel can also support the growth of primary human

Table 2. Possible Mechanisms of Matrigel Enhancement of Tumor Growth

1.	Protective environment
2.	Physical support
3.	Promotion of cell-cell interactions
4.	Stabilization of autocrine and paracrine growth factors and angiogenic factors produced by the tumor cells and the stroma
5.	Laminin-cell interactions
6.	Growth factors and cytokines sequestered in the matrix
7.	Stimulation of inflammation

cancer cells including renal and colon carcinoma cells (Fridman et al., 1991). The latter were obtained from metastatic lesions in the liver, disaggregated with enzymes and the single-cell suspension mixed with either Matrigel or culture media prior to injection into beige/nude mice. In the presence of Matrigel, the cells formed tumors that were 10-fold larger than those formed in the absence of the reconstituted basement membrane (Fridman et al., 1991). Similar results were reported by other investigators. Pretlow and coworkers (1991) showed that human prostatic carcinoma cell lines coinjected with Matrigel formed rapidly proliferating tumors with fewer cells. Most important was their observation that fragments of primary human prostatic carcinomas suspended in Matrigel and transplanted into male nude mice that had not received immunosupresive or hormonal treatments, produced tumors that retained the histopathological characteristics of prostate carcinomas and expressed prostate-specific antigen, acid phosphatase, and hexosaminidase. In light of the low rate of success in obtaining tumors from human prostatic cancer tissues, the results reported by Pretlow and coworkers using Matrigel are of great significance. Similar findings were reported by Passaniti and coworkers (1991) using human prostate carcinoma cell lines coinjected with Matrigel. Besides accelerated tumor growth, these researchers also observed prostate carcinoma cells in the bone marrow of mice bearing subcutaneous tumors, suggestive of an ability of the tumor cells to form distant metastasis in a pattern similar to that observed with prostate tumors. The metastatic prostate cells isolated from the bone marrow were established in culture and found to be more aggressive than the parental cell line. The appearance of bone metastasis from the subcutaneous tumors may be related to the accelerated growth of the primary tumor in response to Matrigel, or due to a direct effect of the matrix on the expression of malignant properties. Our previous results with other tumor cells stimulated by Matrigel indicate that rapid tumor growth does not necessarily result in metastasis formation even from cell lines derived from highly malignant tumors. The diversity of responses of tumor cells to the presence of Matrigel will be as heterogeneous as the populations existent in each particular tumor, and while certain cells will progress into more malignant stages, others may in fact become more differentiated and cease proliferation (Vukicevik et al., 1991). Indeed, we have found that a number of established cell lines and primary human tumor cells do not respond to Matrigel (Fridman et al., 1990; 1991).

A possible involvement of Matrigel in tumor progression has been suggested in a series of studies performed in our laboratory using the murine cell line NIH-3T3 (Fridman et al., 1992). Due to the sensitivity of these cells to become transformed, usually in a single step, the NIH-3T3 cell line has served well to define factors responsible for malignant transformation. The transforming abilities of some members of the large family of oncogene proteins were determined using NIH-3T3 cells as a target (Bishop, 1991). After transfection of a single dominant gene, the NIH-3T3 cells undergo profound changes in morphology, rate of cell proliferation, and tumorigenic potential. Transfection of activated *ras* oncogenes into these cells, for example, induces both tumorigenic and metastatic abilities (Thorgeirsson et al., 1986). Malignant transformation of NIH-3T3 cells can also be achieved by modulating the culture conditions (Rubin and Xu, 1989). Interestingly, the expression of the malignant phenotype is not irreversible, and under the appropriate conditions the cells can return to the nontransformed phenotype (Rubin et al., 1990). Since the transition between a normal and a transformed state could be controlled by the growth conditions, Rubin and Xiu (1989) suggested that the acquisition of a transformed phenotype in NIH-3T3 cells can occur in the absence of conventional genetic alterations. In these cells, however, most of the genetic alterations and qualitative changes involved in tumor progression probably occurred earlier during their establishment in culture. Thus, NIH-3T3 cells reflect a very advanced stage in tumor progression, and exposure to the right stimulus initiates the necessary events leading to their conversion into a malignant cell. We have found that coinjection of nontransformed and nontumorigenic NIH-3T3 cells with Matrigel resulted in tumor formation (Fridman et al., 1992). These tumors grew very slowly and showed the histopathological characteristics of a malignant invasive fibrosarcoma. An extensive degree of vascularization was common, but no distant metastases were observed. The cells isolated from the tumors and maintained in culture showed a transformed morphology with spindle shape cells and absence of contact inhibition. When injected again without Matrigel into athymic mice, the cells formed rapidly, proliferating tumors. If injected intravenously, the transformed NIH-3T3 cells formed many tumors in the surface of the lungs. Thus, the cells isolated from the Matrigel-induced tumors acquired a transformed phenotype and the ability to form experimental metastasis. It is difficult to determine whether the formation of tumors by the NIH-3T3 cells coinjected with Matrigel was a result of a selection of a malignant subpopulation of cells already present in the parental cell line, or to the appearance of newly transformed cells; however, the long latency period of the tumors suggested that a gradual process of selection and adaptation were probably required for the expression of the malignant phenotype. Matrigel acted as an optimum microenvironment for tumor development (see Table 2), and demonstrated that the contact of premalignant cells with the ECM facilitated malignant transformation.

IV. CONCLUSIONS

The complex interactions occurring between cells (tumor and stroma) and the ECM during tumor progression are schematically described in Figure 2. This model ascribes a fundamental role for the ECM in tumor progression. The ECM comprising the basement membrane and the connective tissue can act as a reservoir for growth factors, enzymes, protease inhibitors, cytokines, and other bioactive molecules deposited by both the tumor and the stroma cells. The ECM-bound molecules act upon the cells affecting gene expression. In response, the tumor and stroma cells modify the structure of the ECM through the deposition and degradation of ECM components and secretion of bioactive molecules into the matrix. The modified ECM elicits new cellular responses via specific receptors such as the integrins (Albelda and Buck, 1990). Changes in ECM receptors can also modulate the response of cells to the ECM. There is also a direct interaction between the tumor

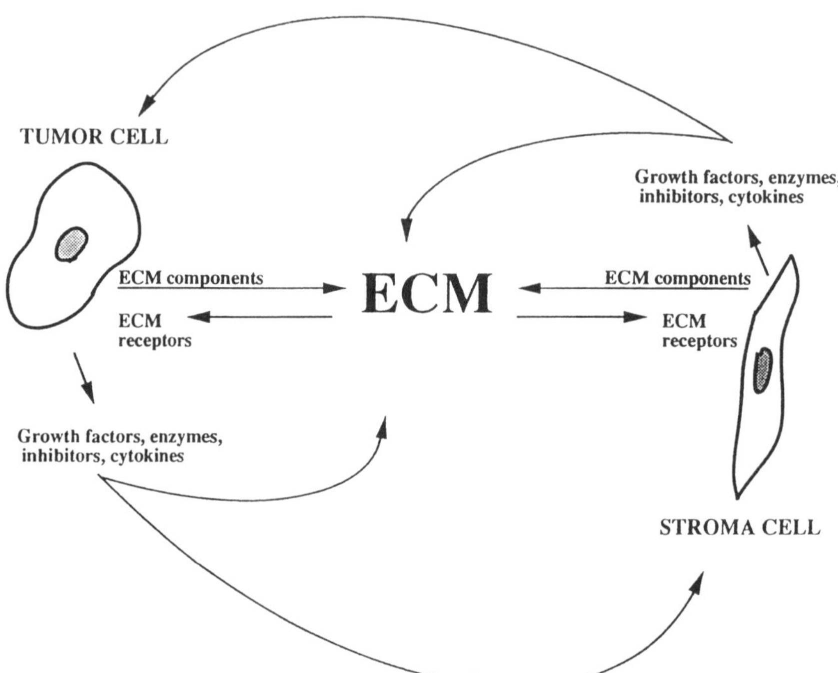

Figure 2. Schematic diagram of the interactions of tumor cells with the ECM and stroma. The ECM includes both the basement membrane produced by the tumor cells and the connective tissue matrix deposited by the stroma cells.

cells and the stroma through the production of diffusible substances and the establishment of cell-cell contacts. This multifactorial process finally concludes with the establishment of a malignant population of cells and the appearance of a cancer growth. The molecular events occurring in response to the ECM are now starting to unveil (DiPersio et al., 1991; Streuli and Bissell, 1991). The elucidation of the mechanism of action of the various ECM receptors, their interactions with the cytoskeleton, and their role in signal transduction will help to define in a more precise manner how signals transmitted by the ECM affect gene expression. Also, the study of the interactions of bioactive molecules with the components of the ECM will provide new insights into the role of the ECM in normal situations and in diseases.

REFERENCES

Albelda, S.M. & Buck, C.A. (1990). Integrins and other cell adhesion molecules. FASEB J. 4, 2868–2880.

Bashkin, P., Doctrow, S., Klagsbrun, M., Svahn, C.M., Folkman, J., & Vlodavsky, I. (1989). Basic fibroblast growth factor binds to subendothelial extracellular matrix and is released by heparitinase and heparin-like molecules. Biochemistry 28, 1737–1743.

Beck, K., Hunter, I., & Engel, J. (1990). Structure and function of laminin: anatomy of a multidomain glycoprotein. FASEB J. 4, 148–160.

Beresford, W.A. (1988). A stromal role in epithelial metaplasia and cancer. The Cancer J. 2, 145–147.

Bishop, J.M. (1991). Molecular themes in oncogenesis. Cell 64, 235–248.

Bissell, M.J., Hall, H.G., & Parry, G. (1982). How does the extracellular matrix direct gene expression? J. Theor. Biol. 99, 31–68.

Brake, D.A., Debouck, C., & Biesecker, G. (1990). Identification of an Arg-Gly-Asp (RGD) cell adhesion site in human immunodeficiency virus type 1 transactivation protein, *tat*. J. Cell Biol. 111, 1275–1281.

Camps, J.L., Chang, S-M., Hsu, T.C., Freeman, M.R., Hong, S-J, Zhau, H.E., von Eschenbach A.C., & Chung, L.W.K. (1990). Fibroblast-mediated acceleration of human epithelial tumor growth *in vivo*. Proc. Natl. Acad. Sci. USA 87, 75–79.

Chiquet-Ehrisman, R., Mackie, E.J., Pearson, C.A., & Sakakura, T. (1986). Tenascin: an extracellular matrix protein involved in tissue interactions during fetal development and oncogenesis. Cell 47, 131–139.

Chiquet-Ehrisman, R., Kalla, P., Pearson, C.A., Beck, K., & Chiquet, M. (1988). Tenascin interferes with fibronectin action. Cell 53, 383–390.

Collier, I.E., Wilhelm, S.M., Eisen, A.Z., Marmer, B.L., Grant, G.E., Seltzer, J.L., Kronberger, A., He, C., Bauer, E.A., & Goldberg, G.I. (1988). H-ras oncogene-transformed human bronchial epithelial cells (TBE-1) secret a single metalloprotease capable of degradaing basement membrane collagen. J. Biol. Chem. 263, 6579–6587.

Cooper, M. and Pinkus, H. (1977). Intrauterine transplantation of rat basal cell carcinoma: a model for reconversion of malignant to benign growth. Cancer Res. 37, 2544–2552.

De Cosse, J., Gossens, C.L., & Kuzma, J.F. (1973). Breast cancer: induction of differentiation by embryonic tissue. Science (Washington, DC) 181, 1057–1058.

DiPersio, C.M., Jackson, D.A., & Zaret, K.S. (1991). The extracellular matrix coordinately modulates liver transcription factors and hepatocyte morphology. Mol. Cell. Biol. 11, 4405–4414.

Donjacour, A.A. & Cunha, G.R. (1991). Stromal regulation of epithelial function. In: Regulatory

Mechanisms in Breast Cancer (Lippman, M. & Dickson, R., Eds.), pp. 335–364. Kluwer Academic Publishers, Boston.

Ekblom, P. (1989). Developmentally regulated conversion of mesenchyme to epithelium. FASEB J. 3, 2141–2150.

Ensoli, B., Barillari, G., Salahuddin, S.Z., Gallo, R.C., and Wong-Staal, F. (1990). Tat protein of HIV-1 stimulates growth of cells derived from Kaposi's sarcoma lesions of AIDS patients. Nature (Lond.) 345, 84–86.

Folkman, J. & Moscona, A. (1978). Role of cell shape in growth control. Nature (Lond.) 273, 345–349.

Fridman, R., Giaccone, G. Kanemoto, T., Martin, G.R., Gazdar, A.F., & Mulshine, J.L. (1990). Reconstituted basement membrane (Matrigel) and laminin can enhance the tumorigenicity and the drug resistance of small lung cancer cell lines. Proc. Natl. Acad. Sci. USA 87, 6698–6702.

Fridman, R., Kibbey, M.C., Royce, L.S., Zain, M., Sweeney, T.M., Jicha, D.L., Yannelli, J.R., Martin, G.R., & Kleinman, H.K. (1991). Enhanced tumor growth of both primary and established human and murine tumor cells in athymic mice after coinjection with Matrigel. J. Natl. Can. Inst. 83, 769–774.

Fridman, R., Sweeney, T.M., Zain, M., Martin, G.R., & Kleinman, H.K. (1992). Malignant transformation of NIH-3T3 cells after coinjection with a reconstituted basement membrane (Matrigel). Int. J. Cancer 51, 740–744.

Fukamashi, I., Mizuno, T., & Kim, Y.S. (1986). Morphogenesis of human colon cancer cells with fetal rat mesenchyme in organ culture. Experientia (Basel) 42, 312–315.

Gleave, M., Hsieh, J-T., Gao, C., von Eschenbach A.C., & Chung, L.W.K. (1991). Acceleration of human prostate cancer growth *in vivo* by factors produced by prostate and bone fibroblasts. Cancer Res. 51, 3653–3761.

Goldscmith-Clermont, P.J., & Janmey, P.A. (1991). Profilin, a weak CAP for actin and RAS. Cell 66, 419–421.

Graf, J., Iwamoto, Y., Sasaki, M., Martin, G.R., Kleinman, H.K., Robey, F.A., & Yamada, Y. (1987). Identification of an amino acid sequence in laminin mediating cell attachment, chemotaxis and receptor binding. Cell 48, 989–996.

Greenburg, G. & Hay, E.D. (1982). Epithelia suspended in collagen gels can loose polarity and express characteristics of migrating mesenchymal cells. J. Cell Biol. 95, 333–339.

Greenburg, G. & Hay, E.D. (1986). Cytodifferentiation and tissue phenotype change during transformation of embryonic lens epithelium to mesenchyme-like cells *in vitro*. Dev. Biol. 115, 363–379.

Greenburg, G. & Hay, E.D. (1988). Cytoskeleton and thyroglobulin expression change during transformation of thyroid epithelium to mesenchyme-like cells. Development 102, 605–622.

Grobstein, C. (1967). Mechanisms of organogenetic tissue interaction. J. Natl. Cancer Ins. Monogr. 26, 279–299.

Hayashi, N. & Cunha, G.R. (1991). Mesenchyme-induced changes in the neoplastic characteristics of the dunning prostatic adenocarcinoma. Cancer Res. 51, 4924–4930.

Hay, D.E. (1990). Role of cell-matrix contacts in cell migration and epithelial-mesenchymal transformation. Cell Differ. Dev. 32, 367–376.

Herlyn, M., Graeven, U., Speicher, D., Sela, B-A, Bennicelli, J.L., Kath, R., & Guerry IV, D. (1991). Characterization of tenascin secreted by human melanoma cells. Cancer Res. 51, 4853–4858.

Inaguma, Y., Kusakabe, M., Mackie, E.J., Pearson, C.A., Chiquet-Ehrisman, R., & Sakakura, T. (1988). Epithelial induction of stromal tenascin in the mouse mammary gland: from embryogenesis to carcinogenesis. Dev. Biol. 128, 245–255.

Iozzo, R.V. (1988). Cell surface heparan sulfate proteoglycan and the neoplastic phenotype. J. Cell. Biochem. 37, 61–78.

Iwamoto, Y., Robey, F.A., Graf, J., Sasaki, M., Kleinman, H.K., Yamada, Y., & Martin, G.R. (1987). YIGSR, a synthetic laminin pentapeptide, inhibits experimental metastasis formation. Science (Washington, DC) 238, 1132–1134.

Janmey, P. & Stossel, T. (1987). Modulation of gelsolin function by phosphatidylinositol 4,5-biphosphate. Nature (Lond.) 325, 362–364.

Kawasaki, K., Namikawa, M., Murakami, T., Mizuta, T., Iwai, Y., Hama, T., & Mayumi, T. (1991). Amino acids and peptides. XIV. Laminin related peptides and their inhibitory effect on experimental metastasis formation. Biochem. Biophys. Res. Comm. 174, 1159–1162.

Kiefer, P., Peters, G., & Dickson, C. (1991). The int-2/fgf-3 oncogene product is secreted and associates with extracellular matrix: implication for cell transformation. Mol. Cell. Biol. 11, 5929–5936.

Kleinman, H.K., McGravey, M.L., Hassell, J.R., Star, V.L., Cannon, F.B., Laurie, G.W., & Martin, G.R. (1986). Basement membrane complexes with biological activity. Biochemistry 25, 312–318.

Kleinman, H.K., Graf, J., Iwamoto, Y., Kitten, G.T., Ogle, R.C., Sasaki, M., Yamada, Y., Martin, G.R., & Luckenbill-Edds, L. (1987). Role of basement membrane in cell differentiation. N.Y. Acad. Sci. 513, 134–145.

Kleinman, H.K., Kibbey, M.C., Schnaper, H.W., Hadley, M.A., Dym, M., & Grant, D.S. (1993). Role of basement membrane in differentiation. In: Molecular Aspects of Basement Membranes (Timpl, R., & Rohrbach, D., Eds.), pp. 161–186. Academic Press, New York.

Labat-Robert, J., Bihari-Varga, M., & Robert, L. (1990). Extracellular matrix. FEBS Lett. 268, 386–393.

Levesque, J-P., Hatzfeld, A., & Hatzfeld, J. (1991). Mitogenic properties of major extracellular proteins. Immunol. Today 12, 258–266.

Liotta, L.A. (1986). Tumor invasion and metastases. Role of the extracellular matrix. Cancer Res. 46, 1–7.

Liotta, L.A., Steeg, P.S., & Stetler-Stevenson, W.G. (1991). Cancer metastasis and angiogenesis: an imbalance of positive and negative regulation. Cell 64, 327–336.

Mann, D.A. & Frankel, D.A. (1991). Endocytosis and targeting of exogenous HIV-l *Tat* protein. EMBO J. 10, 1733–1739.

Mareel, M.M. & Van Roy, F.M. (1986). Are oncogenes involved in invasion and metastasis? Anticancer Res. 6, 419–436.

Martin, G.R. & Timpl, R. (1897). Laminin and other basement membrane components. Ann. Rev. Cell Biol. 3, 57–85.

Matrisian, L.M. (1990). Metalloproteinases and their inhibitors in matrix remodelling. Trends Gen. 6, 121–125.

Mintz, B. & Illmense, K. (1975). Normal genetically mosaic mice produced from malignant teratocarcinoma cells. Proc. Natl. Acad. Sci. USA 72, 3585–3589.

Mirumo, J. & Loskutoff, D.J. (1989). Binding of type I plasminogen activator inhibitor to the extracellular matrix of cultured bovine endothelial cells. J. Biol. Chem. 264, 5058–5063.

Moscatelli, D. & Rifkin, D.B. (1988). Membrane and matrix localization of proteinases: a common theme in tumor cell invasion and angiogenesis. Biochim. Biophys. Acta 948, 67–85.

Murphy-Ulrich, J.E., Schultz-Cherry, S., & Hook, M. (1990). Thrombospondin is associated with growth factor-β; effect on endothelial cell growth. J. Cell Biol. 111, 148a.

Nicolson, G.L. (1987). Tumor cell instability, diversification and progression to the metastatic phenotype: from oncogene to oncofetal expression. Cancer Res. 47, 1473–1487.

Nicolson, G.L. (1991). Molecular mechanisms of cancer metastasis: tumor and host properties and the role of oncogenes and suppressor genes. Current Opinions in Oncology 3, 75–92.

Orkin, R.W., Gehron, P., McGoodwin, E.B., Martin, G.R., Valentine, T., & Swarm, R. (1977). A murine tumor producing a matrix of basement membrane. J. Exp. Med. 145, 204–220.

Panayotou, G., End, P., Aumailley, M., Timpl, R., & Engel, J. (1989). Cell 56, 93–101.

Paralkar, V.M., Vukicevic, S., & Reddi, A.H. (1991). Transforming growth factor β type 1 binds to collagen IV of basement membrane matrix: Implications for development. Dev. Biol. 143, 303–308.

Passaniti, A., Guo, Y., Haney, J., & Long, P.V. (1991). Isolation of highly invasive human prostate carcinoma from the bone marrow of an athymic nude mouse after subcutaneous tumor growth of PC-3 cells. J. Cell Biol. 115, 263a.

Passaniti, A., Taylor, R.M., Pli, R., Guo, Y., Long, P.V., Haney, J.A., Pauly, R.R., Grant, D.S., & Martin, G.R. (1992). A simple, quantitation method for assessing angiogenic and antiangiogenic agents using reconstituted basement membrane, heparin, and fibroblast growth factor. Lab. Invest. 67, 519–528.

Pearson, C.A., Pearson, D., Shibahara, S., Hofsteenge, J., & Chiquet-Ehrismann, R. (1988). Tenascin: cDNA cloning and induction by TGF-β. EMBO J. 7, 2977–2981.

Pierce, G.B., Lewis, S.H., Miller, G.J., Motitz, E., & Miller P. (1979). Tumorigenicity of embryonal carcinoma as an assay to study control of malignancy by the blastocyst. Proc. Natl. Acad. Sci. USA, 76, 6649–6655.

Pretlow, T.G., Delmoro, C.M., Dilley, G.G., Spadafora, C.G., & Pretlow, T.P. (1991). Transplantation of human prostatic carcinoma into nude mice in Matrigel. Cancer Res. 51, 3814–3817.

Rubin, H. & Xiu, K. (1989). Evidence for the progressive and adaptive nature of spontaneous transformation in the NIH 3T3 cell line. Proc. Natl. Acad. Sci. USA 86, 1860–1864.

Rubin, A.L., Arnstein, P., & Rubin, H. (1990). Physiological induction and reversal of focus formation and tumorigenicity in NIH 3T3 cells. Proc. Natl. Acad. Sci. USA 87, 10005–10009.

Ruoslahti, E. & Yamaguchi, Y. (1991). Proteoglycans as modulators of growth factor activities. Cell 64, 867–869.

Sage, E.H., Vernon, R.B., Funk, S.E., Everitt, E.A., & Angello, J. (1989). SPARC, a secreted protein associated with cellular proliferation, inhibits cell spreading *in vitro* and exhibits Ca^{+2}-dependent binding to the extracellular matrix. J. Cell Biol. 109, 341–356.

Sage, E.H. & Bornstein, P. (1991). Extracellular proteins that modulate cell-matrix interactions. J. Biol. Chem. 266, 14831–14834.

Sakamoto, N., Iwahana, M., Tanaka, N.G., & Osada, Y. (1991). Inhibition of angiogenesis and tumor growth by a synthetic laminin peptide CDPGYIGSR-NH_2. Cancer Res. 51, 903–906.

Saksela, O., Moscatelli, D., Sommer, A., & Rifkin, D.B. (1988). Endothelial cell-derived heparan sulfate binds basic fibroblast growth factor and protects it from proteolytic degradation. J. Cell Biol. 107, 743–751.

Seiffert, D., Mimuro, J., Schleef, R.R., & Loskutoff, D.J. (1990). Interactions between type 1 plasminogen activation inhibitor, extracellular matrix and vitronectin. Cell Differen. Develop. 32, 287–292.

ShengLee, Y., Bullard, D.E., Zalutsky, M.R., Coleman, R.E., Wikstrand, C.J., Friedman, H.S., Colapinto, E.V., & Bigner, D.D. (1988). Therapeutic efficacy of antiglioma mesenchymal extracellular matrix ^{131}I-radiolabeled murine monoclonal antibody in a human glioma xenograft model. Cancer Res. 48, 559–566.

Shimizu, Y. & Shaw, S. (1991). Lymphocyte interactions with extracellular matrix. FASEB J. 5, 2292–2299.

Sousa, M, Tilney, N.L., Kupiec-Weglinski, J.W. (1991). Recognition of self within self: specific lymphocyte positioning and the extracellular matrix. Immunol. Today 12, 262–266.

Starkey, J.R. (1990). Cell-matrix interactions during tumor invasion. Cancer and Met. Rev. 9, 113–123.

Streuli, C.H. & Bissell, M. (1991). Mammary epithelial cells, extracellular matrix, and gene expression. In: Regulatory Mechanisms in Breast Cancer (Lippman, M. & Dickson, R., Eds.), pp. 365–381. Kluwer Academic Publishers, Boston.

Sweeney, T.M., Kibbey, M.C., Zain, M., Fridman, R., & Kleinman, H.K. (1991). Basement membrane and the SIKVAV laminin-derived peptide promote tumor growth and metastases. Cancer and Met. Rev. 10, 245–254.

Terranova, V., Aumailley, M., Sultan, L.H., Martin, G.R., & Kleinman, H.K. (1986). Regulation of cell attachment and cell number by fibronectin and laminin. J. Cell. Physiol. 127, 473–479.

Thorgeirson, U.P., Turpeenniemi-Hujanen, T., Williams, J.E., Westin, E.H., Heilman, C.A., Talmadge, J.E., & Liotta, L.A. (1985). NIH/3T3 transfected with human tumor DNA containing activated ras oncogenes express the metastatic phenotype in nude mice. Mol. Cell. Biol. 5, 259–262.

Vlodavsky, I., Lui, G.M., & Gospodarowicz, D. (1980). Morphological appearance, growth behavior and migratory activity of tumor cells maintained on extracellular matrix versus plastic. Cell 19, 607–616.

Vlodavsky, I., Folkman, I., Sullivan, R., Fridman, R., Ishai-Michaeli, R., Sasse, J., & Klagsbrun, M. (1987). Endothelial cell-derived basic fibroblast growth factor: synthesis and deposition into subendothelial extracellular matrix. Proc. Natl. Acad. Sci. USA 84, 2292–2296.

Vlodavsky, I., Bar-Shavit, R., Ishai-Michaeli, R., Bashkin, P., & Fuks, Z. (1991). Extracellular sequestration and release of fibroblast growth factor: a regulatory mechanism? Trends Biochem. Science 16, 268–271.

Vukicevic, S., Somogyi, L., Martinovic, I., Zic, R., Kleinman, H.K., & Marusic, M. (1992). Reconstituted basement membrane (Matrigel) promotes the survival and influences the growth of murine tumors. Int. J. Cancer 50, 791–795.

Weinberg, R.A. (1989). Oncogenes, antioncogenes, and the molecular basis of multistep carcinogenesis. Cancer Res. 49, 3713–3721.

Weinstein, G.D. & Van Scott, E.J. (1965). Autoradiographic analysis of turnover times of normal and psoriatic epidermis. J. Invest. Dermatol. 45, 257–265.

Wilhelm, S.M., Collier, I. E., Marmer, B.L., Eisen, A.Z., Grant, G.A., & Goldberg, G.I. (1989). SV40-transformed human lung fibroblasts secrete a 92-kDa type IV collagenase which is identical to that secreted by normal human macrophages. J. Biol. Chem. 264, 17213–17221, 1989.

Woessner, J.F. (1991). Matrix metalloproteinases and their inhibitors in connective tissue remodeling. FASEB J. 5, 2145–2154.

Yamada, K.M. (1989). Fibronectins: structure, functions, and receptors. Curr. Opin. Cell Biol. 1, 956–963.

Yayon, A., Klagsbrun, M., Esko, J.D., Leder, P., & Ornitz, D.M. (1991). Cell surface, heparin-like molecules are required for binding of fibroblast growth factor to its high affinity receptor. Cell 64, 841–848.

SOME ASPECTS OF INBORN AND ACQUIRED CONNECTIVE TISSUE DISEASES:

A SPECIAL EMPHASIS ON RENAL DISEASE

A. Noël, J.A. Bruijn, E.C. Bergijk, and J.M. Foidart

Advances in Molecular and Cell Biology
Volume 6, pages 271–292.

ISBN: 1-55938-515-4

I. INTRODUCTION

Connective tissue is an ubiquitous component of the organism. Connective tissue cells and extracellular matrix (ECM) play an important role in supporting the tissue architecture and morphological differentiation of cells and organs. Besides the mechanical properties, however, the ECM also influences important biological functions such as cell differentiation, proliferation, migration, and adhesion. These activities are mediated by distinct components. Most of them are multidomain glycoproteins interacting with one another and with cells via specific membrane receptors belonging mostly to the integrin family (Abelda and Buck, 1990; Ruoslahti, 1991; David Cheresh, this volume). The multiple functions of connective tissue are achieved by varying the composition and the proportions as well as the macromolecular organization of the individual components. Synthesis, degradation of ECM molecules, and multimolecular modeling of complex ECM structures are tightly controlled events that guarantee the physiological functions of connective tissues.

Several diseases are due to defects in the primary structure of the individual components of the ECM, or to more complex disturbances in the regulatory mechanisms of connective tissue metabolism. The accumulated information concerning the structure and biological activities of the protein constituents provides new insights about their possible role in inborn and acquired diseases. The innate errors of collagen metabolism are the best characterized model which turned out to be extremely useful in understanding the more complex alterations taking place in acquired diseases. Here our knowledge is much more limited. Intense efforts are still needed to precisely define the molecular events involved in various diseases. Innate disorders of connective are valuable models to understand the function of different structures and the steps of the metabolism of matrix proteins. The elucidation of mutations in collagen genes has already provided significant insight into molecular assembly, chain folding, secretion, processing, and fibril formation. The creation of animal models by the transgenic approach and the identification of natural or introduced mutations in additional collagens will allow a wider understanding of the function of many collagens. Another important area in which there may be extensive therapeutic applications is the characterization and regulation of promoter regions for collagen genes. Such mutations in these regions could explain diseases resulting in abnormal deposition of collagen. Overexpression of collagen genes in various fibroses (cirrhosis, sclerosis, keloids), or underexpression of collagen genes in osteoporosis might also be the consequence of mutations occurring in the regulatory promoter regions of these collagen genes.

Whereas the inborn disorders of collagen metabolism are relatively rare, acquired diseases in which connective tissues play a major role are much more common. Scleroderma, hypertrophic scars, keloids, fibrosis (lung, liver), arthrosis, atherosclerosis, atrophy, wound healing defects, and metastasis of tumors are well known examples of such diseases. In such conditions, the matrix proteins have a normal composition, but their synthesis and/or degradation is altered. The isolation and molecular dissection of the several domains of extracellular matrix glycoproteins allowed the definition of their biological properties. The precise regulation of the genes that encode the collagen chains and large matrix proteins is crucial for normal embryo development. The control by hormones, cytokines, and growth factors on the expression of matrix components and of the enzymes that affect their cross-linking and accumulation or degradation is extensively studied during normal growth and in various acquired diseases. The precise definition of the disturbed regulation of ECM production and remodeling in acquired disorders may provide alternative therapeutic strategies and lead to a better control of the healing process, fibrogenesis, and tumor progression. Furthermore, the action of auto-immune antibodies directed against ECM components and especially basement membrane (BM) can lead to severe diseases often involving the kidney and skin. The complex interplays between ECM and the immune system represent a particular area of intense clinical research. In the second part of this chapter, we will attempt to demonstrate how the extracellular matrix is involved in acquired diseases. We will focus on renal diseases. Indeed, multiple pathogenetic mechanisms involving the glomeruli, the interstitium or both lead to ECM accumulation and renal dysfunction. During glomerular matrix expansion, specialized cellular structures are replaced by fibroblasts, collagen and mesenchymal matrix, with concomitant disruptions of the kidney's functions of filtration, secretion and reabsorption. The pathogenesis involves different types of cells and various growth factors or inflammatory factors released by resident or infiltrating cells such as T cells, neutrophils, monocytes/macrophages and platelets. On the other hand, autoimmune attack against ECM components can also result in severe renal disorders.

This review will analyze the specific involvement of biochemically defined components of the ECM in both inborn and acquired diseases with a particular emphasis on immune-mediated renal injury. Due to the extensive research on ECM structure, function and pathology, no attempt can be made here to summarize comprehensively the entire topic. This paper will first briefly describe collagen-related inborn disorders and then focus on acquired diseases with special attention to the kidney. After a brief description of normal glomerular structure, two fields of renal pathology in which ECM plays a central role will be considered: (1) glomerular ECM expansion; and (2) ECM as target for autoantibodies. Although this review will emphasize glomerulosclerosis, much of the discussion may be applicable to abnormal ECM deposition in general.

II. COLLAGEN IN INBORN DISEASES

The investigation of the inherited matrix disorders has progressed on two parallel fronts. Two decades of protein chemistry have provided detailed background information on the structure of collagens. This has been used to identify the

Table 1. Inborn Diseases Associated to Mutations in Collagen Genes

Collagen Genes	*Diseases*
Type I collagen genes (COL1A1, COL1A2)	Osteogenesis Imperfecta Ehlers-Danlos syndrome Osteoporosis
Type II collagen gene (COL2A1)	Chondrodysplasias Stricker syndrome Achondrogenesis Spondyloephyseal dysplasia Kniest syndrome
Type III collagen gene (COL3A1)	Ehlers-Danlos syndrome Familial aneurysm
Type IV collagen gene (COL4A5)	Alport's syndrome

candidate genes for analysis using DNA markers. On the other hand, segregation analysis has, in turn, sorted out which diseases are caused by the following collagen gene mutations and which are not (Vuorio, 1986; Sykes, 1989) (Table 1).

- Mutations in type I procollagen genes are the best studied and characterized (Prockop, 1985, 1990). They change the primary structure of the protein, but do not decrease expression of the genes. Such mutations thus cause synthesis of a structurally abnormal but partially functional type I procollagen. The deleterious effects of the abnormal procollagen chains are then amplified by three mechanisms in which the presence of a single abnormal pro-α chain in the procollagen molecule can drastically disturb the processing of the molecule of collagen, or the presence of abnormal molecules in fibrils distorts their structure and mechanical properties.
- Mutations that alter the amino acid sequence of the region of a pro-α chain that forms the triple helix interfere with the formation of the collagen. The protein remains unfolded, accumulates in the cells and is then slowly secreted and degraded.
- Mutations that change the structure of the pro-α chains can cause lateral displacement of one pro-α chain. As a consequence, the processing of the N-propeptide by the procollagen-N-peptidase is reduced or abolished. Persistence of the N-propeptide interferes in turn with the aggregation of collagen into fibrils (Hulmes et al., 1989).

The deleterious effects of an abnormal pro-α chain are also amplified when a kink is introduced into the triple helix of the protein (Vogel et al., 1988). The kink in the rigid collagen molecule does not prevent the association of collagen into fibrils. These fibrils are, however, abnormally branched and mechanically dysfunctional.

Studies on brittle bone disease in children, Osteogenesis Imperfect (OI), have provided evidence that over 95% of the patients have mutations in one of the two genes that determine the structure of type I procollagen. More than 70 mutations affecting these genes have been found in OI probands (Kuivaniemi et al., 1991). It has also been demonstrated that some patients with a familial history of osteoporosis have mutations in the type I procollagen genes that are similar but milder in consequence than those present in OI (Constantinou et al., 1990; Spotila et al., 1990). Several forms of cartilage dysfunctions (chondrodysplasias, Stricker syndrome, spondyloepiphyseal dysplasia, spondylometaphyseal dysplasia, achondrogenesis, and the Kniest syndrome) result from mutations in the COL2A1 gene that encodes the chains of type II collagen (Byers, 1990). Some patients with osteoarthritis also have mutations in that gene (Kuivaniemi et al., 1991). Mutations in the human gene that encodes type III procollagen (COL3Al) have been shown to produce some forms of Ehlers–Danlos syndrome (EDS), particularly those with aneurysms of the large vessels (Prockop, 1990; Kuivaniemi et al., 1991) (see Table 1). Mutations in one isoform of type IV collagen, the COL 4A5 gene, are responsible for the X-linked Alport's syndrome, an hereditary glomerular nephritis characterized by a progressive loss of renal function (Barker et al., 1990). No mutations have yet been identified in the majority of the more than 20 genes that encode the chains of other human collagens (type V-XI).

III. ACQUIRED RENAL DISEASES

A. Normal Glomerular Structural Elements

The glomerulus is a network of capillaries containing different cell types separated by extracellular matrices. The endothelial cells line the lumen of glomerular capillaries and large epithelial cells cover the external surface of glomerular basement membrane (GBM). The mesangial cells and resident macrophages represent the cellular element of the interstitial portion of the glomerulus (mesangium). The mesangial cells resemble smooth-muscle cells, presenting contractile properties. The glomerular tuft is surrounded by Bowman's capsule, an extension of the BM of proximal tubule which is covered by a single layer of cells (parietal epithelial cells). The mesangial extracellular matrix and GBM of normal glomerulus contain collagen types IV, V, VI, laminin, fibronectin, entactin, and proteoglycans (including heparan sulfate and chondroitin sulfate) (Fouster et al., 1987; Bruijn et al., 1988a; Klahr et al., 1988; Funabiki et al., 1990) (Figures 1, 2a, 3a; Table 2). These components have been extensively reviewed and it is not our purpose to describe them in detail here (see other chapters of this volume). Fibronectin appears to be

a

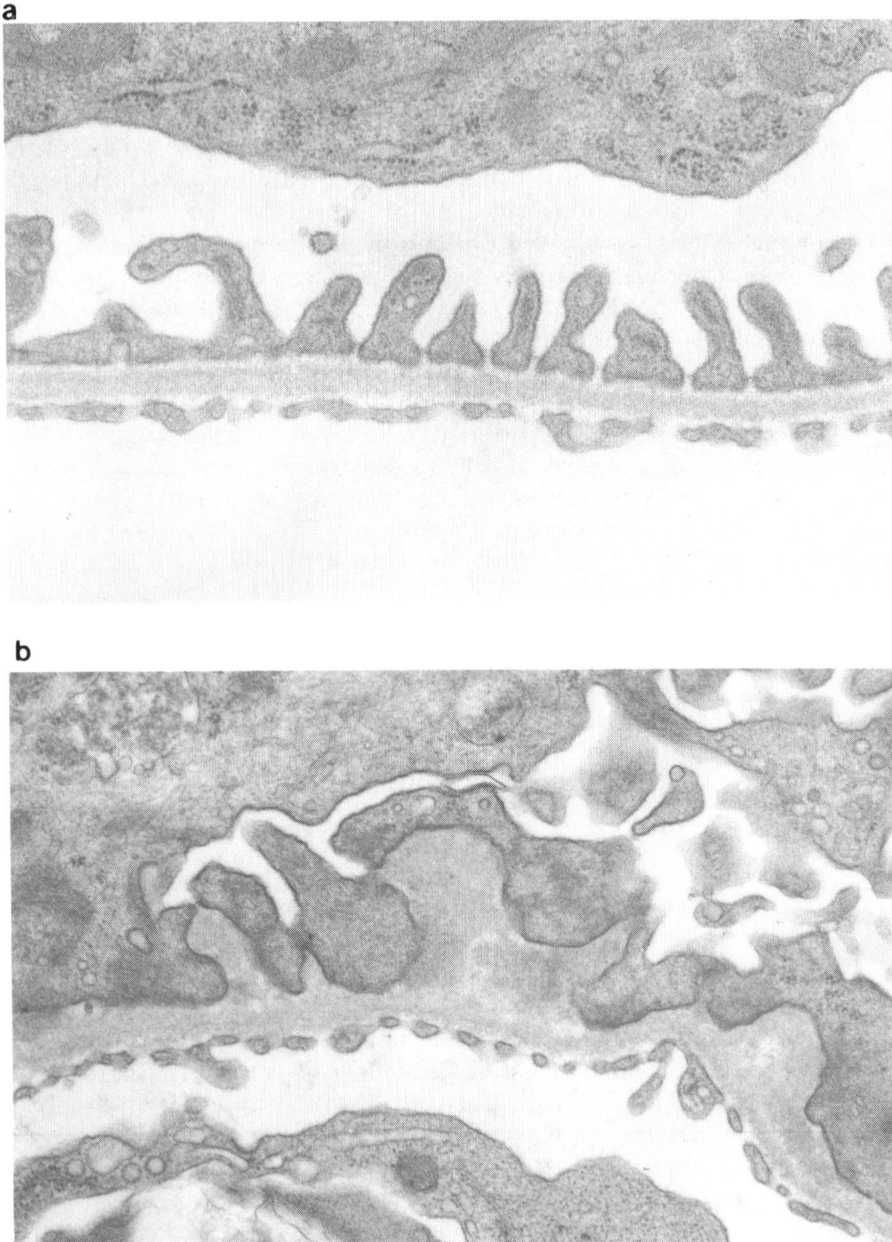

b

Figure 1. (**a**) Electron micrograph of a glomerular capillary wall in the kidney of a normal (C57BL10xDBA/2)F1 mouse. (**b**) Glomerular capillary wall in (C57BL10xDBA/2)F1 mouse kidney suffering from chronic graft-versus-host disease, showing subepithelial electron-dense deposits with spike formations of the glomerular basement membrane (× 30,000).

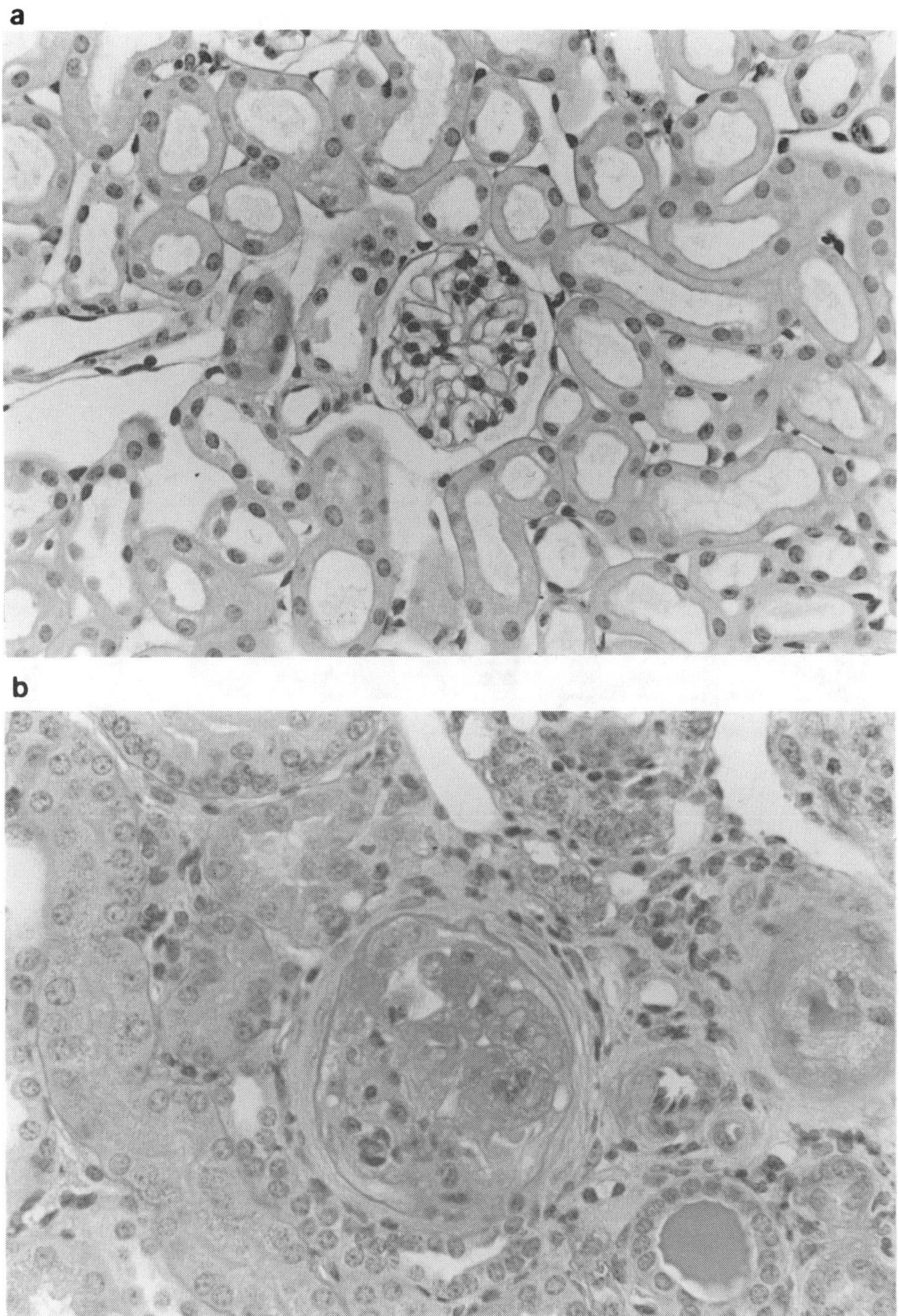

Figure 2. (**a**) Normal mouse glomerulus. (**b**) Light microscopic picture of a glomerulus of a mouse 12 weeks after the induction of chronic graft-versus-host disease, showing global sclerosis (PAS, × 480).

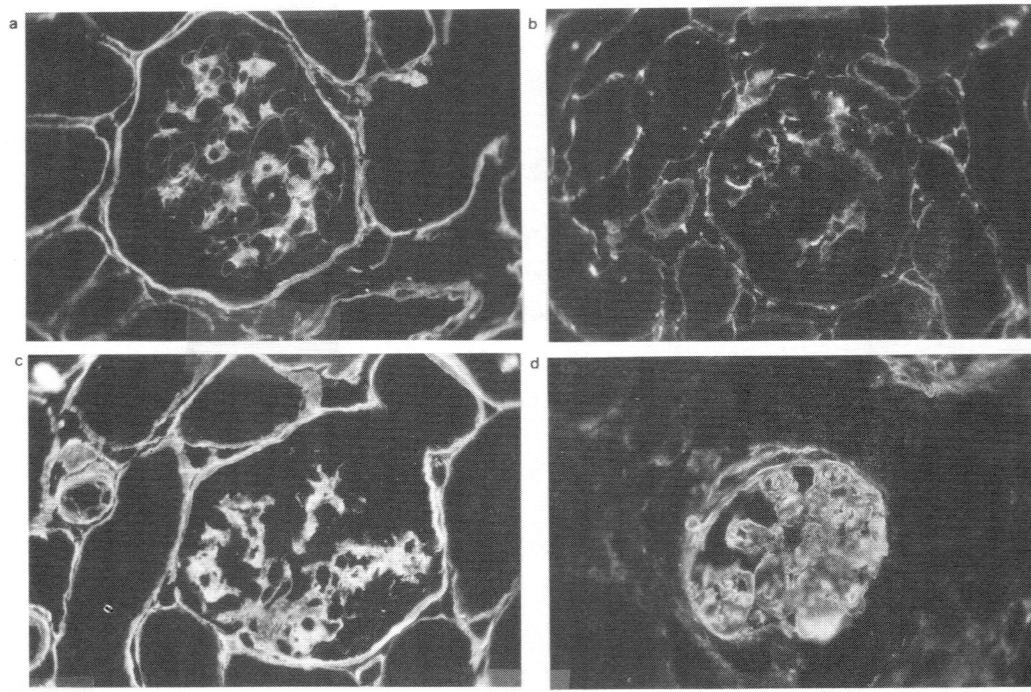

Figure 3. Immunofluorescence pictures showing the molecular composition of expanding extracellular matrix in the glomeruli of mice at different stages after induction of chronic graft-versus-host disease: (**a**) laminin at week 0: normal distribution in mesangium, GBM, Bowman's capsule and tubular BM; (**b**) collagen type III, not normally present, is present in mesangium at week 6; (**c**) collagen type IV strongly present in expanding glomerular matrix at week 6; (**d**) fibronectin diffusely present in sclerotic glomerulus at week 10 (× 480).

Table 2. Localization of ECM Components in the Glomerulus

	Normal Glomerulus	*Pathological Glomerulus*
Collagens		
Type I	–	+ (glomerulosclerosis)
Type III	– (in rat, human)	+ (glomerulosclerosis)
Type IV		
α1 chain	+ (GBM)	
α2 chain	+ (GBM)	
α3 chain	–	+ (GBM in Goodpasture's syndrome)
α4 chain	–	+ (GBM in Goodpasture's syndrome)
α5 chain	–	+ (GBM in Goodpasture's syndrome)
Type V	+ (GBM)	
Type VI	+ (mesangium)	+ (GBM in diabetes)
Fibronectin	+	
Laminin	+ (GBM)	

localized along the GBM, along epithelial cell foot processes, in the mesangium, and in Bowman's capsule. However, its largest concentration is in the mesangium. The interstitial type I collagen is absent from normal glomerulus (Scheinman et al., 1980). Collagen type III, also absent in human and rat glomeruli has, however, been found in normal murine glomeruli (Bergijk et al., 1991, 1992; Downer et al., 1988).

The GBM contains type IV and V collagens. The latter is a minor cell-associated collagen whose structure is similar to the interstitial collagens with a large triple-helical domain (Burgeson et al., 1976; van der Rest et al., in this volume). Type IV collagen is exclusively localized in BM and consists primarily of α1 (IV) and α2 (IV) chains (Timpl and Dziadek, 1986). In contrast to interstitial collagen, its central helical domain presents interruptions in the Gly-X-Y repeat leading to mechanical flexibility and protease sensitivity. The interactions between type IV molecules provide a three-dimensional network forming the structural backbone of the BM upon which the other components (laminin, proteoglycan, entactin) are assembled (Yurchenco and Schittny, 1990). The recent copurification of two distinct type IV collagen α3 (IV) and α4 (IV) chains with the two classical α1 and α2 chains (Butkowski et al., 1987; Saus et al., 1988) indicates that this protein must have several molecular compositions (Hudson et al., 1989). Immunological studies provide evidence that the major component of collagen IV containing the α1 (IV) and α2 (IV) chains is an ubiquitous BM component (Timpl and Dziadek, 1986). However, the presence of other type IV collagen chains has not been ruled out. Organ specificity, at least in the kidney, is suggested by the identification of the Goodpasture's antigen as the α3 chain of collagen IV (Hudson et al., 1989). Similarly, the more recently identified α5 (IV) chain seems to be restricted to the

Table 3. Distribution of the $\beta1$ Integrin Subfamily in Normal Glomerulus

Type	*Structure*	*Ligands Specificity*	*Localization*
VLA-1	$\alpha1\beta1$	COL (I, II, III, IV, VI)	mesangium
VLA 2	a2~ 1	COL (I, IV, VI) LAM	cortical ducts in fetal
VLA-2	$\alpha2\beta1$	COL (I, IV, VI)	cortical ducts in fetal kidney
		LAM	weak in adult
VLA-3	$\alpha3\beta1$	COL (I, III, VI)	mesangium, endothelium, visceral epithelial cells Bowman's epithelium
VLA-4	$\alpha4\beta1$	FN (alt.)	—
VLA-5	$\alpha5\beta1$	FN	mesangium, endothelium
VLA-6	$\alpha6\beta1$	LM	endothelium and podocytes, along GBM

GBM within the kidney (Barker et al., 1990; Hostikka et al., 1990). Since the COL4A5 gene coding for this distinct chain is located in the locus of the Alport's Syndrome, it is believed to be involved in the generation of this X chromosome-linked renal disease (Hostikka et al., 1990).

Endothelial and epithelial cells both participate in the synthesis of GBM. Foidart et al. (1980, 1982), using a culture of glomerular cells, reported that glomerular epithelial cells synthesize type IV collagen, fibronectin, and proteoglycans, while glomerular mesangial cells produce type I, III, and IV collagens, as well as fibronectin and proteoglycans. The latter also synthesize entactin, thrombospondin, and type V and VI collagens (Striker and Striker, 1985; Lovett and Sterzel, 1986; Border, 1988; Mohan et al., 1990).

The interactions between cells and matrix are mediated by specific membrane receptors. Most of them belong to the integrin family which consist of two noncovalently bound protein chains (α and β) (Buck and Horwitz, 1987; Hynes, 1987; Abelda and Buck, 1990; Ruoslahti, 1991; Cheresh, this volume) (Table 3). The α chain is characteristic for each integrin while the β chain is shared by several integrins. Some of the integrins appear to have a unique ligand while others bind to more than one ligand and can exhibit different ligand specificities depending on the cell which expresses them. The Very Late Antigen (VLA), or $\beta1$ subfamily, includes cell membrane receptors for collagen ($\alpha1\beta1$, $\alpha2\beta1$, $\alpha3\beta1$), laminin ($\alpha1\beta1$, $\alpha3\beta1$, $\alpha6\beta1$), and fibronectin ($\alpha3\beta1$, $\alpha4\beta1$, $\alpha5\beta1$). By their interactions with the cytoskeleton, integrins transfer the information from the matrix to the cell. During nephrogenesis, the basal organization of $\beta1$ integrins appears to take place concomitantly with the reorganization of talin, a cytoskeletal protein (Korkhonen et al., 1990a). Through this system, matrix influences cell morphology and cell functions (for a review, see Abrahamson, 1991). The distribution of various integrin subunits in embryonic and adult kidneys has been described (Cosio et al., 1990; Korkhonen et al., 1990a,b) (Table 3).

The carboxy terminal part of the laminin A chain has been suggested to play a role in the morphogenesis of the tubular epithelium of the kidney and its expression during embryogenesis coincides with the onset of tubular morphogenesis (Ekblom et al., 1990). Antibodies directed against laminin A chain or its E8 domain inhibit cell adhesion and interfered with the establishment of epithelial cell polarity during kidney tubule development (Klein et al., 1988). The laminin A chain and the integrin $\alpha6\beta1$ complex, recently identified as the principal receptor for the E8 cell-binding site (Aumailley et al., 1990; Sonnenberg et al., 1990), concomitantly appear in regions where nonpolarized cells convert into polarized epithelial cells in embryonic kidney (Sorokin et al., 1990). The integrin $\beta4$ subunit, which forms a complex with the $\alpha6$ subunit (Hemler et al., 1989), is expressed only in fetal collecting duct, and is absent from adult kidney (Korkonen et al., 1990b). The $\alpha3\beta1$ complex binds fibronectin, laminin, and collagen, and is the predominant integrin in the human glomerulus and in mesangial cells in culture. The strictly basal confinement of both $\alpha3\beta1$ and $\alpha6\beta1$ complexes, in contact with GBM, suggests their role in the adhesion of glomerular cells to BM (Cosio et al., 1990; Korkhonen et al., 1990b). The glomerular mesangium contains $\alpha1\beta1$ (Fradet et al., 1984). The fibronectin-specific receptor $\alpha5\beta1$ is present on both endothelial and mesangial cells (Kerjaschki et al., 1989; Cosio et al., 1990). Although the $\alpha2\beta1$ complex appears to be a dominant endothelial integrin during development, it is only weakly expressed in the adult (Korkhonen et al., 1990b). The $\alpha4\beta1$ integrin recognizing an alternatively spliced CSIII region of fibronectin (Mould et al., 1990) appears to be unexpressed in the normal glomerulus (Cosio et al., 1990).

B. Glomerular Extracellular Matrix Expansion

ECM Metabolism

In pathological conditions, changes in the biosynthesis or metabolism of ECM components can lead to an overproduction of matrix and/or a reduction in the rate of turnover. The intensity of ECM expansion varies considerably. The "minimal-change" disease is defined by the presence of normal or only a slight increase in mesangial matrix or cellularity (Border, 1988). This disorder affects the visceral epithelial cells and is characterized by effacement of foot processes, as revealed by electron microscopic observations. Glomérulosclerosis may be focal and segmental, involving only segments of glomeruli (e.g., Alport's syndrome, IgA nephropathy) (Goldszer et al., 1984). Alternatively, diffuse glomerulosclerosis involving all the glomeruli may occur in vascular diseases, metabolic diseases (e.g., diabetes, amyloidosis), hereditary diseases (e.g., Fabry's disease, familial lecithin-cholesterol acyltransferase deficiency), and immunologic diseases (e.g., systemic lupus erythematosus, Goodpasture's syndrome) (Khlar et al., 1988). In experimental murine chronic graft-versus-host disease (GvHD) (Bruijn et al., 1988a, 1992), electron microscopy reveals the presence of mesangial and subepithelial electron-dense deposits resulting in thickening of the GBM, the formation of "spikes", and expansion of the mesangial matrix (Figure 1). The animals develop glomerular sclerosis after 10 weeks (Figures 2, 3).

The excessive deposition of ECM components may be associated with an alteration in their respective amounts and distribution. In diabetes, thicker basement membranes (GBM and BM in blood vessels) contain a smaller amount of proteoglycans and more laminin and type IV collagen than their normal counterparts (Abrahamson et al., 1986). The type VI collagen, absent in the normal GBM has been detected in human diabetic GBM (Mohan et al., 1990). In passive Heyman's nephritis, sclerotic ECM is composed of glomerular collagen (not interstitial collagen) synthesized by glomerular cells (Adler et al., 1986). On the contrary, collagen types I and III appear in the sclerotic glomeruli of patients with glomerulonephritis (Funabiki et al., 1990) as well as in immunologically mediated renal disorders occurring spontaneously or experimentally induced in mice (Bruijn et al., 1988a; Bergijk et al., 1991; Nakamura et al., 1991).

The absolute quantity of ECM represents a dynamic balance between synthesis and degradation. Studies on the steady-state levels of mRNA coding for ECM molecules demonstrate that the increase of matrix components in experimental murine Graft versus-host disease is preceded by increased levels of their specific mRNA. In this model of lupus nephritis, during the progression of the disease, an enhancement of the amount of all components is observed and begins to occur 6 weeks after GvHD induction (Figure 3). Laminin and type I and IV collagen mRNAs increase 4 weeks after induction of the disease, before light microscopic changes (Bruijn et al., 1990b). Similarly, mRNA levels for type I, III, and IV collagens, laminin A, B1, and B2 chains, and heparan sulfate proteoglycan, increase significantly with the progression of nephritis in NZB/W F1 mice which spontaneously develop the immune disease (Nakamura et al., 1991). The enhanced production of ECM leading to glomerulosclerosis is thus regulated at a transcriptional level (Bruijn et al., 1990b). On the contrary, the development of capillary BM thickening in diabetes mellitus is not associated with an enhancement of the level of mRNAs coding for BM components (Kolbe et al., 1990), and may therefore arise more as a consequence of reduced BM degradation. Indeed, the net amount of ECM is also determined by the activity of highly specific metalloproteinases such as collagenases, as well as by broad spectrum metalloproteinases such as stromelysin (Matrisian, 1990). Neutrophil serine proteinases degrade GBM *in vitro* (Davies et al., 1990). The neutral metalloproteinases are in turn regulated by specific inhibitors. Both of these are secreted by mesangial cells (Davies et al., 1990). The mesangial cells release into the culture medium type IV collagenase, but not interstitial collagenase (Lovett and Sterzel, 1986). This absence of collagenase type I in the glomerulus could partly explain the accumulation in the sclerotic glomerulus of interstitial collagen resistant to collagenase type IV activity. A novel metalloproteinase has been extracted recently from freshly isolated rat glomeruli. This glomerular metalloproteinase (GLOMP), with a molecular mass of approximately

116 to 125 kDa, appears distinct from the previously described metalloproteinase (Shah et al., 1991). Heparinase secreted by platelets can degrade heparan sulfate, resulting in disruption of the selective permeability of GBM.

Matrix protein binding to its cellular receptor could regulate the amount of the ECM components produced by cells. Holderbaum and Ehrhart (1986) have demonstrated the reduction of fibronectin synthesis in cells grown on fibronectin. Our knowledge of the role of glomerular cell matrix receptors in renal physiology and pathology is incomplete. However, data accumulated in other cells and tissue systems suggest a significant role in the maintenance of normal interactions between glomerular cells and matrix.

Role of Intrinsic and Infiltrating Cells

The process of renal sclerosis involves various types of cells including intrinsic glomerular cells and infiltrating cells. The mesangial cells synthesize components of the mesangial matrix. When exposed to diverse stimuli including cytokines (see below) and immune complexes, mesangial cells undergo a profound transformation beginning by an increased proliferative rate and a modification in their morphology (Border, 1988). A quantitative and qualitative modification in protein and collagen synthesis can alter the structure of the ECM. Furthermore, mesangial cells are able to synthesize and secrete a neutral proteinase degrading GBM (Martin et al., 1986). Mesangial cells can generate growth factors (EGF, PDGF) and mediators of inflammation (see below). As described previously, endothelial cells and epithelial cells participate in the elaboration of GBM. In immunologically mediated renal disease, excessive synthesis of ECM components by visceral epithelial cells results in the formation subepithelial BM extensions or "spikes" (Weidner and Lorentz, 1986).

Progressive glomerulosclerosis is associated with the infiltration of the mesangium by fibroblasts and inflammatory cells (macrophages and lymphocytes). Macrophages are known to release enzymes capable of degrading collagen and elastin. However, the contribution of macrophages to sclerosis probably results more from their capacity to regulate the activity of intrinsic renal cells, particularly mesangial cells, than from their direct action on ECM degradation. Furthermore, macrophages may stimulate the proliferation of fibroblasts as well as their production of collagen (Khlar et al., 1988). The role of cytokines secreted by macrophages will be discussed below.

It has been suggested that the appearance of interstitial collagens (types I and III) in the glomerulus in certain disease states is associated with interruptions of Bowman's capsule through which interstitial fibroblasts may migrate into the diseased glomerulus (Foellmer et al., 1986; Bruijn et al., 1988a; Wiggins et al., 1991). Proliferation and migration of fibroblasts are stimulated by macrophages, platelet-derived growth factor (PDGF), fragments of collagen, and possibly other cytokines secreted at the site of inflammatory reaction (Striker et al., 1984; Stoker and Gheraldi, 1991). The interstitium may therefore play a significant role in the development of glomerulosclerosis. Alternatively, glomerular cells may produce type I and III collagens resulting from changes in gene expression (Bruijn et al., 1988a,b).

Role of Growth Factors and Cytokines

Besides producing ECM components and proteinases, glomerular intrinsic cells, as well as infiltrating cells, secrete various biologically active factors able to modulate the growth and biosynthetic activities of the surrounding cells. The cytokines may be produced by lymphocytes, monocytes, or macrophages residing in the renal tissue or outside the kidney. These cytokines thus exert their effects in a endocrine, paracrine, or autocrine fashion. In this regard, interleukin 1 (IL-1) causes pleiotropic effects on nonimmune tissues. Macrophages are the principal source of IL-l, but other cells including mesangial and endothelial cells can also secrete this cytokine (Khlar et al., 1988). Exposure of mesangial cells to interleukin 1 produced by macrophages results in their own production of IL-1 acting in an autocrine fashion (Lovett et al., 1986). In a rat model of lupus nephritis, mRNA coding for IL-1 is increased in mesangial cells (Werber et al., 1987). It affects collagen metabolism either by stimulating or inhibiting its synthesis, depending upon cell types (Matsuchima et al., 1985; Pujol et al., 1985; Kähäri et al., 1987). Torbohm et al. (1989) reported that recombinant IL-1 enhanced type IV collagen synthesis by glomerular epithelial cells, while epidermal growth factor (EGF) displayed an opposite effect (Haralson et al., 1990) (Table 4).

Platelet-derived growth factor (PDGF) synthesized by platelet and glomerular endothelial and mesangial cells is a potent mitogen and growth factor for fibroblasts, mesangial cells, and smooth muscle cells (Ross et al., 1986; Khlar et al., 1988; Schultz et al., 1988). It increases responsiveness of cultured mesangial cells to the mitogenic effect of IL-1 (Lovett et al., 1983; Lovett and Sterzel, 1986). *In vivo*, its effect on mesangial cells could lead to mesangial proliferation and synthesis of matrix. PDGF is also a chemoattractant for fibroblasts and macrophages (Table 4).

Tumor necrosis factor (TNF) is another potent mediator released by resident and infiltrating cells. If infiltrating macrophages represent a potential source of TNF, the mesangial cells themselves could produce this cytokine as well (Baud et al., 1989). The exact effects of TNF *in vivo* remain to be determined. It could participate in the inflammatory process by stimulating the synthesis of prostaglandins and prostacyclin by glomerular cells and macrophages (Bachwich et al., 1986; Baud et al., 1988). TNF also affects surface antigen expression by vascular endothelial cells, the growth of fibroblasts, and various functions of polymorphonuclear leukocytes (Le and Vilcek, 1987) (Table 4).

Transforming growth factor (TGFβ) has widespread effects on ECM (Massagué, 1987; Roberts et al., 1988) (Table 4). Most cells express receptors for TGFβ,

Table 4. Cytokines and Growth Factors

	Source	*Effects On*
IL-1	lymphocytes, macrophages,, monocytes, mesangial cells, endothelial cells, fibroblasts	collagen metabolism, proliferation
PDGF	platelets, macrophages, endothelial and mesangial cells	proliferation protein synthesis, migration of fibroblasts and macrophages
TNF	macrophages, monocytes, mesangial cells	inflammatory process, proliferation of fibroblasts
TGFβ	mesenchymal cells	production of collagen, fibronectin, proteoglycans, collagenase, inhibitors of metalloproteinases

including rat kidney fibroblasts, glomerular mesangial cells, and endothelial and epithelial cells (Wakefield et al., 1987; MacKay et al., 1989). TGFβ enhances the synthesis of collagen, fibronectin, and proteoglycans *in vitro* and *in vivo*, causing excessive deposition of ECM (Ignotz and Massagué, 1986; Roberts et al., 1986). In mesangial cell cultures, it increases the production of two chondroitin/dermatan sulfate proteoglycans, biglycan and decorin, and fibronectin (Border et al., 1990a). In a rat model, glomerulonephritis induced by specific immunologic injury to the mesangial cells is accompanied by a marked increase in the expression of TGFβ. The associated enhancement of proteoglycans and fibronectin was greatly reduced by addition of antiserum raised against a synthetic peptide from TGFβ (Border et al., 1990b; Okuda et al., 1990). The increased deposition of ECM induced by TGFβ may be aided by a TGFβ-mediated decrease in the levels of collagenases and other proteinases, and enhancement of metalloproteinase inhibitors (Edward et al., 1987). Furthermore, TGFβ is also known to influence the extracellular matrix integrin receptors of some cells (Ignotz and Massagué, 1987). By switching the pattern of integrin expression, TGFβ may modulate ECM component deposition and also change the adhesive properties of cells.

Cytokines and growth factors are thus involved in different ways in the glomerulosclerotic process. Besides the direct regulation of ECM metabolism, they influence intraglomerular fibrin deposition and coagulation, the inflammatory process and the vascular permeability (Cotran and Pober, 1989; Bruijn et al., 1992).

C. Extracellular Matrix as Target for Antibodies

Autoimmune attack against ECM components results in acute and severe disorders of the kidney. Although the underlying causes for these diseases remain obscure, they generally are manifested by deposition of antibasement membrane immunoglobulins or complement components (Bohnsack et al., 1985; Kaysen et al., 1986; Andres et al., 1986; Leivo and Engvall, 1986). Among the serious autoimmune diseases, the Goodpasture's syndrome is characterized by glomerulonephritis and alveolar hemorrhage associated with IgG deposition in a linear manner along the GBM (Abrahamson et al., 1986; Bruijn et al., 1988a, 1989a, 1990a). Antisera from Goodpasture's syndrome have been found to bind the noncollagenous NC1 domain of type IV collagen α3 (IV) (Saus et al., 1988). Patients with Alport's syndrome, a hereditary glomerulonephritis, develop anti-GBM antibody disease after renal transplantation (Milliner, 1981; McCoy, 1982). In these patients, both the α3 and α4 chains are absent from the GBM (Gunwar et al., 1990). Ultrastructural defects in the GBM of Alport's patients implicate an altered structural α5 (IV) collagen chain, a specific component of GBM within the kidney (Barker et al., 1990). The detection of mutations in the COL4A5 gene in patients with Alport's syndrome suggests that the Alport's syndrome constitutes the first example of a genetic BM and kidney disease (Zhou et al., 1990). The Goodpasture's and Alport's syndromes are both characterized by the absence of fusion of the epithelial and endothelial BM in glomerular development.

In certain autoimmune disorders, antibodies directed against ECM components play a role in glomerular involvement. Antibodies against laminin and type IV collagen have been found in sera from patients with poststreptococcal glomerulonephritis (Kefalides et al., 1985), and in sera from rats with mercuric chloride-induced glomerulonephritis (Bellon et al., 1982). This is associated with the appearance of subepithelial electron-dense complexes in the kidney. In an early stage of experimental Trypanosomiasis, a linear binding of antibodies directed against laminin and type IV collagen was visualized along the GBM. Later, this changes to a more granular distribution with subendothelial electron-dense complex formation (Bruijn et al., 1990a). In Heymann's nephritis, a 330-kDa cell surface glycoprotein was identified as the major nephritogenic antigen involved (Kerjaschki and Farquhar, 1982). Other antigen-antibody systems, including 90- and 110-kDa antigens and laminin, have been suggested to play a pathogenetic role in Heymann's nephritis. Antibodies against GBM components are thus involved in formation of large electron-dense deposits, situated either subepithelially or subendothelially. They are subsequently translocated into the mesangium by several possible mechanisms discussed previously (for a review, see Bruijn et al., 1992).

IV. CONCLUDING REMARKS

In the last decade, ECM has been extensively studied and found to play a major role as a mechanical support for tissues and to influence important biological functions. The ECM is thus involved in a variety of biological processes and diseases. Extensive work has characterized mutations in collagen genes in patients with connective disorders. This research has helped our understanding of the normal biochemistry of ECM proteins and will further aid in elucidating the complex alterations occurring during the progression of acquired connective tissue diseases. The abnormal metabolism of ECM leads to various disorders. In this review, we attempted to provide updated information on the ECM deposition occurring in renal pathology controlled by several cytokins and growth factors, and involving several cell types. Much remains unclear about the pathogenetic mechanisms leading to renal disorders. A better understanding of these complex processes will improve diagnosis and treatment in patients.

REFERENCES

Albelda, S.M. & Buck, C.A. (1990). Integrins and other cell adhesion molecules. FASEB J. 4, 2868–2880.

Abrahamson, D.R. (1986). Recent studies on the structure and pathology of basement membranes. J. Pathol. 149, 257–278.

Abrahamson, D.R. (1991). Glomerulogenesis in the developing kidney. Seminars of Nephrology 11, 375–389.

Adler, S., Striker, L.J., Striker, G.E., Perkinson, D.T., Hibbert, J., & Couser, W.G. (1986). Studies of progressive glomerular sclerosis in the rat. Am. J. Pathol. 123, 553.

Andres, G., Brentjens, J.R., Caldwell, P.R.B., Camussi, G., & Matsuo, S. (1986). Biology of disease. Formation of immune deposits and disease. Lab. Invest. 55, 510–520.

Aumailley, M., Timpl, R., & Sonnenberg, A. (1990). Antibody to integrin alpha-6-subunit specifically inhibits cell-binding to laminin fragment 8. Exp. Cell Res. 188, 55–60.

Bachwich, P.R., Chensue, S.W., Larrick, J.W., & Kunkel, S.L. (1986). Tumor necrosis factor stimulates interleukin-I and prostaglandin E_2 production by resting macrophages. Biochem. Biophys. Res. Comm. 136, 94–101.

Barker, D.F., Hostikka, S.L., Zhou, J., Chow, L.T., Oliphant, A.R., Gerken, S.C., Gregory, M.C., Skolnick, M.H., Atkin, C.L., & Tryggvason, K. (1990). Identification of mutations in the COL 4A5 collagen gene in Alport's syndrome. Science 248, 1224–1226.

Baud, L., Oudinet, J.P., Bens, M., Noe, L., Peraldi, M.N., Rondeau, E., Etienne, J., & Ardaillou, R. (1989). Production of tumor necrosis factor by rat mesangial cells in response to bacterial lipopolysaccharide. Kidney Int. 35, 1111–1118.

Baud, L., Perez, J., Friedlander, G., & Ardaillou, R. (1988). Tumor necrosis factor stimulates prostaglandin production and cyclic AMP levels in rat cultured mesangial cells. FEBS Lett. 239, 50–54.

Bellon, B., Capron, M., Druet, E., Verroust, P., Vial, M.C., Sapin, C., Girard, J.F., Foidart, J.M., Mahieu, P., & Druet, P. (1982). Mercuric chloride induced autoimmune disease in Brown-Norway rats: sequential search for anti-basement membrane antibodies and circulating immune complexes. Eur. J. Clin. Invest. 12, 127–133.

Bergijk, E.C., Munaut, C., Baelde, J.J., Prins, F., Foidart, J.M., Hoedemaeker, P.J., & Bruijn, J.A. (1991). A histological study of the extracellular matrix during the development of glomerulosclerosis in murine chronic graft-versus-host disease. Kidney Int., abstract, 40, 344.

Bergijk, E.C., Munaut, C., Baelde, J.J., Prins, F., Foidart, J.M., Hoedemaeker, P.J., & Bruijn, J.A. (1992). A histological study of the extracellular matrix during the development of glomerulosclerosis in murine chronic graft-versus-host disease. Am. J. Pathol. 140, 1147–1156.

Bohnsack, J.F., Tenner, A.J., Laurie, G.W., Kleinman, H.K., Martin, G.R., & Brown, E.J. (1985). The C_{1q} subunit of the first component of complement binds to laminin: a mechanism for the deposition and retention of immune complexes in basement membrane. Proc. Natl. Acad. Sci. USA 82, 3824–3828.

Border, W.A. (1988). Distinguishing minimal change from mesangial disorders. Kidney Int. 34, 419–434.

Border, W.A., Okuda, S., Languino, L.R., & Ruoslahti, E. (1990a). Transforming growth factor-β regulates production of proteoglycans by mesangial cells. Kidney Int. 37, 689–695.

Border, W.A., Okuda, S., Languino, L.R., Sporn, M.B., & Ruoslahti, E. (1990b). Suppression of experimental glomerulonephritis by antiserum against transforming growth factor β1. Nature 346, 371–374.

Bruijn, J.A., Hoedemaeker, Ph.J., & Fleuren, G.J. (1989a). Biology of disease. The pathogenesis of anti-basement membrane glomerulopathy and immune-complex glomerulonephritis: dichotomy dissolved. Lab. Invest. 61, 480–488.

Bruijn, J.A., Hogendoorn, P.C.W., Corver, W.E., van den Broek, L.J.C.M., Hoedemaeker, Ph.J., & Fleuren, G.J. (1990a). Pathogenesis of experimental lupus nephritis: a role for anti-basement membrane and anti-tubular brush border antibodies in murine chronic graft-versus-host disease. Clin. Exp. Immunol. 79, 115–128.

Bruijn, J.A., Hogendoorn, P.C.W., Hoedemaeker, Ph.J., & Fleuren, G.J. (1988a). The extracellular matrix in pathology. A review. J. Lab. Clin. Med. 111, 140–148.

Bruijn, J.A., Munaut, C., Baelde, J.J., & Foidart, J.M. (1990b). Extracellular matrix (ECM) expansion and mRNA levels in the development of experimental glomerular sclerosis (abstract). Kidney Int. 37, 409.

Bruijn, J.A., van Elven, E.H., Hogendoorn, P.C.W., Corver, W.E., Hoedemaeker, Ph.J., & Fleuren, G.J. (1988b). Murine chronic graft-versus-host disease as a model for lupus nephritis. Am. J. Pathol. 130, 639.

Bruijn, J.A. et al. (1992). Induction and progression of experimental lupus nephritis: exploration of a pathogenetic pathway. Kidney Int. 44: 5–13.

Buck, C.A., & & Horwitz, A.F. (1987). Cell surface receptors for extracellular matrix molecules. Annu. Rev. Cell Biol. 3, 179–205.

Burgeson, R.E., Eladli, F.A., Kaitila, I.I., & Hollister, D.W. (1976). Fetal membrane collagens. Identification of a new collagen alpha-chains. Proc. Natl. Acad. Sci. USA 73, 2579–2583.

Butkowski, R.J., Langeveld, J.P.M., Wieslander, J., Hamilton, J., & Hudson, B.G. (1987). Localization of the Goodpasture epitope to a novel chain of basement membrane collagen. J. Biol. Chem. 262, 7874–7877.

Byers, P.H. (1990). Mutations in collagen genes: biochemical and phenotypic consequences. In: Extracellular Matrix Genes (Sandell, L.J. & Boyd, C.D., Eds.). pp. 251–265.

Constantinou, C.D., Pack, M., & Prockop, D.J. (1990). A mutation in the type I procollagen gene on chromosome 17q21.31-q22.05 or 7q21.3-q22.1 that decreases the thermal stability of the protein in a woman with ankylosing spondylitis and osteoponia. Cytogenet. Cell Genet. 51, 979.

Cosio, F.G., Sedmak, D.D., & Nahman, J.R. (1990). Cellular receptors for matrix proteins in normal kidney and human mesangial cells. Kidney Int. 38, 886–895.

Cotran, R.S. & Pober, J.S. (1989). In: Progress in Immunology (Melchers, F. et al., Eds.), pp. 747–753. Springer Verlag, Berlin.

Davies, M., Coles, G.A., Thomas, G.T., Martin, J., & Lovett, D.H. (1990). Proteinases and the glomerulus: their role in glomerular diseases. Klin. Wochenschr. 16, 1145–1149.

Downer, G., Phan, S.H., & Wiggins, E.C. (1988). Analysis of renal fibrosis in a rabbit model of crescentic nephritis. J. Clin. Invest. 82, 998–1006.

Edwards, D.R., Murphy, G., Reynolds, J.J., Whitman, S.E., Docherty, A.J.P., Angel, P., & Keath, J.K. (1987). Transforming growth factor beta modulates the expression of collagenase and metalloproteinase inhibitor. EMBO J. 6, 1899–1904.

Ekblom, M., Klein, G., Mugraver, G., Fecker, R., Deutzmann, R., Timpl, R., & Ekblom, P. (1990). Transient and locally restricted expression of laminin A chain mRNA by developing epithelial cells during kidney organogenesis. Cell 60, 337–346.

Foidart, J.M., Foidart, J.B., & Mahieu, P.R. (1980). Synthesis of collagen and fibronectin by glomerular cells in culture. 1st International Symposium on the Glomerular Basement Membrane, Vienna. Renal Physiol. 3, 193 (abstract).

Foidart, J.M., Foidart, J.B., Van Cauwenberghe, D., Decoux, M., Stul, M., Lenaers, A., Niebes, P., & Mahieu, P.R. (1982). In: Biochemistry (Morel, F., Ed.). INSERM Symposium n° 21. Elsevier Biomedical, New York.

Foellmer, H.G., Bernd-Sterzel, R., & Kashgarian, M. (1986). Progressive glomerular sclerosis in experimental antiglomerular basement membrane glomerulonephritis. Am. J. Kidney Diseases 7, 5–11.

Fradet, Y., Cordon-Cardo, C., Thomson, T., Daly, M.E., Whitmore, M.F., Lloyd, K.O., Melamed, M.R., & Old, L.J. (1984). Cell surface antigens of human bladder cancer defined by mouse monoclonal antibodies. Proc. Natl. Acad. Sci. USA 81, 224–228.

Fouster, L.S., Michael, A.F., Kleppel, M.M., Vemier, R.L., & Fish, A.J. (1987). Nephritogenicity in sheep of the NC 1 domain of type IV collagen from human glomerular basement membrane (hGBM). Kidney Int. 31, 319.

Funabiki, K., Horikoshi, S., Tomino, Y., Nagai, Y., & Koide, H. (1990). Immunohistochemical analysis of extracellular components in the glomerular sclerosis of patients with glomerulonephritis. Clin. Nephrol. 34, 239–246.

Goldszer, A.C., Sweet, J., & Cotran, R.S.(1984). Focal segmental glomerulosclerosis. Annu. Rev. Med. 35, 429–449.

Gunwar, S., Saus, J., Noelken, M.E., & Hudson, B.G. (1990). Glomerular basement membrane. Identification of a fourth chain, α4, of type IV collagen. J. Biol. Chem. 265, 5466–5469.

Haralson, M.A., Dimari, S.J., Hoover, R.L., & Harris, R.C. (1 990). Epidermal growth factor suppresses collagen biosynthesis in retinoic acid-treated cultured rat kidney mesangial cells (abstract). Kidney Int., 37, 195.

Hemler, M.E., Crouse, C., & Sonnenberg, A. (1989). Association of the VLA α6 subunit with a novel protein. J. Biol. Chem. 264, 6529–6536.

Holderbaum, D., & Ehrhart, L.A. (1986). Substratum influence on collagen and fibronectin biosynthesis by arterial smooth muscle cells *in vitro*. J. Cell. Physiol. 126, 216–224.

Hostikka, S.L., Eddy, R.L., Byers, M.G., Höyhtyä, M., Shows, T.B., & Tryggvason, K. (1990). Identification of a distinct type IV collagen α chain with restricted kidney distribution and assignment of its gene to the locus of X chromosome-linked Alport syndrome. Proc. Natl. Acad. Sci. USA 87, 1606–1610.

Hudson, B.G., Wieslander, J., Wisdom, B.J., & Noelken, M.E. (1989). Biology of disease. Goodpasture syndrome: molecular architecture and function of basement membrane antigen. Lab. Invest. 61, 256–269.

Hulmes, D.J. S., Mould, A.P., Kadler, K.E., Chapman, J.A., & Prockop, D.J. (1989). Procollagen processing control of type I collagen fibril assembly. In: Springer Series in Biophysics (Bayley, P.M., Ed.), Vol. 3, pp. 292–301. Springer, Berlin.

Hynes, R.O. (1987). Integrins, a family of cell surface receptors. Cell 48, 549–555.

Ignotz, R.A. & Massagué, J. (1986). Transforming growth factor beta stimulates the expression of fibronectin and collagen and their incorporation into the extracellular matrix. J. Biol. Chem. 261, 4337–4345.

Ignotz, R.A. & Massagué, J. (1987). Cell adhesion protein receptors as targets for transforming growth factor beta action. Cell 51, 189–197.

Kähäri, V.M., Heino, J., & Vuorio, E. (1987). Interleukin-1 increases collagen production and mRNA levels in cultured skin fibroblasts. Biochim. Biophys. Acta 929, 142–147.

Kaysen, G.A., Myers, B.D., Couser, W.G., Rabl-kin, R., & Felts, J.M. (1986). Biology of disease: mechanisms and consequences of proteinuria. Lab. Invest. 54, 479–498.

Kefalides, N.A., Howard, P., & Ohno, N. (1985). In: Basement Membranes (Shibata, S., Ed.), pp. 73–87. Elsevier Science Publishers B.V., Amsterdam.

Kerjaschki, D., Ojha, P.P., Susani, M., Horvat, R., Binder, S., Hovorka, A., Hillemanns, P., & Pytel, A. (1989). A β1-integrin receptor for fibronectin in human kidney glomeruli. Am. J. Pathol. 134, 481–489.

Kerjaschki, D. & Farquhar, M.G. (1982). The pathogenic antigen of Heymann nephritis is a membrane glycoprotein of the renal proximal tubule brush border. Proc. Natl. Acad. Sci. USA, 79, 5557–5561.

Khlar, S., Schreiner, G., & Ichikawa, I. (1988). Mechanisms of diseases. N. Engl. J. Med. 318, 1657–1666.

Klein, G., Langegger, M., Timpl, R., & Ekblom, P.(1988). Role of laminin A chain in the development of epithelial cell polarity. Cell 55, 331–341.

Kolbe, M., Kaufman, J.L., Friedman, J., Dinerstein, Ch., Mackenzie, J.W., & Boyd, Ch.D. (1990). Changes in steady-state levels of mRNAs coding for type IV collagen, laminin and fibronectin following capillary basement membrane thickening in human adult onset diabetes. Connective Tissue Res. 25, 77–85.

Korkhonen, M., Ylänne, J., Laitinen, L., & Virtanen, I. (1990a). Distribution of β_1 and β_3 integrins in human fetal and adult kidney. Lab. Invest. 62, 616–625.

Korkhonen, M., Ylänne, J., Laitinen, L., & Virtanen, I. (1990b). The α_1-α_6 subunits of integrins are characteristically expressed in distinct segments of developing and adult human nephron. J. Cell Biol. 111, 1245–1254.

Kuivaniemi, H., Tromp, G., & Prockop, D.J. (1991). Mutations in collagen genes: causes of rare and some common diseases in humans. FASEB J. 5, 2052–2060.

Le, J. & Vilcek, J. (1987). Tumor necrosis factor and interleukin I: cytokines with multiple overlapping biological activities. Lab. Invest. 56, 234–248.

Leivo, I. & Engvall, E. (1986). C3d fragment of complement interact with laminin and binds to basement membranes of glomerulus and trophoblast. J. Cell Biol., 103, 1091–1099.

Lovett, D.H. & Sterzel, R.B. (1986). Cell culture approaches to the analysis of glomerular inflammation. Kidney Int. 30, 246–254.

Lovett, J., Ryan, J.L., & Sterzel, R.B. (1983). Stimulation of rat mesangial cell proliferation by macrophage interleukin 1. J. Immunol. 136, 2830–2836.

Lovett, J., Szamel, M., Ryan, J.L., Sterzel, R.B., Gemsa, D., & Resch, K. (1986). Interleukin 1 and the glomerular mesangium. I: Purification and characterization of a mesangial cell-derived autogrowth factor. J. Immunol. 136, 3700–3705.

McCoy, R.C., Johnson, H.K., Stone, W.J., & Wilson, C.B. (1982). Absence of nephritogenic glomerular basement membrane antigen(s) in some patients with hereditary nephritis. Kidney Int. 21, 642.

MacKay, K., Striker, L.J., Tauffer, J.W., Doi, T., Agodoa, L.Y., & Striker, G.E. (1989). Transforming growth factor-β. Murine glomerular receptors and responses of isolated glomerular cells. J. Clin. Invest. 83, 1160–1167.

Martin, J., Lovett, D.H., Gemsa, D., Sterzel, R.B., & Davies, M. (1986). Enhancement of glomerular mesangial cell neutral proteinase secretion by macrophages. Role of interleukin 1. J. Immunol. 137, 525–529.

Massagué, J. (1987). The TGF-β family of growth and differentiation factors. Cell 49, 437–438.

Matrisian, L.M. (1990). Metalloproteinases and their inhibitors in matrix remodeling. Trends in Genetics 6, 121–125.

Matsushima, K., Bano, M., Kidwell, W.R., & Oppenheim, J.J. (1985). Interleukin 1 increases collagen type IV production by murine mammary epithelial cells. J. Immunol. 134, 904–909.

Milliner, D.S., Pierides, A.M., & Holley, K.E. (1982). Renal transplantation in Alport's syndrome. Anti-glomerular basement membrane glomerulonephritis in the allograft. Mayo Clin. Proc. 57, 35–40.

Mohan, P.S., Carter, W.G., & Spiro, R.G. (1990). Occurrence of type IV collagen in extracellular matrix of renal glomeruli and its increase in diabetes. Diabetes 39, 31–37.

Mould, A.P., Wheldon, L.A., Komoriya, A., Wayner, E.A., Yamada, K.M., & Humphries, M.J. (1990). Affinity chromatographic isolation of the melanoma adhesion receptor for the IIICS region of fibronectin and its identification as the integrin $\alpha_4\beta_1$. J. Biol. Chem. 265, 4020–4024.

Nakamura, T., Ebihara, I., Shirato, I., Tomino, Y., & Koide, H. (1991). Increased steady-state levels of mRNA coding for extracellular matrix components in kidneys of NZB/W F1 mice. Am. J. Pathol. 139, 437–450.

Okuda, S., Languino, L.R., Ruoslahti, E., & Border, W.A. (1990). Elevated expression of transforming growth factor β and proteoglycan production in experimental glomerulonephritis. J. Clin. Invest. 86, 453–462.

Prockop, D.J. (1985). Mutations in collagen genes. J. Clin. Invest., 75, 783–787.

Prockop, D.J. (1990). Mutations in type I procollagen genes: an explanation for brittle bones and a paradigm for other diseases of connective tissue. Annals of Rheumatic Diseases 49, 440–444.

Pujol, J.P., Penfornis, H., Arenzana-Seisdedos, F., Bocquet, J., Farjanel, J., Rattner, A., Brisset, M., Virelizier, L., Beliard, R., & Loyau, G. (1985). Mononuclear cell-mediated modulation of synovial cell metabolism. I: Collagen synthesis. Exp. Cell Res. 158, 63–74.

Roberts, A.B., Flanders, K.C., Kondaiah, P., Thompson, N.L., Obberghen-Schilling, E.V., Wakefield, L., Rossi, P., de Crombrugghe, B., Heine, U., & Sporn, M.B. (1988). Transforming growth factor β: biochemistry and roles in embryogenesis, tissue repair and remodeling, and carcinogenesis. Recent Prog. Horm. Res. 44, 157–197.

Roberts, A.B., Sporn, M.B., Assoian, R.K., Smith, J.M., Roche, N.S., Wakefield, L.M., Ui, H., Liotta, L.A., Falanga, V., Ehrl, J.H., & Fauci, A.S. (1986). Transforming growth factor type-β. Rapid induction of fibrosis and angiogenesis in vivo and stimulation of collagen formation *in vitro*. Proc. Natl. Acad. Sci. USA 83, 4167–4171.

Ross, R., Raines, E.W., & Bowen-Popes, D.F. (1986). The biology of platelet-derived growth factor. Cell 46, 155–169.

Ruoslahti, E. (1991). Integrins. J. Clin. Invest. 87, 1–5.

Saus, J., Wieslander, J., Langeveld, J.P.M., Quinones, S., & Hudson, B.G. (1988). Identification of Goodpasture antigen as the α_3(IV) chain of collagen IV. J. Biol. Chem. 263, 13374–13380.

Scheinman, J.I., Foidart, J.M., & Mickael, A. (1980). The immunohistology of glomerular antigens. V: The collagenous antigens of the glomerulus. Lab. Invest. 43, 373–381.

Shah, S., Nguyen, H., Cortez, S., & Baricos, W. (1991). A novel metalloproteinase present in freshly isolated rat glomeruli. Am. J. Physiol. 260, 555–561.

Shultz, P.J., Dicorleto, P.E., Silver, B.J., & Abboud, H.E. (1988). Mesangial cells expressed PDGF, mRNAs and proliferate in response to PDGF. Am. J. Physiol. 255, F674–F684.

Sonnenberg, A., Linders, C.J.T., Modderman, P.W., Damsky, C.H., Aumailley, M., & Timpl, R. (1990). Integrin recognition of different cell binding fragments of laminin (P1, E3 and E8) and evidence that α_6/β_1 but not α_6/β_4 functions as a specific receptor for E8. J. Cell Biol. 110, 2145–2155.

Sorokin, L., Sonnenberg, A., Aumailley, M., Timpl, R., & Ekblom, P. (1990). Recognition of the laminin E8 cell-binding site by an integrin possessing the α_6 subunit is essential for epithelial polarization in developing kidney tubules. J. Cell Biol. 111, 1265–1273.

Spotila, L., Constantinou, C.D., Sereda, L., Riggs, L. & Prockop, D.J. (1990). Substitution of serine for gly alpha 2-661 in the gene for type I procollagen (COL1A2) as a cause of post-menopausal osteoporosis. Am. J. Hum. Genet., 47, A237.

Stoker, M. & Gheraldi, E. (1991). Regulation of cell movement: the mitogenic cytokines. Biochim. Biophys. Acta 1072, 81–102.

Striker, G.E. & Striker, L.J. (1985). Biology of disease. Glomerular cell culture. Lab. Invest. 53, 123–128.

Striker, L.M., Killen, P.D., Chi, E., & Striker, G.E. (1984). The composition of glomerulosclerosis and membranoproliferative glomerulosclerosis. Lab. Invest. 51, 181–192.

Sykes, B. (1989). Inherited collagen disorders. Mol. Biol. Med. 6, 19–26.

Timpl, R., & Dziadek, M. (1986). Structure, development and molecular pathology of basement membranes. Int. Rev. Exp. Pathol. 29, 119p.

Torbohm, I., Berger, B., Schönermark, M., von Kempis, J., Rother, K., & Hänsch, G.M. (1989). Modulation of collagen synthesis in human glomerular epithelial cells by interleukin 1. Exp. Immunol. 75, 427–431.

Vogel, B.E., Doelz, R., Kadler, K.E., Hojima, Y, Engel, J., & Prockop, D.J. (1988). A substitution of cystein for glycine 748 of the alpha 1 chain to cystein produces a kink at this site in the procollagen I molecule and an altered N-proteinase cleavage site over 225 nm away. J. Biol. Chem. 263, 19249–19255.

Vuorio, E. (1986). Connective tissue diseases: mutations of collagen genes. Annals of Clinical Res. 18, 234–241.

Wakefield, L.M., Smith, S.M., Masui, T., Harris, C.C., & Sporn, M.B. (1987). Distribution and modulation of the cellular receptor for transforming growth factor-beta. J. Cell Biol. 105, 965–975.

Wayner, E.A., Garcia-Pardo, A., Humphries, M.J., McDonald, J.A., & Carter, W.G. (1989). Identification and characterization of the T lymphocyte adhesion receptor for an alternative cell attachment domain (CS-1) in plasma fibronectin. J. Cell Biol. 109, 1321–1330.

Weidner, N. & Lorentz, W.B. (1986). Scanning electron microscopy of the acellular glomerular and tubular basement membrane in lupus nephritis. Am. J. Clin. Pathol., 85, 135–145.

Werber, H.I., Emancipator, S.N., Tykocinski, M.L., & Sedor, J.R. (1987). The interleukin-1 gene is expressed by rat glomerular mesangial cells and is augmented in immune complex glomerulonephritis. J. Immunol. 138, 3207–3212.

Wiggins, R.C., Goyal, M., Merritt, S.E., & Killen, P.D. (1991). Localization of alpha 2(I) collagen mRNA by *in situ* hybridization to the periarterial and periglomerular regions in anti-GBM crescentic nephritis in the rabbit. J. Am. Soc. Nephrol. 2, 568.

Yurchenko, P.D. & Schittny, J.C. (1990). Molecular architecture of basement membranes. FASEB J. 4, 1577–1590.

Zhou, J., Hostikka, S.L., Barker, D.F., Chow, L.T., Gregory, M.C., Atkin, C.L., & Tryggvason, K. (1990). Cloning of the human α5(IV) collagen gene and characterization of mutations in Alport syndrome. Fifth International Symposium on Basement Membranes, Oulu, Finland. Abstract 3.11.

Index

Advances in Molecular and Cell Biology

Edited by **E. Edward Bittar,** *Department of Physiology, University of Wisconsin-Madison*

Cell biology is a rapidly developing discipline which brings together many of the branches of the biological sciences that were separate in the past. The interrelations between cell structure and function at the molecular and subcellular level are the central connecting theme of this JAI series. This particular series also deals with new techniques. "As the twentieth century nears its close," write Kenneth Miller, "the development fo an enormous range of tools and techniques, some physical, some chemical, some biological, has changed the situation forever. Cell biology today crosses the boundary, links the molecule with the organelle, associates the cellular response with the larger organism." Most of the contributors are cell and molecular biologists who are currently engaged in research and have made important contributions to our present understanding of cell function. Each article is meant to be an essay rather than review. And each volume is written along thematic lines. The series should thus act as a source of information and ideas for junior and senior scientists, as well as teachers and graduate students of the biological sciences.

Volume 1, 1987, 200 pp. $90.25
ISBN 0-89232-792-8

CONTENTS: Introduction, *Kenneth R. Miller.* **The Analysis of Fractured Membrane "Halves",** *Knute A. Fisher, University of California, San Francisco.* **The Brush Border: A Model For Structure, Biochemistry, Mobility, and Assembly of the Cytoskeleton,** *David R. Burgess, University of Miami.* **The Lysosomal Concept: From Isolated Particles To Acid Hydrolase Compartments of The Cell,** *Phyllis M. Novikoff, Yeshiva University.* **Glycoprotein Dynamics in the Chlamydomonas Flagellar Membrane,** *Robert A. Bloodgood, University of Virginia.* **A Structural Analysis of One Photosynthetic Membrane: Rhodopseudomonas Viridis,** *Kenneth R. Miller, Brown University.* **Molecular Cytochemistry of Freeze-Fractured Cells: Freeze-Etching, Fracture-Label, Fracture-Permeation, and Label-Fracture,** *Pedro Pinto da Silva, National Cancer Institute, Frederick, Maryland.*

Volume 2, 1989, 307 pp. $90.25
ISBN 0-89232-886-X

CONTENTS: Introduction to Volume 2: Surpassing the Boundaries, *Kenneth R. Miller.* **New Trends in Understanding the Regulation of Red Cell Membrane Skeleton Organization,**

Carl Cohen, St. Elizabeth's Hospital. **The Structure and Function of Cell Adhesion Molecules,** *Bruce Cunningham, Rockefeller University.* **Hexabrachion Protein (Tenascin, Cytotactin, Brachionectin) in Connective Tissues, Embryonic Brain and Tumors,** *Harold P. Erickson, and Virginia A. Lightner, Duke University Medical Center.* **Cell-Substrate Adhesion: Induction of Cell Spreading and Apical/Basal Plasma Membrane Polarity,** *Bruce S. Jacobson, University of Massachusetts, Amherst.* **X-Ray Diffraction Studies of Gap Junction Structure,** *Lee Makowski, Columbia University School of Medicine.* **Processing of Endocytosed Material,** *Robert F. Murphy, Carneige Mellon University.* **The Cytoskeleton of the Blood Platelet: A Dynamic Structure,** *Vivian Nachmias and Ken-ichi Yoshida, University of Pennsylvania.* **Physiological Electric Fields Can Influence Cell Mobility, Growth, and Polarity,** *Richard Nuccitelli, University of California, Davis.* **Mapping Detailed Shape and Specific Loci of Macromolecules by Brightfield, Darkfield and Immunoelectron Microscopy,** *Henry S. Slayter, Dana-Farber Cancer Institute.* **Recent Advances in the Study of Mitochrondria in Living Cells,** *James R. Wong and Lan Bo Chen, Dana-Farber Cancer Institute and Harvard Medical School.* **DNA-Protein Interactions at Telomeres in Ciliated Protozoans,** *Daniel E. Gottschling and Virginia Zakian, Fred Hutchinson Cancer Research Center.*

Volume 3, 1990, 275 pp. $90.25
ISBN 1-55938-013-6

CONTENTS: The *Dictyostelium Discoideum* Plasma Membrane: A Model System for the Study of Actin-Membrane Interactions, *Elizabeth J. Luna, Linda J. Wuestehube, Hilary M. Ingalls and Catherine P. Chia, Worcester Foundation for Experimental Biology. Tektins and Microtubules, R.W. Linck, University of Minnesota Medical School.* **The Elusive Organization of the Spindle and the Kinetochere Fiber: A Conceptual Retrospective,** *Andrew S. Bajer, University of Oregon.* **Talin: Biochemistry and Cell Biology,** *Keith Burridge and Leslie Molony, University of North Carolina, Medical School.* **Digital Imaging Flourescence Microscopy: Statistical Analysis of Photobleaching and Passive Cellular Uptake Processing,** *Zeljko Jericevic, B. Wiese, R. Homan, J. Bryan and L.C. Smith, Baylor College of Medicine.* **Clathrin Assembly Proteins and the Organization of the Coated Membrane,** *James H. Keen, Temple University of Medicine.* **Control of the Mast Cell Secretion,** *David Lagunoff, St. Louis University Medical School.* **Pattern Formation: The Differentiation of Pigment Cells from Embryonic Neural Cells,** *Sally K. Frost, University of Kansas.* **Experimental Analysis of Centrosome Reproduction in Echinoderm Eggs,** *Greenfield Sluder, Worcester Foundation for Experimental Biology.* **Multiple Pathways of Protein Secretion in Exocrine Cells,** *J. David Castle, University of Virginia Medical School.*

JAI PRESS